The Electronic Packaging Series

Series Editor: Michael Pecht, University of Maryland

Published

Long-Term Non-Operating Reliability of Electronic Products
Judy Pecht and Michael Pecht

Advanced Routing of Electronic Modules
Michael Pecht and Yeun Tsun Wong

High Temperature Electronics
F. Patrick McCluskey, Richard Grzybowski, and Thomas Podlesak

Forthcoming

Estimating the Influence of Temperature on Microelectronic Device Reliability
Pradeep Lall, Michael Pecht, and Edward Hakim

HIGH TEMPERATURE ELECTRONICS

Edited by

F. Patrick McCluskey

CALCE Research Center
University of Maryland
College Park, MD

Richard Grzybowski

United Technologies Research Center
East Hartford, CT

Thomas Podlesak

U.S. Army Research Lab
Ft. Monmouth, NJ

CRC Press
Boca Raton New York London Tokyo

Acquiring Editor: *Norm Stanton*
Senior Project Editor: *Susan Fox*
Cover Design: *Jonathan Pennell*
PrePress: *Carlos Esser and Kevin Luong*
Marketing Manager: *Susie Carlisle*
Direct Marketing Manager: *Becky McEldowney*

Library of Congress Cataloging-in-Publication Data

High temperature electronics / edited by F. Patrick McCluskey, Richard
Grzybowski, Thomas Podlesak.
p. cm.
Includes bibliographical references and index.
ISBN 0-8493-9623-9 (alk. paper)
1. Electronic apparatus and appliances—Thermal properties.
2. Electronic packaging. 3. Heat resistant materials.
I. McCluskey, F. Patrick. II. Grzybowski, Richard. III. Podlesak, Thomas.
TK7870.25.H54 1996 96-36956
621.381—dc21 CIP

International Standard Book Number 0-8493-9623-9
Library of Congress Card Number 96-36956
Printed in the United States of America 1 2 3 4 5 6 7 8 9 0
Printed on acid-free paper

Foreword

The development of electronics that can operate at elevated temperatures is a critical technology for the production of distributed control systems, smart sensors, remote actuators, and other next generation products. In addition, it has the potential to significantly reduce the cost of electronic systems used to monitor automotive, aerospace, and chemical process environments. The importance of this technology is evidenced by the increasing number of programs being instituted by avionics and automotive electronics companies and their suppliers.

While advances in the design and manufacture of electronic components have now made it possible to design electronic systems that will operate reliably above the traditional temperature limit of 125°C, successful system development efforts hinge on a firm understanding of the fundamentals of semiconductor physics and device processing, materials selection, package design, and thermal management, together with a knowledge of the intended application environments. This information, while critical to practicing engineers and technologists who are increasingly being called upon to design systems for elevated temperature use, has never before been available in a single text.

This book brings together this essential information and presents it for the first time in a unified way. Students will find its detailed level of explanation and coverage of the fundamental scientific concepts useful, while managers and policymakers will value it for its excellent overview of the key issues involved in extreme temperature electronics and its identification of those areas where further development is needed. Finally, packaging and device engineers and technologists should consider it required reading for its coverage of the techniques and tradeoffs involved in materials selection, design, and thermal management and its presentation of best design practices using actual fielded systems as examples.

Michael G. Pecht
Professor and Director
CALCE Electronic Packaging Research Center
University of Maryland

Preface

The development of electronics that operate at highly elevated temperatures has been identified as a critical technology for the next century. U. S. Air Force interest in this area is focused on the more electric airplane (MEA) and more electric engine (MEE) initiatives, while the Army Research Laboratory and the U. S. Army Tank-Automotive Command are conducting similar programs for ground vehicles. Commercial aircraft and automotive manufacturers are pursuing similar initiatives in the private sector.

Limiting temperature to below 125°C has become a severe design constraint. It hinders the development of distributed control systems, smart sensors, and remote actuators, and increases the cost of electronic systems used to monitor such systems as automotive engines, anti-lock brakes, aircraft engines, aerospace propulsion systems, and chemical processes. The costs are a result of the additional size, weight, expense, and reliability risks related to remote placement or cooling of these electronic assemblies.

Motivation

As part of its research charter, the CALCE Electronics Packaging Research Center has surveyed the available literature, held discussions with numerous professionals and high temperature device manufacturers, and conducted research on the use of electronic hardware at elevated temperatures. This has led to the conclusion that recent advances in the design and manufacture of electronic components have made it possible to develop electronic systems that will operate reliably above the currently accepted temperature limits. However, successful development hinges on a firm understanding of the issues related to materials selection and compatibility, thermal management, and reliability assessment. A knowledge of the effects of temperature on performance and reliability and of the strategies and designs that accommodate higher temperatures is needed for engineers to make informed design tradeoffs. Currently no book provides this essential information.

What This Book Is About

This book presents a comprehensive, critical review of the state of the art in high temperature electronics, providing a systematic and scientific exploration of the issues involved in designing and specifying components, developing packaging techniques, designing for particular applications, and testing the final system. Experts from the field of high temperature electronics have contributed to nine chapters covering topics ranging from semiconductor device selection to testing and final assembly.

Chapter 1 introduces the field of high temperature electronics, explaining its current importance, listing potential applications, and presenting the technical challenges that must be addressed for successful high temperature products to be developed. The next three chapters focus on the selection of components for use in elevated temperature electronic systems. Chapter 2 discusses the fundamental temperature limits for the use of silicon devices, presents design guidelines and technologies to extend the useful temperature range of silicon, and provides data on the use of commercially available silicon devices at elevated temperatures. Chapter 3 presents the current state of the art in wide bandgap semiconductor devices used as alternatives to silicon at extreme temperatures. The technical concerns that limit their performance are discussed together with the status of research efforts to remove these limitations. Chapter 4 focuses on high temperature passive components. Special attention is devoted to leakage currents and capacitance shifts in capacitors, since these factors present the most severe most severely limit the use temperature of passive components.

Once components that can operate at elevated temperatures are chosen, packaging and assembly schemes must be developed. The next four chapters discuss materials selection, design, and thermal management in packaging devices for high temperature use. A systems approach is taken with chapters addressing concerns at the package, board, and module/box level. Chapters 5 and 6 present materials selection criteria for hardware intended for use at elevated temperatures. The maximum use temperatures of materials at each level of electronic packaging are provided, along with the temperature dependence of key mechanical, electrical, and thermal properties. Chapter 7 describes novel cooling, layout, and thermal management techniques that can be used to lower the system temperature. Chapter 8 discusses typical applications requiring electronic systems that operate at elevated temperatures, and the design approaches currently being employed for these applications. Aerospace, automotive, and well logging applications are covered. Finally, Chapter 9 presents procedures for accelerated testing of high temperature electronics, including the design of test fixtures.

Who This Book Is For

Since this book takes a scientific approach to analyzing and addressing the issues involved in high temperature electronics, which include designing and specifying components, developing packaging techniques, designing for particular applications, and testing the final system, it is designed primarily to be a textbook for graduate courses. However, it is equally valuable as a design guide and handbook for engineering, manufacturing, and science professionals. In addition, the book is organized to be easily tailored as a resource for undergraduate, community college, and continuing education courses.

Acknowledgments

In order to create a book presenting the most comprehensive and up-to-date review of the state of the art in high temperature electronics, many experts in many different areas of the field must be called upon to participate.

Apart from the listed contributors, the editors would particularly like to thank Rich Bauernschub, Sanir Patel, and Jong Kim for their contributions to this book. In addition,

Richard R. Siergiej, the author of section 3.3 is indebted to his colleagues at Northrop-Grumman Science and Technology Center for major contributions, particularly A. K. Agarwal, G. Augustine, V. Balakrishna, D. L. Barrett, C. D. Brandt, A.A. Burk, Jr., R. C. Clarke, R. C. Glass, H. M. Hobgood, R. H. Hopkins, L. B. Rowland, T. J. Smith, and S. Sriram. Chris Carlin wishes to express his appreciation to John Gerstle, Tom Odell and Don Huling for their review and comments on the text and to the Boeing Company and NASA (contract NASA-19360) for their support of the work. Kitt Reinhardt wishes to express thanks to Joseph Weimer and Steve Przybylko of the Air Force Wright Laboratory, Dayton OH, for their technical support.

Thanks also to Lloyd Condra, Bryan Buchanan, Chris Carlin, Karl Eisinger, John Fink, Jeff Green, Greg Kromholtz, Gary Miller, Tom Torri, and Kevin Wodkins who helped begin the efforts at the CALCE Electronic Packaging Research Center in the area of high temperature electronics and provided guidance and representative electronic modules for study over the last two years. A special thanks to all the member companies of the CALCE Electronic Packaging Research Center who have supported and reviewed this book: AlliedSignal Aerospace, AMP, Army Materiel Systems Analysis Activity, Bellcore, Boeing, Cetar, Collins, Crane-Eldec, Daewoo, Defence Research Agency (U.K.), Delco Electronics, Department of Defense (U.S.), DY4 Systems, ERS, GEC-Marconi, Honeywell, Litton, Lockheed-Martin, LZR Electronics, McDonnell-Douglas, Ministry of Defense (U.K.), NASA, National Science Foundation, Naval Air Warfare Center, Northrop-Grumman, Philips, Rockwell, Saturn, Sonix, Sonoscan, State of Maryland, Texas Instruments, TRW, and United Technologies. Thanks also to DARPA who funded this book through the TRP for Workforce Retraining in Manufacturing Science and Engineering of Cost Effective Electronics.

The editors are also indebted to our English reviewer, Dr. Lesley Northup and our publisher, Norm Stanton, for ensuring a uniform style throughout the book, and to the University of Maryland CALCE Electronic Packaging Research Center staff: Joanyuan Lee, Amal Hammad, Ioulia Fedotova, Lakshmi Sundararaman, Erica Du, and especially to Iuliana Roshou Bordelon.

Lastly, a very special thanks goes to Dr. Michael Pecht, Director of the CALCE Electronic Packaging Research Center, who initiated the center's efforts in high temperature electronics, and without whose contributions, assistance, direction, encouragement, and support this book would never have been completed.

Contributors

Phil Brusius has been a reliability engineer at Honeywell's Solid State Electronic Center for 17 years. During this time he has worked on all aspects of microelectronic reliability, including device wearout studies such as electromigration, hot carrier effects, and time dependent dielectric breakdown; accelerated life tests and their correlation to defect phenomena; reliability modeling; packaging; and high temperature microelectronics. He has authored papers in all of these areas. He also worked as a Hybrid Packaging Engineer for Harry Diamond Laboratories for 5 years and as a Failure Analysis Engineer for Sperry Support Services and Goddard Space Flight Center for 5 years. He has a B.S.E.E. and M.S.E.E. from University of Wisconsin, Madison.

Chris M. Carlin is the Propulsion Systems Manager, High Speed Civil Transport at Boeing Commercial Airplane Company. Dr. Carlin is responsible for the entire propulsion system for the Boeing high speed civil transport airplane, including the design of the electronic control system. A graduate of MIT, he has over 30 years experience in the development of aerospace related flight and propulsion control systems. He is a recognized expert on the design of avionic systems for use in high temperature environments.

Philip Dreike is Senior Member of Technical Staff at Sandia National Laboratories. He received a B.S. degree in Physics and an M.S. degree in Electrical Engineering from Stanford University in 1974, and M.S. and Ph.D. degrees from Cornell University in Applied Physics in 1976 and 1980 respectively. He joined Sandia National Laboratories Target Experiments Division in 1981, and worked in the area of Inertial Confinement Fusion from 1981 until 1991. He worked in the Reliability Physics Department from 1991 until 1994 on integrated circuit reliability in hostile environments, particularly those involving high temperature. He is presently working in the area of hardware/software co-design in the Power Electronics and Custom Controllers Department. He is a member of the IEEE and the American Physical Society.

Darrel R. Frear is a Senior Member of the Technical Staff in the Mechanical and Corrosion Metallurgy Department, and is a member of the Center for Solder Science and Technology, at Sandia National Laboratories. He received his A.B. from Dartmouth College in 1982, his M.S. (1984) and Ph.D. (1987) in Materials Science from the University of California, Berkeley. His area of research is the characterization and modeling of the microstructure/mechanical properties relationships and thermomechanical fatigue behavior of solder alloys and other interconnections for electronic applications. Darrel is an internationally recognized expert in the fatigue behavior of solders. Darrel has co-edited three books that describe the state-of-the-art of solders and over 60 publications. Darrel has also worked in the area of aluminum thin films and electromigration behavior. He is a member of ASM International, and is on the board of directors of TMS.

Catherine Gallagher has been extensively involved for the last five years in the development and statistical optimization of the ORMET line of electrically conductive materials for printed wiring boards and other electronic packaging substrates. Her contributions have included the design, synthesis, and characterization of the polymer-flux systems in the ORMET products by employing HPLC, GPC, IR, GC, TMA, TGA and DSC techniques. She has extensively characterized the powder metallurgy for packing density, flow, percolation, oxidation levels, alloying and phase transition effects by the use of DSC, TGA, and sedigraph. She has had extensive experience modifying the rheology, flow and wetting characteristics to meet specific applications including screen-printing, stenciling and filling of photo-imaged grooves, and engineering the chemistry to successfully couple the metal matrix to the adhesive to form a true composite material. She is well versed in investigating compatibility of various packaging materials with respect to coefficient of thermal expansion, glass transition temperature, thermal decomposition temperature and moisture absorption. She has also had experience in patenting R&D materials, transferring them into engineering, and transferring technology to licensees. Ms. Gallagher earned her bachelor of science in organic chemistry from the University of California, Los Angeles. She has over 5 publications related to this field.

George Goetz is a Senior MTS in the Microelectronics and Technology Center of Allied Signal He has conducted studies on the nature and characteristics of surfaces with applications ranging from microelectronics to spark plugs. He holds several patents in the area of reactions at solid surfaces, and currently is involved in developing thin film conductors for high temperature microelectronics.

Richard Grzybowski received his Ph.D. in electromagnetics from the University of CT in 1993 and is a senior Research Scientist at the United Technologies Research Center. His areas of expertise include electromagnetics, device physics and high temperature electronics and packaging. He is a senior member of the IEEE and has published papers extensively on the subject of harsh environment device characterization, modeling and packaging. He has been with UTRC since 1981 and has over ten years experience in the field of high temperature electronics circuit design and packaging.

John Hearn is a Senior Development Engineer with Delco Electronics Corporation. He has been working in technology development, design and packaging of semiconductors and electronic circuits for forty years. Thirty four have been with Delco Electronics developing discrete high power and high voltage silicon devices; and developing hybrid circuit technologies and packaging directed toward severe environments such as under the hood of an automobile. Technologies he has pioneered range from small sensors to large micro-processor based engine control modules. His present interests include developing a better understanding of both the underhood environment, and the thermal-mechanical fatigue mechanisms in packages. He is a member of the International Society of Hybrid Microelectronics (ISHM).

Y.K. Joshi is an Associate Professor of Mechanical Engineering and a faculty member at the CALCE Electronics Packaging Research Center at the University of Maryland at College Park. His research and instructional activities are in the area of thermal engineering. Dr. Joshi's academic experience also includes seven years on the faculty at the Naval Postgraduate School

(NPS), Monterey, California. Prior to joining NPS, he spent one year as Advanced Mechanical Engineer in the semiconductor packaging industry, investigating heat transfer problems during semiconductor assembly. Dr. Joshi serves as an Associate Editor of the ASME Transactions, Journal of Electronic Packaging.

Mike Liu has been with Honeywell since 1968 and is the principal investigator for the "Advanced chemically treated oxide development" program sponsored by the Naval Research Laboratory. He holds an Ph.D. in Electrical Engineering from University in Minnesota. Since 1990 Mike has been leading the process development of various SOI materials for radiation hardened and high temperature CMOS SRAM and ASIC applications. He won a Sweatt Award (Honeywell's highest award for engineers) for his development of production worthy SOI materials in 1995. Dr. Liu has been granted 14 U.S. patents on semiconductor devices; he is the author or co-author of over 80 technical papers; he is the author of three chapters on "Pyroelectric Coefficients" in the Landolt-Bernstein New Series; and he is co-editor of two IEEE Proceedings of the International Symposium of Applications of Ferroelectrics.

Sam Liu is a Materials Research Engineer at the University of Dayton Research Institute. Dr. Liu received his B.S. and M.S. degree from the University of Science and Technology, Beijing in 1967 and 1980 respectively. He received his Ph.D. from the University of Dayton in 1989. Dr. Liu has contributed to the development of new magnetic and electronic materials. His current research interests, include high performance rare-earth permanent magnet materials, contact metallizations and metal-insulator-semiconductor (MIS) devices based on wide-band gap semiconductors.

Goran Matijasevic is Senior Member of Technical Staff at Toranaga Technologies, Inc., where beside being the manager of the R&D department, he is responsible for developing new metal technologies for conductive adhesive compositions as well as other electronics materials and processes. He has been the principal investigator on one Phase II and three Phase I SBIR Projects. Prior to joining Toranaga, he was a Post-Doctoral Researcher at the University of California, Irvine where he investigated solder attachment for microelectronic packaging. He received his Ph.D. and his M.S. in Electrical and Computer Engineering from University of California, Irvine in 1991 and 1985, respectively, and his bachelor of engineering degree in Materials Science from the University of Belgrade in 1984. He co-wrote book chapters in Microelectronic Packaging, Van Nostrand Reinhold, New York, 1993, and "Chip Attachment," Chip-on-Board (COB) Technology, Van Nostrand Reinhold, New York, 1994. He has over 20 journal and conference publications in the field of electronic packaging materials and processes.

Patrick McCluskey is a Research Scientist at the CALCE Electronic Packaging Research Center at the University of Maryland. He received a Ph.D. in Materials Science from Lehigh University in 1991 and is currently directing research in the areas of high temperature electronics, reliability assessment of microelectronic packages, and qualification of plastic encapsulated microelectronics. He has published in the areas of thin film growth, surface analysis, and microelectronic reliability. Prior to joining CALCE EPRC, he was a materials technologist in the electronic products division of W.L. Gore and Associates, Inc.

Mark Nelms received the B.E.E. and M.S. degrees in electrical engineering from Auburn University in 1980 and 1982, respectively. He received the Ph.D. degree in electrical engineering from Virginia Polytechnic Institute and State University in 1987. He is currently an Associate Professor in the Electrical Engineering Department at Auburn University. His research interests are in the areas of power electronics, energy conversion, and power systems.

Thomas Podlesak is an Electronics Engineer for the U.S. Army Research Laboratory at Fort Monmouth, NJ. He received the B.S. in Physics from St. John's University, NY and the M.S. in Engineering Science and the Ph.D. in Electrical Engineering from the State University of New York at Buffalo. After working in industry, he took his present position where he specializes in research in pulse power and power conditioning. Dr. Podlesak co-authored "Power Electronics for the All-Electric Tank", a landmark study on the application of power electronics to a major Army system for the 21st Century. He served as Technical Chairman of the First Workshop on High Temperature Power Electronics for Vehicles, held at Fort Monmouth in April of 1995. Dr. Podlesak's major interest in high temperature electronics is their application in power conditioning systems for transportation applications.

Kitt C. Reinhardt received B.S. and M.S. degrees in Electrical Engineering from the State University of New York at Buffalo in 1988 and a Ph.D. degree in Engineering Physics from the Air Force Institute of Technology in 1994. His major area of scholastic study was semiconductor device physics and fabrication. Since graduation his activities have included the fabrication and analysis of Si, InP, GaAs and GaInP solar cells, GaAs microwave MESFETs, and SiC p/n and Schottky diodes. He has also conducted technical trade-studies concerning Air Force power system needs, including analyses of satellite space power subsystems, the impact of wide-bandgap power electronics on the More Electric Aircraft (MEA) and the effect of power system performance on solar-powered unmanned aerial vehicles (UAVS). He has authored or co-authored 5 journal articles and 22 conference papers.

Gerald Servais is a staff Research Engineer at Delco Electronics. He has worked for more than 35 years with General Motors, the last 23 years at Delco Electronics. At Delco, he spent more than 13 years in electrical part failure analysis (discrete devices to microprocessors) and for the last 10 years he has been supervisor of Digital Parts Group and the Parts Research & Test Development Group. He has published more than 10 technical papers and has been involved with various technical organizations including EIA JEDEC, and IPC. He is one of the founders of the CDF Automotive Electronics Council (developing automotive standards for the qualification of integrated circuits, discrete semiconductor devices, and passive components). He has a B.S. in Aeronautics from St. Louis University (Parks College) and MS in Mechanical Engineering (Material Science) from Marquette University.

Richard R. Siergiej is presently a senior engineer/scientist at the Northrup Grumman Science and Technology Center in Pittsburgh (formerly Westinghouse Science and Technology Center). His present interests include all aspects of prototyping novel SiC devices. He has been actively involved in the development of the SiC static induction transistor. He also supports other SiC device development programs from DC to large signal rf modeling. Dr. Siergiej received his Ph.D. in electrical engineering from Lehigh University, where he studied the role quantization effects in silicon have on the interpretation of parameters describing Si/SiO_2 trapping kinetics.

Thomas Torri is working in the Advanced Technology Center with responsibilities for Product Technical Development. He has worked for General Motors for 36 years. The last nine years have been at Delco Electronics in advanced product development and in product reliability. He holds a B.S. in Electrical Engineering from the University of Notre Dame and a M.S. in Electrical Engineering from the University of Michigan. He is a member of the Society of Automotive Engineering (SAE) and is a Registered Professional Engineer.

Thomas Zipperian is the Manager of the Compound Semiconductor Technology Department at Sandia National Laboratories. He received his B.S. degree in Electrical Engineering from Montana State University in 1975 and the M.S. and Ph.D. degrees in Electrical Engineering from the University of Minnesota in 1978 and 1980, respectively. He joined Sandia National Laboratories in Albuquerque in August, 1980. Since that time he has been engaged in research on a variety of physics, materials and device topics in compound semiconductor technology. Subjects specifically of interest include studies of strained-layer superlattice and strained quantum well materials and devices. He has either authored or co-authored over 130 technical articles in these fields. At present, he continues studies in compound semiconductor areas as well as initiating new materials and device projects in thallium-based high-temperature superconducting thin films. He is a member of Tau Beta Pi, the IEEE and the American Vacuum Society.

To

Monica McCluskey
Donna, Cassandra, and Nerissa Grzybowski,
and
Mary Podlesak

for their support and encouragement
throughout this project

Contents

Chapter 1

OVERVIEW OF HIGH TEMPERATURE ELECTRONICS

1.1 What is High Temperature Electronics?

Commercial and military aircraft that can fly at greater than twice the speed of sound with improved safety, reliability, and maintainability, automobiles with longer lifetimes and greater fuel economy, chemical processes with ultra-precise control and minimal waste — these are just a few of the future products which are possible with the use of elevated temperature electronics. The ability to use electronic systems at elevated temperatures will not only make new products possible but it will decrease the cost and increase the reliability of current products by removing the need for large, heavy, complex cooling systems and the cabling and interconnections required for remote placement of the electronics.

However, there are many technical challenges involved in developing electronic systems that will operate at elevated temperatures. These range from proper IC design, the appropriate use of passives, and the development of robust packaging structures to the use of the latest thermal management techniques. Before attempting to describe all these issues of elevated temperature electronic design, however, we need to define elevated temperature electronics and put it in a historical context.

The 1960s saw the gradual displacement of the vacuum tube from its predominance in electronic products. Tubes were slowly replaced by new devices, first referred to as "solid-state," but later called semiconductors. Pundits of the time extolled the virtues of these devices, which eliminated the negative features of the old-fashioned vacuum tube, such as high interelectrode voltage drop and the problematic need for tubes to be heated to high temperatures in order to function. The latter shortcoming necessitated stocking filaments and expending power for heating. The reference to electronic equipment as "heaters" was as literal as it was figurative. One of the last remaining tube manufacturers in the U.S. has no heating system. The heat from the tubes undergoing qualification testing is more than sufficient to heat a large factory.

Semiconductors were to change all that. It is ironic, however, that the early semiconductors, particularly those with higher power ratings, generated considerable heat, even without a heater circuit, due to their own operating characteristics. Worse, this heat tended to destroy the devices. The vacuum tube functioned, by and large, quite well at higher temperatures; not so the semiconductor.

Various methods were used to alleviate this problem. Heat sinks were added to absorb the excess heat generated, but greater power ratings required ever larger heat removal systems. In effect, the inability of these devices to function at elevated temperatures has restricted them to relatively benign operating environments. Although this is not an issue in most cases, it can seriously limit the ability to operate electronic systems in some applications where there is a critical need (for example, sensors in the combustion chamber of a jet engine).

What is needed is a new class of electronic components capable of operating at higher-than-normal temperatures and the associated packaging to support them. These components include both semiconductors and passive components. In the past, associated components and packaging were not issues, since if the semiconductors could not survive, it made no sense to develop related components that could. Today, however, their development is a rapidly growing field of technology. This work will examine all aspects of making high temperature electronics a reality, from the basic issues of device physics and material science to manufacturing and relevant applications.

Currently, there is no agreement as to just what is "high temperature electronics," or, better stated, what constitutes high temperature? Different experts will cite ranges to define the realm of high temperature electronics. These definitions are particularly troublesome when they imply that devices operating at temperatures lower than one expert's range are not high temperature devices, while another's range, lower than the first, places them in the high temperature electronics category.

For example, the sensor in the combustion chamber of a jet engine may experience temperatures of 300°C or higher. This is definitely a high temperature environment. However, so is the case of power semiconductors in the electric propulsion system of a military ground vehicle, such as a main battle tank, even though these devices may operate at temperatures no greater than 200°C, with the engine lubricant oil being used as a coolant.

The best definition for "high temperature electronics" is electronics operating at temperatures in excess of those normally encountered by conventional, silicon-based semiconductors or their auxiliary components. These temperature limits have been set by convention at 70 to 85°C, with some special military systems being operated at a maximum temperature of 125°C. Even this definition has its shortcomings. For the lower levels of the elevated temperature regime, 125 to 200°C, it is often possible to use "conventional" silicon semiconductors by modifying the device design or packaging, or by appropriately derating various device properties. This is a completely acceptable procedure in some cases, especially in commercial applications where there is little additional burden. For high performance systems, such as those used by the military, however, this is unacceptable and a new approach is necessary. Similarly for passive components, the temperature limitations may be related to the choice of a packaging material or manufacturing process or they may be more fundamental requiring a totally new design.

It should be apparent that the realm of "high temperature electronics" covers a wide range of applications and must satisfy different criteria. The main objective of this book is to explore these criteria and the issues in each case. The first task is to describe briefly the applications of high temperature electronics, and how each will affect the realization of the technology.

1.2 Applications

High temperature electronics are essential for harsh environments in which the use of conventional electronics is impractical, such as under the hood of an automobile or in exposed areas of aircraft. Conventional applications in factories, consumer electronics, and laboratories, by and large, are well served by conventional electronics, a multi-billion dollar industry, with considerable investment in manufacturing and service facilities, which would need to be duplicated or modified to make a transition to high temperature devices. This transition, in most cases, is unwarranted.

Commercial and military ground and air vehicles stand the most to gain from high temperature electronics. Marine vessels are not expected to be a large consumer of high temperature devices. Conventional devices should be more than adequate, even when substantial active cooling is required. This is because once one gets past the corrosion problem, a ship is literally floating in one of the best coolants.

Air and ground vehicles, on the other hand, are ideal candidates for utilization of high temperature electronics. They have both regions of high temperature and limitations on their ability to provide cooling. This second point has been the impetus, to a great extent, the development of high temperature electronics.

Air and ground vehicles have severe limitations on the amount of weight and volume they can devote to auxiliary systems, such as electronics cooling. Although the devices producing the reject heat are, generally speaking, small and lightweight, and it is generally easy to remove the heat from the device generating it, disposing of that heat is a somewhat daunting enterprise. The only available medium for accepting the rejected heat is often ambient air. If natural or forced convection air cooling is not practical due to temperature differential, gas volume or ambient dust conditions, liquid cooling, either single or two phase, is necessary. Liquid to air heat exchangers are notoriously large, making severe demands on allowable weight and volume from the vehicles. Such systems also require pumps or compressors, coolant piping, and controls, which make additional demands on weight and volume.

A common misconception is that high temperature electronics do not require cooling. Although adiabatic operation is possible, to a limited extent, continuous operation will require cooling, and more complex cooling than has been required in the past [Mahefky 1994]. High temperature devices do not require as much cooling, or coolants at low temperatures, but require cooling just the same. This apparent advantage quickly becomes a disadvantage, when the coolant and the apparatus needed to handle the coolant are forced to operate at a temperature higher than standard, forcing the use of new materials, also capable of high temperature operation.

Since air and ground vehicles are largely propelled by heat engines, there are hot zones on the vehicle. To operate close to these hot zones is very advantageous, both from the standpoint of measuring operating parameters, and from the standpoint of mounting electronic control elements in critical locations, such as on the engine. This situation is encountered not only in military applications but commercial ones as well. It is often the case that the environment under the hood of the standard automobile requires components built to more stringent requirements than those of military specification components.

This ability to operate in demanding places leads to the second major application of high temperature electronics, that of hostile environments. There are places where it would be quite advantageous to make measurements, even though these locations exhibit quite high temperatures. Beside the previously mentioned combustion processes, there are other such places of high temperature on the earth or, more properly, under the earth. Deep wells for the exploitation of petroleum and other minerals, and geothermal application create a small but vital niche market for high temperature electronics. It will be shown how data must be gathered from deep in the earth, reliably and accurately, in locations where temperatures will reach several hundred degrees Celsius.

1.3 Technical Challenges

As has been shown above, there is a need to produce high temperature electronic components to fill the demands of a number of electronic systems with significant potential for technological and market growth. However, the establishment of a high temperature electronic industry, comparable to the present conventional electronic industry, will require answering many technical challenges.

Comparison to the conventional electronic industry is a very apt one. The principal problems of the current high temperature electronics industry closely parallel the conventional electronics industry in the 1960s. Similar problems presented themselves and these challenges were met. A similar strategy will need to employed to establish a high temperature electronics industry.

The first challenge could not be more basic. The semiconductor materials required to produce high temperature semiconductor components are in short supply or not of sufficient purity to produce high quality devices. While conventional silicon devices can be used to 200°C, higher temperatures to 300°C require the use of silicon-on-insulator (SOI) technology. This technology extends the useful range of silicon by isolating devices on the IC dielectrically rather than by means of reverse biased junctions. As a result, the structure is immune to the problems of leakage and latch-up at these junctions. However, these wafers are difficult to make and only a handful of companies manufacture them. Above 300°C, even SOI technology is not sufficient to stop leakage from rendering the silicon devices unusable. At these temperatures, a wide bandgap semiconductor is needed. The most commonly used material, silicon carbide, when it is formed into wafers from which components are made, is beset with a problem known as micropipes. Extremely small voids, microns or fractions of microns in diameter, form throughout the wafer structure and disrupt the ability of the material to withstand voltage. This is a critical shortcoming for components that perform switching functions, such as diodes, transistors, and thyristors. The problem is most serious for components designed to handle high power, since they tend to be not only high voltage devices but high current devices. This requires relatively large semiconductor devices, increasing the probability of a micropipe being present, thereby rendering the device useless. Progress is being made to reduce these defects allowing for the growth of better material.

The technical challenge of producing satisfactory semiconductor devices does not end with the production of a satisfactory semiconductor wafer. The next step, SiC device processing, has developed into a sometimes daunting task, definitely different from that of processing conventional silicon, as silicon carbide is most commonly used as an abrasive. New techniques are being developed to form semiconductor structures in such a material.

After processing, devices must be placed in a package in order to be used in a circuit. Conventional techniques are often not suitable for this application. At elevated temperatures, common plastic encapsulants begin to decompose. New die attach materials and solders must be developed which can withstand the temperatures without melting or decomposing as well. Packages are also susceptible to intermetallic growth at the wirebonds at these temperatures. For standard wire and metallization materials, this can result in voiding, reduced bond strength, increased resistance, and eventually bond failure. Metallizations must also be tailored for high temperature use so as not to fail by electromigration. The use of new materials and fabrication techniques are currently being investigated for each of these packaging elements. Research is also being conducted on high temperature resins and inks to improve the mechanical, dielectric, and adhesion strength of organic printed wiring boards at elevated temperatures. High temperature solders that are lead-free and have good fatigue resistance at elevated temperatures are also being developed.

Finally, new materials and designs are needed for passive components. One of the most difficult technological barriers to the development of high temperature electronic systems is the need to create compact, thermally stable, high energy density capacitors. Both the capacitance and the dissipation factor of any capacitive component will often change significantly with increasing temperature. However, this is especially true for the large devices of several microfarads or more which will be required for power conditioning associated with electric motors and power switching applications.

Fundamental properties of traditional ceramic dielectric materials dictate that stability of capacitance with respect to temperature and voltage must be sacrificed to achieve large values of the dielectric constant. Temperature compensating capacitors, such as C0G or NP0, are highly stable with respect to temperature with a predictable temperature coefficient of capacitance, and few adverse effects of aging. However, they are usually made from mixtures of titanates with relatively low dielectric constants. General purpose ceramic capacitors, such as X7R, are made of barium titanate, and are ferro-electric with large dielectric constants, but as such they exhibit wide variations in capacitance with increases in temperature, particularly as the dielectric constant is increased. In addition, the leakage currents become unacceptably high at elevated temperatures making it more difficult for the capacitor to hold a charge. Capacitor manufacturers are actively involved in searching for materials that will provide increased capacitance and increased temperature stability. Glasses, glass-ceramics, and even high temperature polymers such as acrylates, polyimides, and fluoropolymers are being investigated as potential dielectrics.

1.4 Summary

There is a definite need for high temperature electronics, driven by the requirements of systems operating under harsh conditions. To achieve a viable industry capable of producing components to meet these needs, technical challenges largely in the area of materials and the integration of these materials into a device structure must be concurrently addressed. The technical challenges, though substantial, may be overshadowed by the perception of a limited market for such high temperature devices. Here again, lessons may be drawn from the experience of the conventional electronics market. The size of the electronics industry, in its current form, could not have been foreseen at the beginning of the development of modern semiconductors. What led to the development of the current large scale commercial industry was the creation of low volume, high performance products for the military, government, and private sectors. Now there is considerable support for high temperature electronics in the various military and civilian agencies of the U.S. government. These applications will be extensively discussed in this work. The consensus of experts in the field is that a military market is developing which is leading to a civilian market, a civilian market that will challenge, in some instances, the current low temperature electronic devices and thus penetrate into applications where conventional devices are currently used.

The balance of this work will address the critical issues, from the applications for high temperature electronics, the production of satisfactory materials, and the fabrication of components, both active and passive to the issue of thermal management at higher temperature. The work will conclude with a discussion of accelerated testing of these components. In a sense, there is no limit in sight for the operating temperature of semiconductors, and maximum high temperature for high temperature electronics has yet to be defined.

Chapter 2

SELECTION AND USE OF SILICON DEVICES AT HIGH TEMPERATURES

2.1 Basic Principles of Semiconductor Junctions

This section reviews important electrical properties of semiconductor materials and pn junctions with special emphasis on the temperature dependencies of these properties. We assume that the reader has some familiarity with elementary semiconductor device physics, but that it will be useful to focus on how important parameters vary with temperature for different semiconductors. The reader is referred to a text such as the ones by Streetman [1990] or Sze [1981] for detailed analyses.

Ordinarily, circuits must operate over a temperature range; high-temperature electronics must function not only at high temperatures but, for many applications, at very low temperatures as well. Variations with temperature in semiconductor properties can cause changes in circuit performance. As the temperature increases, these changes include increased reverse leakage current and decreased forward voltage drops in pn junctions, decreased conductivities, increased gain in bipolar junction transistors (BJTs), and decreased threshold voltages in metal oxide semiconductor field effect transistors (MOSFETs). At sufficiently high temperatures the pn junctions themselves cease to function.

Reductions in maximum operating voltages are an indirect result of high-temperature operation. Depending upon the temperature range over which the circuit must operate, these changes may range from minor to catastrophic. The variations may be dealt with by circuit design techniques using conventional silicon devices over narrower ranges with modestly high maximum temperatures (perhaps -55 to 160°C). For wide temperature ranges with very high operating temperatures (for example, -30 to 600°C in jet engine applications), it may be necessary to use devices manufactured from exotic semiconductors such as silicon carbide (SiC).

A fundamental property of crystalline solids is that electrons populate *energy bands*, which are separated by energy *bandgaps*, as shown in Figure 2.1. In this model, an insulator is characterized by a full *valence* band, separated from a *conduction* band by a large bandgap. An alternate picture is that all of the electrons are tightly bound in chemical bonds, so that several electron volts of energy are required to free an electron. In a metal, the bandgap will be very small, or zero, and electrons are free to move in the conduction band; not all of the electrons are tightly bound in chemical bonds. In a semiconductor, the bandgap is small enough (1 to 3 eV) so that electrons may be thermally excited from the valence band to the conduction band, or introduced into the conduction band by *doping* the semiconductor with impurities.

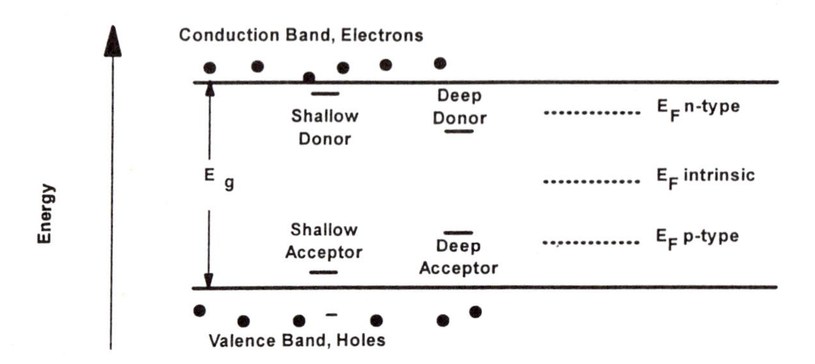

Figure 2.1 Sketch showing the relationship between the energy levels of the valence and conduction bands, the deep and shallow donor and acceptor levels, and the Fermi levels of n-type, intrinsic, and p-type semiconductors.

The energy bands can be more completely characterized by plotting them as functions not only of electron energy, but of momentum as well, as shown in Figure 2.2. In a *direct* bandgap semiconductor, a valence band energy maximum and conduction band minimum have the same momentum value. In an *indirect* bandgap semiconductor they have different momentum values. The size of the bandgap plays an important role in high temperature performance of semiconductors, with the direct/indirect distinction playing an important role as well. In thermal equilibrium, electron hole pairs are formed by thermal dissociation, and recombine by a variety of processes. In a direct semiconductor, such as gallium arsenide (GaAs), radiative recombination is possible between holes and electrons with equal momentum, a two-particle interaction. Recombination times are of the order of 10^{-9} s. In an indirect semiconductor, recombination must take place by more complicated processes, such as capture by a trap and subsequent recombination. Carrier lifetimes in silicon are of the order of 10^{-5} s.

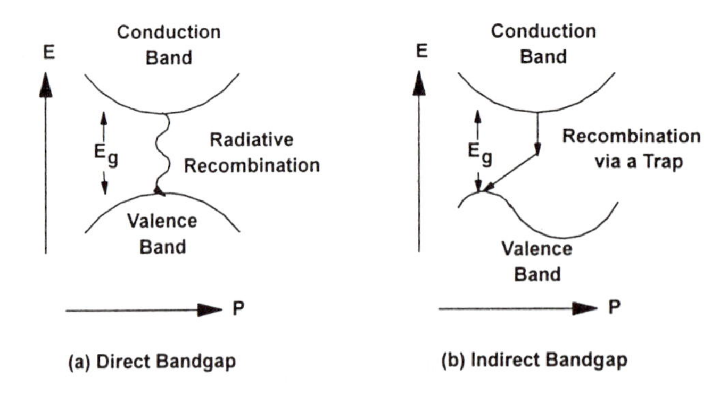

Figure 2.2 (a) Direct and (b) indirect bandgaps in energy-momentum space, showing how recombination takes place by different processes. Direct recombination is much more rapid.

The properties of various semiconductors are shown in Table 2.1 Silicon, germanium, and diamond are elemental semiconductors and SiC is a compound semiconductor from column IV of the periodic table. Binary and ternary compound semiconductors such as GaAs, gallium phosphide (GaP), indium phosphide (InP), gallium aluminum arsenide (GaAlAs), and gallium indium phosphide (GaInP) are formed from elements in columns III and V of the periodic table. At room temperature, the bandgaps range from about 1.1 eV for Si and 1.34 eV for GaAs, to about 2.9 eV for SiC, all at room temperature. Bandgap is also a decreasing function of temperature, as shown in Figure 2.3 for Si and GaAs [Muller and Kamins 1986].

Table 2.1 Properties of various semiconductors [Muller and Kamins 1986, Zipperian 1986]

	Si	GaAs	$Al_{.5}Ga_{.5}As$	GaP	SiC	Diamond
Bandgap (eV)	1.1	1.3	1.9	2.34	2.9	5.5
Direct / Indirect	I	D		I	I	
Breakdown Field (kV/cm)	250	300	500	--	2500	10,000
Thermal Conductivity (W/cm·K)	1.5	0.5	0.1	0.8	4.9	20
Electron Mobility, 25°C $cm^2/v\text{-}s$	1400	4000	3000	350	250	2200
Hole Mobility, 25°C $cm^2/v\text{-}s$	600	400	_	100	50	1600
Melting Temp. (°C)	1412	1238	1304	1470	2700	--
Max. Temp. for 10 V Reverse Breakdown (°C)	300	700		>1000	>1000	
Temperature for 0.1 A/cm^2 Leakage (°C)[4]	250	350	475	550	>900	

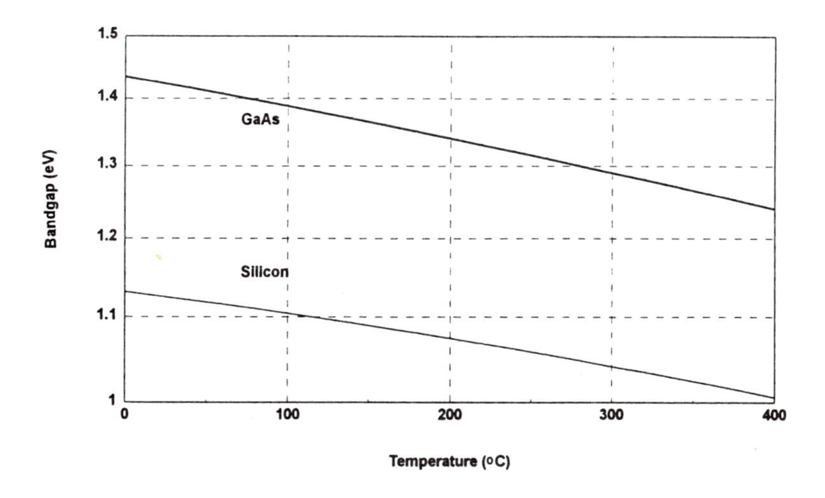

Figure 2.3 Temperature dependence of bandgap energy in Si and GaAs.

Conductivity in a semiconductor is provided by introducing electrical carriers into the valence or conduction bands — electrons in the conduction band or *holes* in the valence band. Physically, a hole is the absence of an electron, or a missing bond, but it behaves as a positively charged carrier. A pure semiconductor is called *intrinsic*. In an intrinsic semiconductor, free charge is produced solely by the thermal excitation of electrons into the conduction band. An equal number of holes are left behind in the valence band. The *intrinsic carrier density*, n_i, of a semiconductor depends on temperature and the bandgap of the semiconductor, as given by Equation 2.1 [Streetman 1990, Sze 1981],

$$n_i = \left(\frac{2\pi kT}{h^2} \right)^{3/2} (m_{dh} m_{de})^{3/4} e^{-E_g/2kT} = AT^{3/2} e^{-E_g/2kT} \qquad (2.1)$$

where k and h are Planck's and Boltzmann's constants, T is the absolute temperature, m_{de} and m_{dh} are the electron and hole effective masses, and E_g is the bandgap. (Discussion of effective masses is beyond the scope of this) The intrinsic carrier density vs. temperature is plotted in Figure 2.4 for Si, GaAs, GaP, and SiC. (The temperature dependence of E_g is taken into account for Si, GaAs, and GaP, but neglected for SiC.) The strong dependence of n_i on the bandgap is apparent, and its significance for high-temperature operation will become clear in the next few paragraphs.

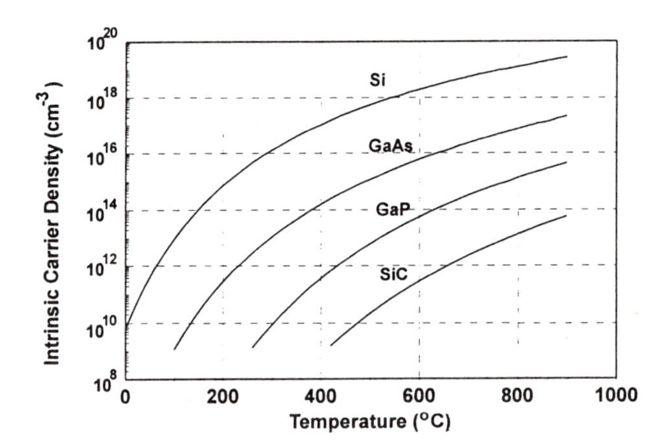

Figure 2.4 Intrinsic carrier density for Si, GaAs, GaP [Thurmond 1975], and SiC (estimated). The variation in bandgap with temperature is taken into account for Si, GaAs, and GaP.

Carriers may also be introduced by means of impurities or *dopants*. If a column V atom such as aluminum (Al) is substituted for a silicon atom in a silicon crystal, there is an unbonded electron. Similarly, if a column III atom, such as boron (B), is substituted there is a missing bond in the crystal lattice, leaving a *hole.* The binding energy of the electron or hole can be characterized by an energy level in the band diagram, as shown in Figure 2.1. If the energy level is close to the conduction band for a column-V impurity, the electron is thermally ionized, contributing a carrier to the conduction band. Holes may be similarly ionized and move into the valence band. These dopants are called *donors* (contributing electrons, with a density usually called N_d) or *acceptors* (accepting electrons or contributing holes, with a density usually called N_a). The doped material is said to be *n-type* if it has excess electrons and *p-type* if it has excess holes. Ionization energies of typical dopants in Si or III-V compounds are of the order of 0.01 eV, so they are readily ionized at room temperature, which is about 0.025 eV. However, dopants in SiC have ionization energies of more than 0.1 eV, so they are only partially ionized at room temperature, leading to variations in carrier density with temperature.

In a doped semiconductor, there are both *majority* and *minority* carriers. In n-type material, electrons are the majority carrier and holes are the minority carrier; the reverse is true in p-type material. Their respective densities are related by a mass-action equation:

$$n_e n_h = n_i^2 \tag{2.2}$$

where n_e and n_h are the electron and hole densities. Majority carrier densities are far larger than minority carrier densities in equilibrium.

The electronic properties of semiconductor devices depend on dopants. Dopant densities can be limited by their solid solubility limits. Solid density is of the order of 10^{22} cm^{-3}, and the solubility limits tend to be in the range of 10^{18} to 10^{20} cm^{-3}. Comparing these values with the intrinsic carrier densities plotted in Figure 2.4, it becomes clear that for sufficiently high temperatures, dopant effects will be swamped by the intrinsic carrier density. This shows the potential advantage of wide-bandgap semiconductors for very high temperature operation.

A final important parameter is the value of an energy level called the *Fermi level.* For an intrinsic semiconductor, the Fermi level lies about midway between the valence and conduction bands, with a value, E_i. For n-type material it is closer to the conduction band, and for p-type material it is closer to the valence band, as shown in Figure 2.1 The value of the Fermi level is related to the dopant density:

$$E_F = E_i \pm kT\ln\frac{N_a/d}{n_i} = E_i + \frac{E_g(T)}{2} + kT\ln\frac{N_a/d}{AT^{3/2}} \tag{2.3}$$

where the "+" sign is for n-type and the "-" sign for p-type doping; A is from Equation 2.1. This temperature dependence is important for field effect transistor (FET) operation.

The most basic device is a diode or pn junction. An abrupt pn junction is shown in Figure 2.5. The left and right sides of the junction are uniformly doped n-type and p-type, with densities N_d and N_a. The carriers are mobile, so the dopant density gradient across the junction leads to diffusion of carriers, which in turn causes the buildup of *built-in electric fields* and a *contact potential.* In equilibrium, the *diffusion currents* of electrons to the p-side and holes to the n-side are countered by *drift currents* of electrons and holes back to the n-side and p-sides. Note that majority carriers on each side of the junction are the source of diffusion current, while minority carriers are the source of charge for drift current. While the majority carrier densities are largely fixed by the dopant densities, Equations 2.1 and 2.2 show that the minority carrier density varies exponentially with temperature. Electrical behavior that depends on minority carriers varies significantly with temperature. Arguments can be made to show how the drift and diffusion currents are balanced when the contact potential is given by:

$$\psi_0 = \frac{kT}{q}\ln\frac{N_a N_d}{n_i^2} = \frac{kT}{q}\ln\frac{N_a N_d}{A^2 T^3} + \frac{E_g}{q} \qquad (2.4)$$

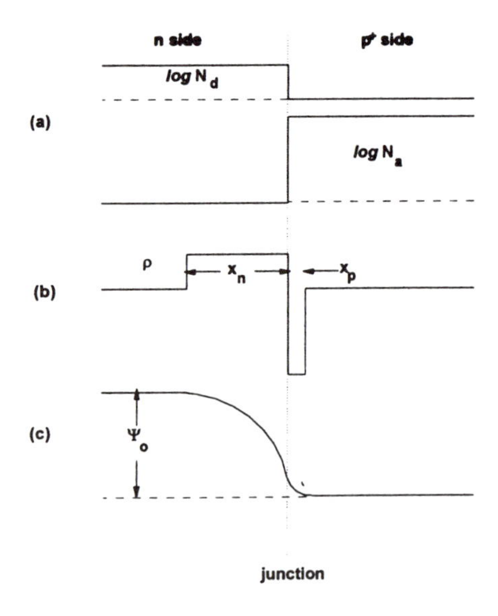

Figure 2.5 An abrupt pn junction. (a) Dopant density vs. position for heavy (p^+) p-doping. (b) Depletion region. (c)Built-in potential.

This shows that the contact potential increases approximately linearly with temperature, and linearly with semiconductor bandgap. The contact potential "sweeps" mobile carriers out of a *depletion region*, leaving bare ionic charge, as shown in Figure 2.5b. When the junction is placed in a circuit, the voltage applied to the junction may add to or subtract from the contact potential. In the case of reverse bias, it adds to the contact potential, and in forward bias it subtracts from it. The width of the depletion region on each side of the junction is given by:

$$x_{p/n} = \left\{ \frac{2\epsilon(\psi_0 - V)}{q} \left[\frac{N_{d/a}}{N_a(N_a + N_d)} \right] \right\}^{1/2} \tag{2.5}$$

where $x_{p/n}$ is the width on the p- or n-side of the junction, V is the applied voltage defined positive when the drop is from the p-side to the n-side (forward bias), $N_{d/a}$ is the corresponding donor or acceptor density, and ϵ is the dielectric constant.

The width of the depletion region on the n-type side of the junction helps determine the reverse breakdown voltage of the junction, which indirectly affects high-temperature device operation in an important way. Reverse breakdown is caused by an electron avalanche breakdown on the n-side (p-material does not break down this way). An avalanche occurs when the electric field is so high that electrons can gain an energy, E_g, between lattice collisions, and then ionize electrons from the valence to the conduction band. For high enough fields, this process exponentiates into avalanche breakdown. Maximizing the breakdown voltage is done by heavily doping the p-side, and doping the n-side of the junction as lightly as possible to maximize x_n and minimize the E-field. However, at higher temperatures, the lowest useful dopant density is bounded from below by the increased intrinsic carrier density.

Breakdown voltage vs. dopant density is plotted for Si, GaAs, GaP, and SiC in Figure 2.6. An estimate of the maximum operating temperature for a semiconductor material can be made by first requiring a minimum reverse breakdown voltage, and then assuming (somewhat arbitrarily) that the dopant density must be at least ten times larger than the intrinsic density of the semiconductor to have a functional junction. For example, for a silicon diode to have a 10 V breakdown, the n-side may be doped no more heavily than 1.4×10^{17} cm^{-3}, as shown in Figure 2.6. The maximum operating temperature for a 10 V silicon technology is the temperature for which the intrinsic density is 1.4×10^{16} cm^{-3}, or 300°C. Similar estimates of maximum operating temperature are made in Table 2.1 for GaAs, GaP, and SiC. This effect is important in the choice of material for very high temperature operations, because almost every circuit depends upon reverse-biased junctions to limit currents or for device isolation. This effect is particularly important for power devices such as may be required for mechanical actuators.

Forward and reverse current density in the junction are given by the diode equation, 2.6a. Some high-temperature effects can be illustrated by casting the simplest form of Equation 2.6a in somewhat unconventional forms, 2.6b and 2.6c:

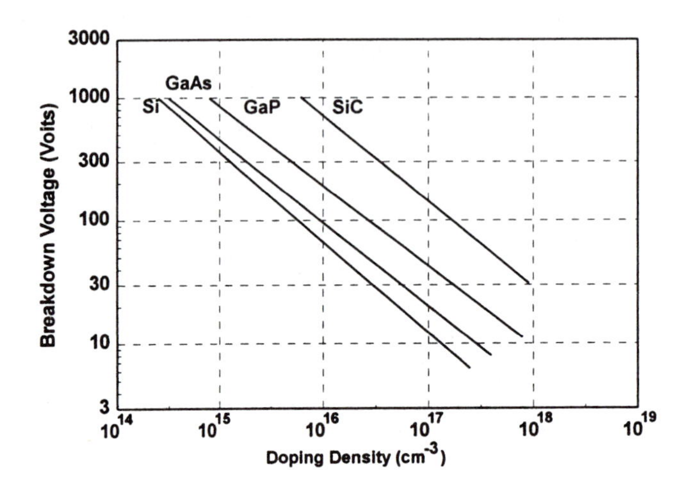

Figure 2.6 Estimates of the reverse breakdown voltage in various semiconductors as functions of n-type doping in abrupt p^+n junctions.

$$j = J_0\left(e^{qV/kT} - 1\right) = q\left(\frac{D_h}{L_h}p_n + \frac{D_e}{L_e}n_p\right)\left(e^{qV/kT} - 1\right) \tag{2.6a}$$

$$\begin{aligned} j &= qAT^3\left(\frac{D_h}{L_hN_d} + \frac{D_e}{L_eN_a}\right)e^{-E_g/kT}\left(e^{qV/kT} - 1\right) \\ &\approx qAT^3\left(\frac{D_h}{L_hN_d} + \frac{D_e}{L_eN_a}\right)e^{(qV-E_g)/kT} \qquad qV/kT \gg 1 \end{aligned} \tag{2.6b}$$

$$j = qAT^3\left(\frac{L_h p_n}{\tau_h} + \frac{L_e n_p}{\tau_e}\right) e^{-E_g/kT}\left(e^{qV/kT} - 1\right) \qquad qV/kT \ll -1 \qquad (2.6c)$$
$$\approx qAT^3\left(\frac{L_h p_n}{\tau_h} + \frac{L_e n_p}{\tau_e}\right) e^{-(E_g/kT)}$$

where j is the current density, $D_{h/e}$ are hole/electron diffusion constants, $L_{h/e}$ are hole/electron diffusion lengths, $\tau_{h/e}$ are hole/electron recombination times, p_n is the hole density on the n-side of the junction, and n_p is the electron density on the p-side of the junction. Equation 2.6b is obtained from Equation 2.6a using Equations 2.1 and 2.2. Equation 2.6c is obtained by using the relationship $L = (D\tau)^{1/2}$.

Analysis of Equation 2.6b shows two things about behavior under forward bias as operating temperature increases. First, for semiconductors with different bandgaps, the tendency will be for the forward voltage drop to be larger in the larger bandgap material for the same current density. A strategy for coping with high-temperature effects by using wide bandgap semiconductors will lead to greater power dissipation. Second, for a given forward voltage ($qV/kT \gg 1$), the junction current increases with increasing temperature. While this may be slightly counter-intuitive, given the exponential form of the equation, physical intuition supports this because forward current is due to diffusion of energetic majority carriers over the (depressed) built-in potential, whose density increases with temperature. It is fairly straightforward to show that:

$$\frac{\partial V}{\partial T} = \frac{(qV - E_g) - 3kT}{qT} + \frac{1}{q}\frac{dE_g}{dT} \qquad (2.7)$$

$$\frac{\partial j}{\partial T} = \left(\frac{3}{T} + \frac{E_g - qV}{kT^2}\right) j \qquad (2.8)$$

For silicon junctions, the decrease in forward voltage can be readily estimated from Equation 2.7 to be about 2 mV/°C : $qV\text{-}E_g \sim 400$ mV, $3kT \sim 100$ mV, and dE_g/dT is about 0.3 mV/°C from Figure 2.3. Similarly, the coefficient of j on the right hand side of Equation 2.8 is clearly positive.

Equation 2.6c illustrates three points about how reverse bias leakage currents ($qV/kT \ll -1$) vary with temperature and bandgap. First, reverse current increases more than exponentially with temperature. Not only does this mean that the junction will block currents poorly at high temperatures, but the junction temperature may also undergo thermal runaway due to internal power dissipation. Second, the reverse current is lower for larger bandgap semiconductors, due to the lower carrier density. Third, and least well-recognized, is that, due to the dependence on the recombination time, direct bandgap semiconductors tend to have larger leakage currents than indirect bandgap semiconductors with similar bandgaps. At the level of detail that this calculation has been performed, similar performance could be expected for Si and GaAs junctions, despite the difference in their bandgaps.

The increase of leakage current with operating temperature is probably the most important cause of circuit failure at elevated temperatures. Table 2.1 shows temperatures measured by Zipperian [1986] at which a reverse leakage current density of 0.1 Amp/cm^2 is reached for p^+n junctions in several semiconductors. (The notation p^+ means heavily doped p-type.) Essential functions performed by reverse-biased junctions are device isolation and rectification. Loss of isolation or rectification will cause circuits to fail at elevated temperatures. For a given semiconductor material, design strategies should be chosen to minimize leakage currents and their impact. An example of this is the use of silicon-on-insulator (SOI) integrated circuits, as described in Section 2.4. Semiconductors with wider bandgaps than Si, such as GaAs and other III-V compounds and SiC, have been investigated for applications above 200°C. Simple GaAs devices have been found to perform little better than Si devices at high temperature partly because GaAs has a direct bandgap and partly because better design strategies can be employed with Si.

The final effect considered in this section is the change in bulk resistivity of semiconductors with temperature. This has an effect on parasitic resistances in devices, and can therefore affect circuit operation. The two main contributions to the bulk resistance of a semiconductor are phonon (lattice vibration) and impurity scattering (mostly by dopants) of the charge carriers. Phonon scattering increases with temperature as $T^{3/2}$, while impurity scattering decreases as $T^{-3/2}$. The importance of these effects are device- and application-dependent.

2.2 Silicon Bipolar Transistors

This section considers the operation of BJTs, emphasizing their temperature-dependent aspects, and briefly reviews work from the literature addressing BJT use at very high temperatures.

The basic structure of a BJT, in this case a pnp transistor, is shown in Figure 2.7. (A pnp

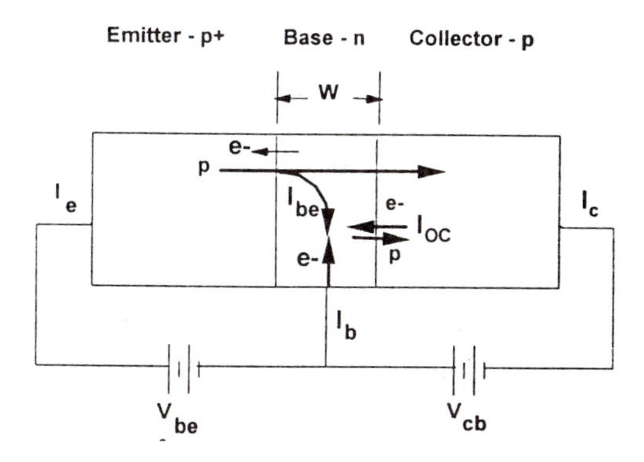

Figure 2.7 A pnp transistor biased in normal mode, showing the principal components of current. The width of the base region is W.

transistor is usually chosen for this discussion to avoid the different issue of how base current changes from electron to hole carriers at the base contact. However, the operation of an npn transistor can be understood by simply reversing junction polarities and interchanging the roles of electrons and holes.) The transistor has a p^+ doped emitter and collector separated by a "thin" n-type base with width $W << L_h$. It is biased in a common base configuration. The base-emitter junction is forward biased, and the base-collector junction is reverse biased. The different components of electron and hole current flow are shown by the arrows. Forward bias drives a large hole current into the base from the emitter. Because the base thickness is much less than a hole diffusion length, a fraction, the *current transfer ratio*, $\alpha \cong 1$, of the hole current reaches the base-collector depletion region. The holes are then collected into the collector by the built-in potential of the reverse-biased junction. So, $I_c \approx I_e$. However, a fraction, $1-\alpha$, of the holes recombine with electrons in the base, leading to a base current, I_{be}. A second component is supplied to the base current by I_{OC}, the thermally generated current in the reverse-biased junction.

Perhaps the most important property of a transistor is that the base current can control the emitter and collector currents. A simple physical model shows this. If the base is uniformly doped, it must have a zero electric field between the junctions. This means that the number of holes, Q, in transit across the base at any one time must be neutralized by electrons supplied mostly from the base electrode. If the supply of electrons is restricted, for example by a bias network, the hole injection rate must adjust to match it. The average hole transit time across the base is τ_t. The average electron life-time is the recombination time, τ_e. Then, neglecting the reverse saturation current from the collector and the electron current from the base to emitter, the ratio of the collector current to the base current is given by:

$$\frac{I_c}{I_b} = \beta = \frac{\tau_e}{\tau_t} \tag{2.9}$$

and β is called the *forward current gain*. More complete calculations of β retain an important term containing the recombination time.

This simple model shows the origin of four important temperature effects in BJTs. First, because recombination time is temperature-dependent, the forward current gain depends upon temperature. An example of how the gain varies with temperature is shown in Figure 2.8. Second, I_c vs. V_{be} is also a function of temperature, as shown in the last section. This affects both the DC bias points of the transistor and its transconductance. Third, the reverse leakage current, I_{cb}, increases with temperature and may also affect bias points, in effect reducing the base current available to control the collector current. And, fourth, the resistivity of Si itself is temperature-dependent.

Some of these effects are evident in a small signal model of a transistor (see, for example, Gray and Meyer [1993]), shown in Figure 2.9. The transconductance, g_m, has explicit and implicit temperature dependence:

$$g_m = \frac{\partial I_c(T)}{\partial V_{be}} \approx \frac{\partial I_e(T)}{\partial V_{be}} = \frac{qI_e(T)}{kT} \tag{2.10}$$

The input resistance, $r_\pi = \beta / g_m$, and $r_\mu = \beta\, r_o$ are likewise temperature dependent; r_o is the large-signal output resistance, given by the slope of I_C-V_{CE}.

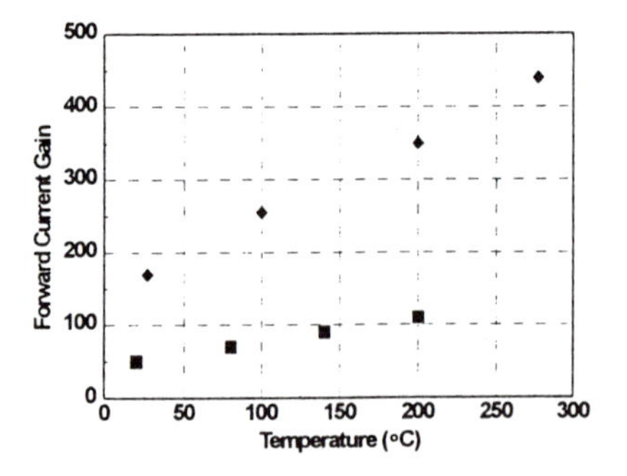

Figure 2.8 Variation of forward current gain with temperature for a 2N6023 power transistor (squares, from [Bromstead 1991]) and for a BiCMOS wafer (diamonds, [Dickman et al.1994]).

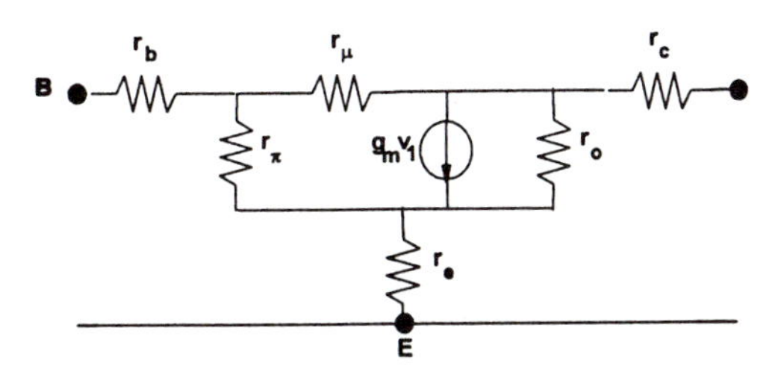

Figure 2.9 Small signal transistor model showing resistive parasitics and other temperature dependent elements.

Plots of I_C-V_{CE} from an npn BJT taken at 300 °K (27°C) and 550 °K (277 °C) are shown in Figure 2.10; degradation of the transistor characteristics is evident. The sharp knee in the curve and linearity above the knee is lost at high temperatures. The output resistance, r_o, is no longer a constant over a wide range of conditions.

Parameter variations such as these make it difficult to design circuits that operate over a wide temperature range. In integrated circuits, leakage through the substrate between elements must also be taken into account. Some of these additional leakage paths can be controlled using "dielectrically isolated" technologies, some of which are described in Sections 2.4, 2.6, and 2.7. Conventional bipolar circuits can be expected to fail at temperatures somewhat above 150°C, and almost certainly below 200°C (see, for example, work by Palmer and Heckman [1978]).

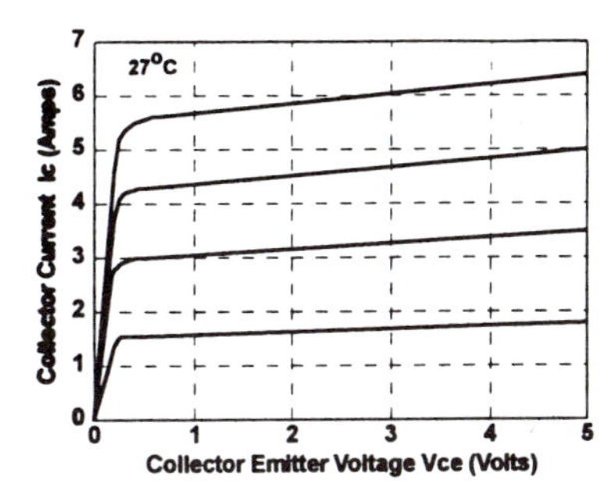

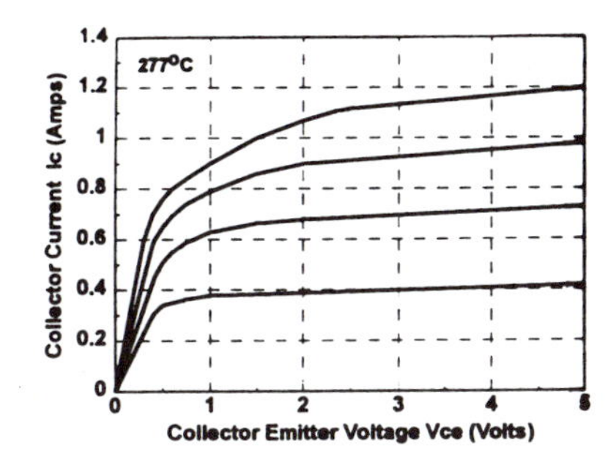

Figure 2.10 V_{CE} -I_C characteristics for a BiCMOS transistor at 27 °C (left) and at 277 °C (right). Notice the loss of linearity in the saturation region, and the dramatic reduction in I_C [Dickmen et al. 1994].

There has been little work to develop high-temperature-capable analog circuitry using bipolar technology. One notable advance, however, is a quad op-amp integrated circuit developed by Beasom and Patterson [1982] which operated successfully over a temperature range of 0-300°C. They chose a dielectrically isolated process to minimize leakage between elements. Special transistor layouts were required, which helped minimize the effect of surface leakage. Leakage currents (I_{CB} and I_{CES}) varied by five orders of magnitude; V_{BE} varied by 0.4 volts, β varied by a factor of two, and the values of diffused resistors (doped silicon) doubled from 0 to 300°C. Considerable attention was paid to compensating for these variations. Because of the large leakage currents, dc bias levels were made large. It was found that changes in V_{BE} could be offset using changes in diffused resistors.

Digital bipolar circuitry for high-temperature use has also received only limited attention, perhaps due to the dominance of CMOS circuitry. Prince et al. [1980] evaluated the high-temperature performance of TTL small scale integrated circuits. They found that standard-process TTL functioned to temperatures of 250°C, although with reduced switching speed, noise margin, and output current. The failure was due to a decrease in the high-level output voltage. Dielectrically isolated TTL suffered similar degradations, but remained functional to 325°C. Sunayama et al. [1991] describes an integrated injection logic (IIL) inverter circuit that was functional to 385°C. Migitaka and Kurachi [1994] describe IIL ring oscillators that functioned to 454°C.

2.3 Silicon MOSFETs

The most widely used integrated circuits are based on complementary metal oxide semiconductor (CMOS) technology. Discrete MOSFETs are also widely used for many applications. This section reviews MOSFET operation, emphasizing temperature-dependent aspects. Use of MOS and CMOS devices for high-temperature applications has been studied intensively, and some important findings from the literature are reviewed here.

A simplified drawing of *enhancement mode* n-channel and p-channel MOSFETs (nMOSFET and pMOSFET) is shown in Figure 2.11. For the nMOSFET, n-type *source* and *drain* regions are implanted or diffused directly into a lightly doped p-type substrate. The source is typically grounded, and a positive voltage is applied to the drain; the drain-substrate junction is reverse-biased. A silicon dioxide *gate oxide* is grown over the surface of the silicon between the gate and drain, and a conducting *gate* electrode is deposited on top of the gate oxide. The gate, gate oxide, and underlying silicon form a capacitor structure. Applying a positive voltage (depositing positive charge) on the gate draws electrons to the surface of the silicon. When the voltage exceeds a threshold voltage, V_T, it induces an n-type *channel* between the source and drain. Current can then flow in the field-induced n-type silicon between the source and the drain. The conductance of the channel is controlled by the gate voltage, V_G. Operation of a p-channel MOSFET is similar, with dopings reversed, and with the gate and drain voltage polarities reversed. In order to have both nMOS and pMOS devices on the same substrate, one or the other must be placed in a *well* with the appropriate doping polarity. In our example, the pMOSFET is in an n-well, an n-doped region. The substrate is typically grounded, but bias, V_{body}, is sometimes applied.

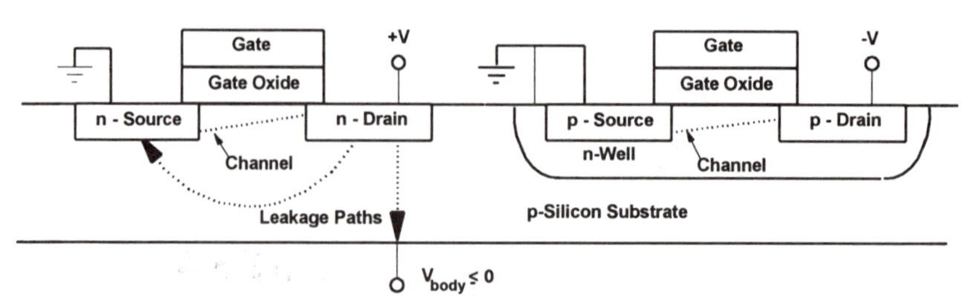

Figure 2.11 Simplified sketch on n-channel and p-channel MOSFETs on a p-type silicon wafer. Note leakage paths shown with dashed arrows at the reverse-biased junctions.

There are also so-called depletion mode devices, which are turned off by applying a gate voltage. In a depletion mode device, the channel is lightly doped, and the carriers are forced out by the gate potential.

Expressions for V_T and for the current-voltage characteristics of MOS transistors are derived in textbooks such as those by Streetman [1990] and Sze [1981] but the derivations are beyond the scope of this book. V_T is given by:

$$V_T = \Theta_{ms}(T) - \frac{Q_i}{C_i} - \frac{Q_d(T)}{C_i} + 2\Phi_F(T) \tag{2.11}$$

where

$$\Phi_F = E_F - E_i = \frac{E_g(T)}{2} + kT\ln\frac{N_a}{AT^{3/2}} \tag{2.12}$$

and θ_{ms} is the metal semiconductor work function, C_i is the capacitance associated with the gate oxide, Q_i is the charge on the gate capacitor, and Q_d is the charge moved in the channel region. Equation 2.12 is largely from Section 2.1. Numerically, a typical value of dV_T/dT in a silicon MOSFET is -2 *mV/°C* to -4 mV/°C, depending upon the thickness of the gate oxide and doping. The small-signal current-voltage relationship is given by:

$$I_D \approx K\mu_c(V_G - V_T)V_D, \qquad V_D \ll V_G - V_T \tag{2.13}$$

Also,

$$I_D \approx \frac{1}{2} K \mu_c (V_G - V_T)^2 \qquad V_D \gg V_G - V_T \tag{2.14}$$

where I_D is the drain current; V_G, V_T, and V_D are the gate, threshold and drain voltages; μ_c is the mobility of charge carriers in the channel; and K is a largely geometrical factor. The channel mobility is different from the bulk mobility because the carriers interact strongly with the silicon-to-silicon-dioxide interface. Like the bulk mobility, the channel mobility decreases with temperature. From Equation 2.13, note that I_D is linear in either V_G or V_D if the other is held constant, for "small" values of V_D. From Equation 2.14, note that I_D saturates for "large" V_D, and that it decreases with temperature through the temperature dependence of μ_c. The simplest small-signal model of a MOSFET, shown in Figure 2.12, displays its principal temperature dependence, which is in the transconductance:

$$g_m = \frac{\partial I_D}{\partial V_G} = k \mu_c(T) \tag{2.15}$$

MOSFETs also suffer from two significant parasitic leakage current paths, as shown by the dashed arrows in Figure 2.11. The drain is reverse-biased with respect to the substrate. Reverse leakage currents flow into the substrate, where they may go to either the grounded source or the substrate supply. They are the main source of *subthreshold* current in MOSFETs. These leakage currents increase exponentially with temperature, as discussed in Section 2.1. At high enough temperatures, the leakage currents become so large that I_D is no longer practically controlled by the gate.

The characteristic curve for a discrete MOSFET (Motorola 2N4351) [1981] is shown in Figure 2.13. It is difficult to identify V_T from a characteristic curve, but, as a practical matter, it is often defined as the gate voltage for which I_D is 1% of I_{sat}. Using this definition, the shift in V_T is quite large, about -10 mV/°C over the 25 to 250°C temperature range. The leakage currents rise dramatically with temperature, and the saturation current and transconductance decrease significantly.

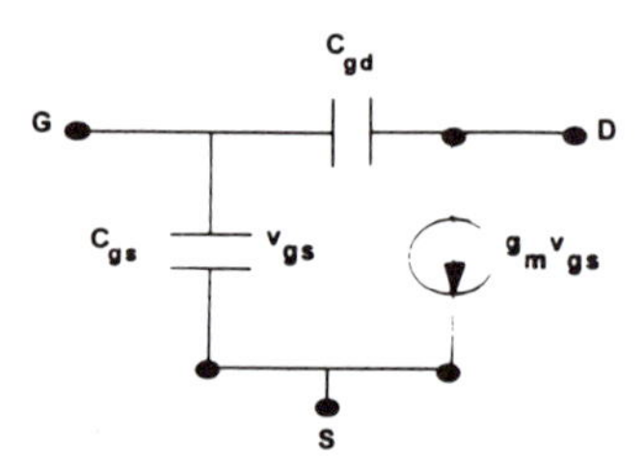

Figure 2.12 Simplest small-signal MOSFET model, showing temperature dependence of g_m.

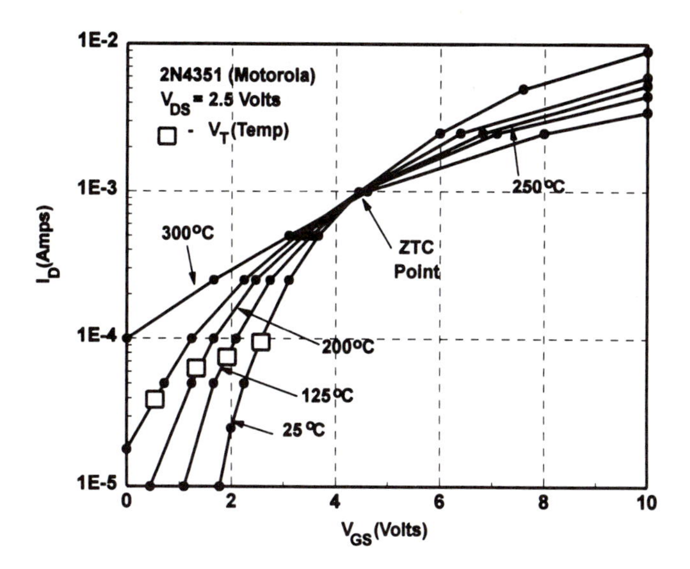

Figure 2.13 Drain current vs. gate voltage with constant drain voltage for a Motorola 2N4351. V_T, as defined in the text, is shown for each curve with an open square. Note the decrease in V_T, the increase in leakage current, and the decrease in saturation current with temperature. The zero temperature coefficient point is about 5V [Veneruso 1981].

The curves also show the zero temperature coefficient (ZTC) point, an important property of MOSFETs for high-temperature operation of some analog circuits. For $V_{GS} = V_{ZTC}$, I_D is independent of temperature. This allows the use of a bias point that does not vary with temperature. Such a property is rare in semiconductors. There is no similar property for BJTs.

In integrated circuits, parasitic current paths also exist between devices. A particularly dangerous path can be seen in Figure 2.11 between the source of the pMOSFET and the drain of the adjacent nMOSFET. The source, well, substrate, and drain form a pnpn structure, which is the basic structure of a thyristor. Under normal conditions, currents are blocked by reverse-biased junctions in this path. Under some abnormal conditions, all three junctions can become forward biased. The conventional analysis of how this occurs is illustrated using Figure 2.14. The four regions are decomposed into coupled npn and pnp transistors. The current transfer efficiencies are α_1 and α_2, and the reverse collector currents are I_{CO1} and I_{CO2}. Then:

$$I_{C1} = \alpha_1 I + I_{CO1} = I_{B2} \tag{2.16}$$

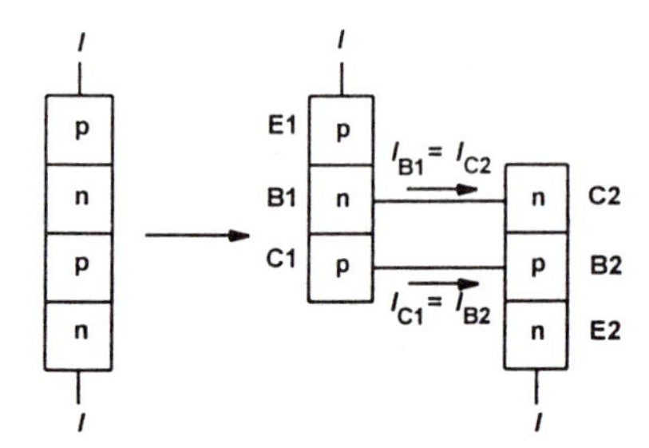

Figure 2.14 Model of thyristor current flow by decomposition into coupled transistors.

$$I_{C2} = \alpha_2 I + I_{CO2} = I_{B1} \qquad (2.17)$$

Noting that $I = I_{C1}+I_{C2}$, solving for I yields:

$$I = \frac{I_{CO1} + I_{CO2}}{1 - (\alpha_1 + \alpha_2)} \qquad (2.18)$$

This shows how the thyristor current can become large if the current transport through the middle n and p regions is efficient enough. This is important at high-temperatures because α depends directly on carrier recombination times, which increase with temperature. Latchup can be dealt with by proper design layout rules [Ambaum 1994] which keep α small; conventional CMOS devices may have an inadequate design margin for high-temperature operation. The various forms of dielectric isolation also reduce latchup danger. One form fills trenches around the FETs with a dielectric, either grown or deposited silicon dioxide. A second important form of isolation uses a SOI substrate, as shown in Figure 2.15. The dielectric layer under the FETs greatly reduces drain-source leakage and latchup paths. SOI technology is discussed in more detail in the next section.

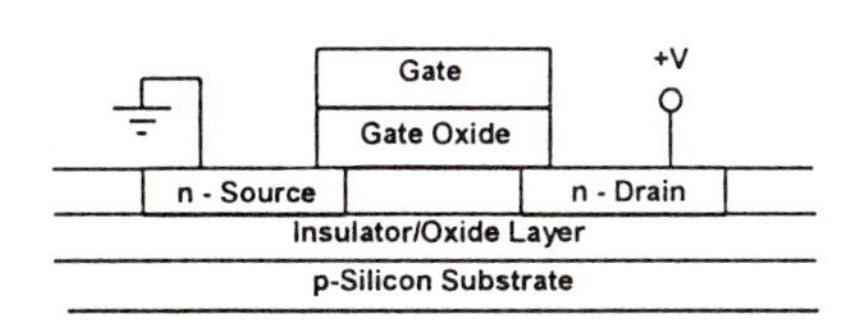

Figure 2.15 Sketch of a MOSFET in an SOI substrate. Leakage currents and latchup paths are largely eliminated.

The use of MOS technologies at temperatures above 125°C has received significant attention [Shoucair 1991]. Commercial MOS device technologies can be used successfully at temperatures significantly above 125°C, as described later in this chapter. Typically, a conventional CMOS device is functional to about 160°C. Dielectrically isolated technologies that are not specially designed for high-temperature operation are often functional to 200°C; and specially designed SOI may function to 300°C or more. There are reports of SOI CMOS that is functional to more than 500°C [1994].

There has also been recent work on power devices called MOS controlled thyristors (MCTs) which are switched using an MOS effect. Temple and co-workers at Harris Power R&D have developed MCTs which show good voltage standoff to 250 °C [Arthur 1992]. Designs for power ICs for operation up to 200°C have also been considered [Marshall 1992].

2.4 Silicon-On-Insulator Technology

Silicon integrated circuits are typically designed and processed to perform reliably at temperatures up to 70°C, for industrial applications, and 125 C for military applications. Above these design limits two things may happen: (1) the semiconductor characteristics can change sufficiently, that specified circuit parameters will not be met, or (2) permanent changes may occur that cause the circuit to cease to be functional. For example, an NMOS threshold voltage will decrease with temperature until the transistor is always on and excessive leakage current flows, compromising the functionality of the circuit. Similarly, the electromigration process increases with temperature and can result in open or short circuits if design limits are exceeded. These problems can be minimized through careful design and process schemes, but often at the expense of other desired characteristics and only to a certain limit.

The fundamental limitation of silicon integrated circuits is the excessive generation of carriers by the thermal energy. As mentioned in the previous sections, the number of carriers increases exponentially with temperature, as given by the intrinsic carrier density [Grove 1967]:

$$n_i(T) = C\,T^{3/2}\exp(-E_g/2kT),cm^{-3} \qquad (2.19)$$

where C is a constant, T is the temperature in ^{o}K, k is Boltzmann's constant, and E_g is the energy bandgap, which is temperature-dependent and given for Si by [Sze 1981]

$$E_g = 1.17 - 4.7x10^{-4}\, T^2/(T + 636), eV \tag{2.20}$$

At a sufficient temperature, thermally generated carriers can exceed doping levels until the semiconductor junction is no longer operative. This temperature is the upper limit for the specific semiconducting material. For silicon devices, the practical temperature limit is on the order of 400°C [Dreike et al. 1994], although the theoretical limit is higher. Below this threshold the excess carriers result in increased junction leakage current, which alters circuit operation. Increasing the doping density decreases this leakage current somewhat; however, increasing the doping decreases the breakdown voltage. Thus, there may need to be a trade-off between high-temperature performance and the breakdown voltage.

This leakage current is the most significant problem for high-temperature silicon integrated-circuit operation. The leakage current for a silicon n^+p junction is given by [Grove 1967]

$$I(T) = qA\sqrt{\frac{(D_n/\tau_n)n_i^2(T)}{N_a}} + \frac{qAn_i(T)W}{t_e}, ampere \tag{2.21}$$

where A is the area of the junction in cm^2, q is the electronic charge, D_n is the electron diffusion constant, τ_n is the electron lifetime in p-type silicon, W is the depletion width, and $\tau_e = (\tau_n + \tau_p)/2$ is the effective lifetime of the thermal generation process in the depletion region. From room temperature to about 125°C, the leakage current is dominated by the second term — the generation process. Above 150°C, the leakage current is dominated by the first term — the diffusion process. Since I_α a n_i^2 (T) is in that range, the leakage current increases exponentially with temperature.

From Equations 2.19 and 2.21 it is clear that one means of reducing the leakage current is to use higher bandgap material, such as SiC, diamond, or GaAs. Another way to reduce the leakage current is to decrease the junction area, which is in the numerator of Equation 2.21. An effective means of doing this, especially for high-temperature operation, is through the use of SOI technology. An example of the leakage current dependence on temperature, comparing a silicon NMOS and an SOI NMOS transistor, is shown in Figure 2.16 [Brusius et al. 1994]. Leakage currents at high temperatures can be two to four orders of magnitude lower for SOI devices than for bulk devices, depending upon the device design [Colinge 1991]. This increase is often approximated by doubling the current for every 10°C temperature rise for the bulk silicon n+p junction.

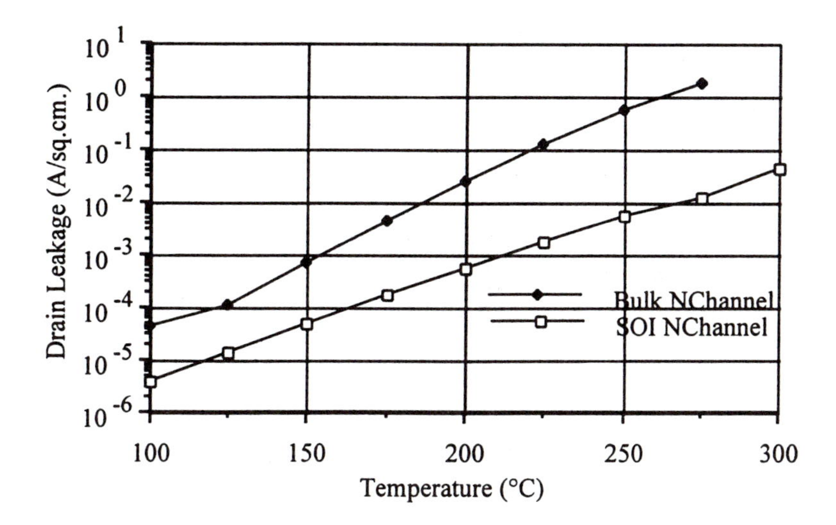

Figure 2.16 Leakage current vs. temperature for bulk and SOI NMOS transistors.

Because of the low leakage current, engineers have successfully designed and fabricated digital and analog circuits for high-temperature operation. SOI CMOS 4K SRAMs have been demonstrated to be functional at 300°C [Eggermont et al. 1994]. SOI CMOS operational amplifiers have been evaluated to 300°C [Grove 1967]. Device testing has demonstrated predictable behavior up to 500°C [Tyson and Grzybowski 1994].

The use of SOI technology offers other features that are advantageous for high-temperature operation, as well as for overall electrical performance. In addition, SOI integrated circuit processing can use the same equipment and processes that conventional silicon integrated circuits use, and is thus highly integratable and manufacturable.

2.4.1 Fundamentals of SOI technology

The SOI wafer contains a thin silicon layer over a thick buried oxide (BOX) layer on a bulk silicon substrate. This structure can be used with dielectric isolation (a CVD-oxide refill trench) in a standard double-metal CMOS process, as shown in Figure 2.17 (not to scale). It is clear that the isolation junction, such as the N-well in Figure 2.17, is eliminated and the junction area is greatly reduced when the junction bottoms out to the buried oxide. For the SOI CMOS transistors shown, the dimensions are: gate oxide ≈ 0.035 μm, top silicon ≈ 0.3 μm, and buried oxide ≈ 0.4 μm for a 10V linear application. Thinner gate oxides, such as .012 μm or .015 μm, have been used for 5 volt applications. The top silicon thickness can range from 0.05 μm to 2.0 μm, and the buried oxide thickness can range from 0.08 μm to 4 μm in practice, depending on the system requirements. A top silicon thickness of less than 0.1 μm is often used for fully depleted devices. For low-voltage applications, the thin buried oxide could be less than 0.2 μm. Not shown in Figure 2.17 is the lateral body tie/contact for each SOI transistor in a partially depleted transistor. The leakage current at high temperature would be reduced by two orders of magnitude as the silicon thickness is reduced in SOI, as shown in Figure 2.16 above 250 °C.

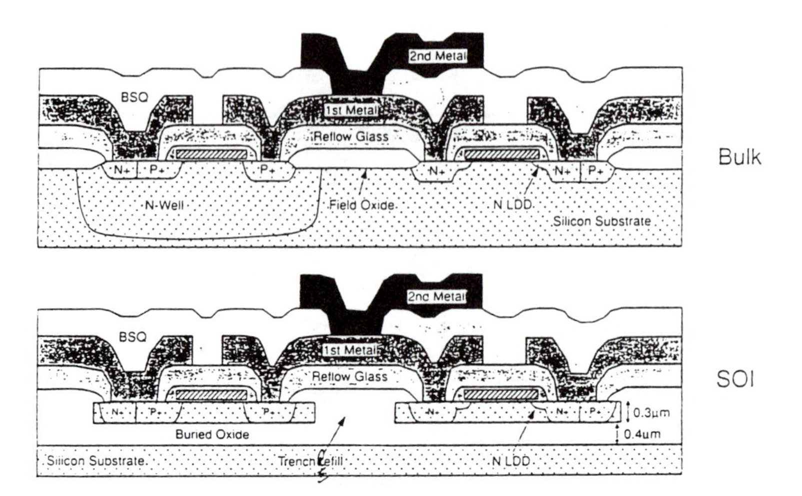

Figure 2.17 Cross-section of bulk and SOI CMOS devices.

A number of methods can provide the dielectric isolation to form SOI wafers. The earliest work on SOI circuits was performed in the 1970s with silicon-on-sapphire (SOS) devices, in which an epitaxial layer of silicon is grown on a single-crystal sapphire (Al_2O_3) substrate. The two materials are not particularly compatible because the crystallographic lattice spacing is very different. SOS circuits were useful for military and radiation applications, but widespread application was limited by the cost of the sapphire substrate and the relatively high level of defect density in the upper silicon material. The defect levels impacted the yield and performance of large, complex integrated circuits.

To overcome these material limitations, new processes have been developed using silicon as the substrate material. Perhaps the most common approach is the use of the two-stage SIMOX (separation by implantation of oxygen) process to form the silicon-on-insulator. The two stages of the process are (1) the high-energy implantation (~1.8×10^{18} / cm^2 at 200 keV) of oxygen ions into the silicon at ~600°C, and (2) a high-temperature annealing process on the order of 6 h at 1300 to 1350°C to form the SOI structure. The high-temperature processing also reduces the dislocation defects by orders of magnitude and mitigates oxygen precipitates in the top silicon layer. As might be expected, higher doses produce thicker buried oxides, and higher energies make the BOX layer deeper and the top silicon layer thicker. A relatively thick 400 nm BOX layer under ~200 nm of silicon is a common configuration. The acceptance of SIMOX has been limited by cost and defects. The cost is due the million-dollar price tag of the implanting equipment and the long implantation time, perhaps 5 hours a batch. Defect levels are decreasing with processing enhancements, but are still on the order of a few hundred per cm^2 [Auberton et al. 1995, Dance 1993].

A newer approach to making SOI material uses bonded and etched-back SOI (BESOI) wafers. In this structure, two wafers, at least one of which has been thermally oxidized, are bonded together using a thermal process. The bonding strength is increased by an anneal above 800°C. One of the wafers is then thinned by etching and polishing to form the top silicon surface. Because the oxide is formed thermally, the defect levels are low, often <.01/cm^2. The difficult part of this process is controlling the top silicon thickness; control to better than ±0.01 μm has been reported [Yallup 1993]. This process is limited because the top silicon only covers the substrate to within about 3 mm of the edge. This shortcoming can mean a loss of as much as 10% of the usable silicon area on a 100 mm wafer.

SOI wafers have been prepared by other methods for a specific application or research. Zone-melting recrystallization (ZMR) is commonly performed by depositing an amorphous or polycrystalline silicon film on an insulator. The film is then recrystallized with a laser, electron beam, or other heating source. SOI wafers can also be obtained using an epitaxial lateral overgrowth (ELO) technique, in which narrow lines are etched in an oxidized silicon wafer to serve as seed areas for vertical and lateral epitaxial growth. The resulting material is chemically and mechanically polished to provide SOI areas up to 10 μm wide. Silicon-on-quartz (SOQ) wafers have been fabricated by bonding a silicon wafer with an etch stop layer to a quartz wafer at room temperature, and then thinning it via mechanically polishing and chemical etching [Sarma and Liu 1994].

Because the quality of the SOI materials varies, several non-destructive techniques have been developed to qualify a batch of wafers. Key silicon and buried oxide (BOX) parameters, such as thickness, composition, and uniformity, are often measured by spectroscopic ellipsometer. The dielectric properties of the oxide can also be obtained with the ellipsometer [Hovel et al. 1993]. Defect densities are often measured by etching to reveal dislocations, using automatic defect-counting equipment. This type of process can be utilized just after the oxygen implantation and after the high-temperature anneal with an automated commercial optical-defect inspection system based on low incident-angle light-scattering to eliminate interference for the BOX [Yue et al. 1994]. A qualitative view of the material surfaces can be obtained by focusing laser light incident on the wafer and observing the emitted luminescence, along with its position on the wafer. This photoluminescence (PL) scanning can also characterize thickness uniformities, defect densities in SIMOX, and voids in BESOI. The electrical quality of the silicon can be evaluated through surface photovoltage response or a microwave detection technique. The former measures the effective minority carrier lifetime through the response to a UV laser; the latter observes the conductivity modulation when a high-intensity UV light is pulsed onto the surface to measure the large-signal lifetime.

2.4.2 High-temperature SOI device performance

SOI wafers, processes, devices, and designs have been optimized to perform at temperatures over 300°C. The design of SOI devices must account for various factors, including the effects of the buried oxide and control of the snapback voltage with a body tie. The buried oxide has breakdown characteristics different than the thermal oxide. It may also introduce different leakage mechanisms in PMOS devices, where as much as 1 mA of current may flow between the source and the drain with enough back bias [Yallup 1993]. NMOS devices do not have such conduction [Collinge 1989].

A negative characteristic of NMOS SOI devices that is related to the floating substrate is the snapback effect [Ochoa et al. 1983], in which the drain-to-source breakdown voltage is less than the drain-to-body breakdown voltage. This effect, which results from the hole flow that forward-biases the body-to-source diode, has been modeled [Huang and Kueng 1991] and unfortunately for high-temperature purposes it is relatively independent of temperature. In measurements on devices similar to those shown in Figures 2.18 and 2.19, the snapback voltage only changed 0.1 volt over the temperature range from 25 to 350°C.

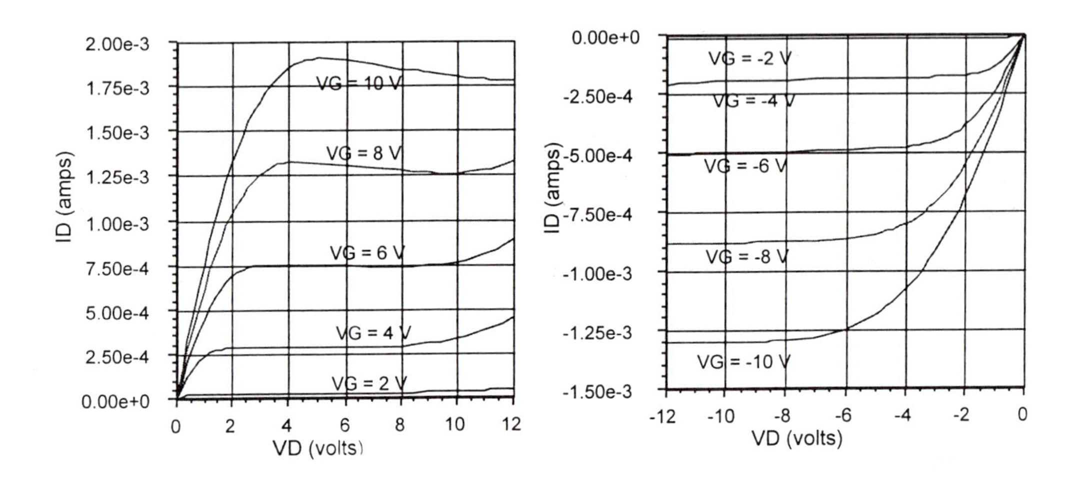

Figure 2.18 Characteristics at 25°C of 16/5 NMOS (with negative conductance) and 15/3 PMOS transistors.

SOI MOS transistors have increased drive current if the channel body is left floating. This condition may cause a kink in the saturation region of the I_{DS}-vs.-V_{DS} characteristics, as well as improved subthreshold slope characteristics. It could also result in snapback and latch-up [Chen et al. 1988]. These effects, good and bad, can be mitigated by tying the body to the source potential. Smart body contacts [Matloubian 1989] have been devised that clamp the body potential when the transistor is off and allow the body to float when the transistor is on; this provides higher current gain, while eliminating the snapback and latch-up potential.

DC transistor measurements of SOI NMOS transistors may not accurately portray the high-speed switching characteristics in a digital circuit because of the thermal insulation of the device from the substrate. The current through the channel may cause self-heating and exhibit a negative conductance at high gate biases in n-channel transistors [McDaid et al. 1989], as shown in Figure 2.18 for T = 25°C. This negative conductance will not affect the digital circuit function but may affect the analog operation. Device characteristics without self-heating have been measured with a pulsed I-V measurement system [Jenkins et al. 1994] saturated drain currents were 20% higher than static currents. As the temperature increases, the negative conductance decreases at high gate bias voltage, and may vanish completely, as shown in Figure 2.19 for T = 350°C.

Each SOI transistor has two channels — a top channel and a back channel. They are decoupled in partially depleted transistors and coupled in fully depleted transistors. Partially depleted transistors have a neutral region between the top and back channels so that they have thicker silicon and heavier doping than fully depleted transistors. In this section partially depleted devices are emphasized because fully depleted devices are still under development.

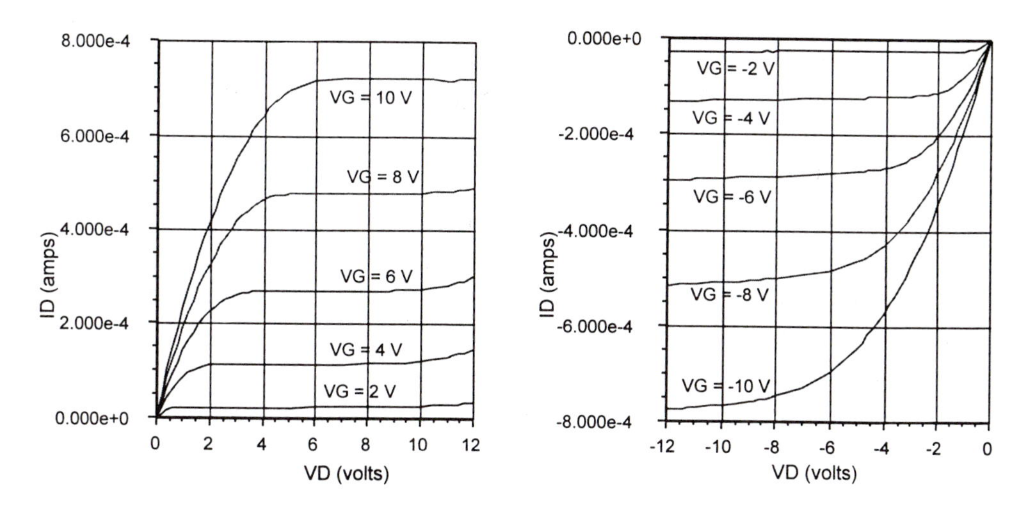

Figure 2.19 Characteristics at 350°C of 16/5 NMOS and 15/3 PMOS transistors.

Since the NMOS and PMOS threshold voltages move closer to zero with increasing temperature, the ambient V_t must be high enough to allow non-zero values at the maximum operating temperature. A zero threshold voltage would have substantial current flowing in the off state. Plots of threshold voltages of the top channel and back channel of PMOS and NMOS transistors, with respect to temperatures up to 350°C, are shown in Figure 2.20 and Figure 2.21.

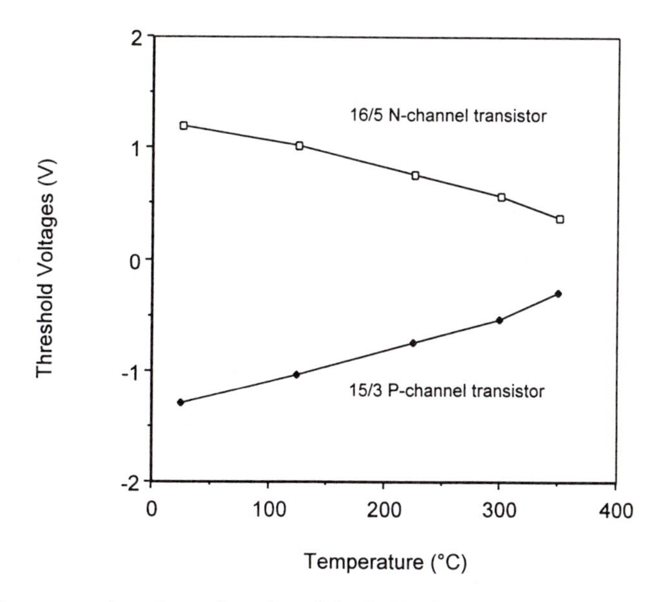

Figure 2.20 Temperature dependence of top-channel threshold voltages of 16/5 NMOS and 15/3 PMOS transistors.

The change is predictable and has a slope of approximately -2.5 mV/°C and 3.0 mV/°C for NMOS and PMOS devices, respectively. Others have reported values of -2.0 mV/°C and 2.75 mV/°C [Grzybowski and Tyson 1993] and ± 1 to 2 mV/°C [Krull and Lee 1988], where it was demonstrated that the slope of the threshold temperature characteristics for SOI devices is the same as that for bulk devices. Measurements on the devices used in Figure 2.22 showed that the change in subthreshold slopes (measured at Vdd = 0.1 V) was +1.73 and -2.2 mV/dec/°C for PMOS and NMOS transistors, respectively. The back-channel threshold voltage temperature coefficients were measured at -82 (NMOS) and +75 (PMOS) mV/°C. The temperature dependence of the subthreshold slope can be described by

$$S_{mm} = \left(\frac{kT}{q}\right) \ln\left(1 + \frac{C_{Dm}}{C_{OXm}} + \frac{C_{itm}}{C_{OXm}}\right), m = 1,2 \tag{2.22}$$

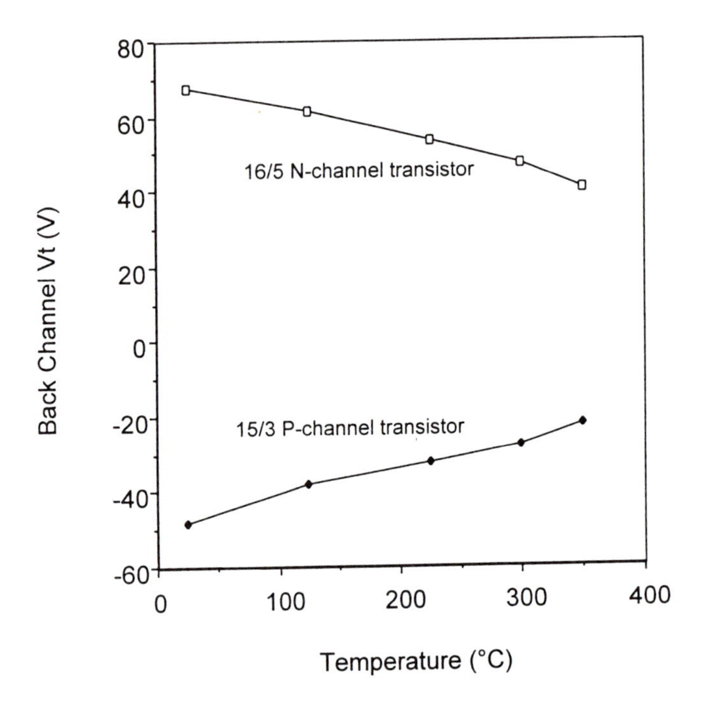

Figure 2.21 Temperature dependence of back-channel threshold voltages of 16/5 NMOS and 15/3 PMOS transistors.

where m = 1 and 2 refers to the top and back interface, respectively. C_{Dm} is the depletion capacitance and C_{itm} is the capacitance due to the interface charge. The slope measured at 2 temperatures is shown in Figure 2.22.

The temperature coefficient of the threshold voltages can be accounted for by [Colinge 1991]

$$\frac{dV_{tm}}{dT} = \frac{d\phi_f}{dT}[1 + \frac{q}{C_{oxm}}\sqrt{\frac{\epsilon_{Si}\epsilon_0 N_a}{kT\ln(N_a/n_i)}}], V/^0K \tag{2.23}$$

where ϵ_{Si} and ϵ_0 are the permittivity of the silicon and the vacuum, respectively; N_a is the doping concentration; and ϕ_f is the Fermi potential, given by $kT\ ln(N_a/n_i)/q$. Here, m = 1 or 2 for V_{tm} and C_{oxm}, and refers to the top gate or back gate properties, respectively. For example, V_{t1} is the top-gate threshold voltage and V_{t2} is the back-gate threshold voltage. $C_{ox1} = \epsilon_{ox}/t_{ox1}$ is the per unit area top-gate capacitance and $C_{ox2} = \epsilon_{ox}/t_{ox2}$ is the per unit area top-gate capacitance. It is evident from Equation 2.23 that a higher temperature coefficient of the back-channel threshold voltage would be expected because the back channel involves a thicker buried oxide than the thin gate oxide. The derivative of f_f with respect to temperature is given by

$$\frac{d\phi_f}{dT} = \frac{k}{q}[\ln(N_a/n_i) - \frac{3}{2} - \frac{E_g}{2kT}], V/^0K \tag{2.24}$$

For the very thin, fully depleted SOI MOS transistor, the dV_t/dT can be made to approach df_t/dT. This type of device is currently under development and should have the least temperature variation to date, as described by Equation 2.24.

Although the SOI approach has some disadvantages, such as higher material cost, availability, higher defect density, and higher thermal resistance than bulk silicon, SOI devices offer some performance advantages over silicon devices. Besides improved radiation immunity, SOI devices are immune to latch-up and may have higher packing densities, simpler processing, lower parasitic capacitance, and low voltage operation. The SOI approach also provides a number of advantages for low-power applications. The power consumption of an IC is proportional to CV^2f, where C is the total capacitance, V the operating voltage, and f the frequency. Since the junction area and capacitance of an SOI device can be reduced significantly, a total capacitance reduction of 15 to 20% can be achieved, depending on the circuit [Auberton et al. 1994].

One of the earliest applications of SOI was radiation-hardened devices. SOI is less sensitive to transient radiation phenomena because the photocurrents generated are dependent upon collection volume. Since SOI devices have less volume, the photocurrents produced by heavy ions are smaller. The lateral body tie contact, which helps to reduce edge leakage, is effective for this purpose and renders improved resistance to high cumulative doses and single event upset (SEU) [Auberton et al. 1995].

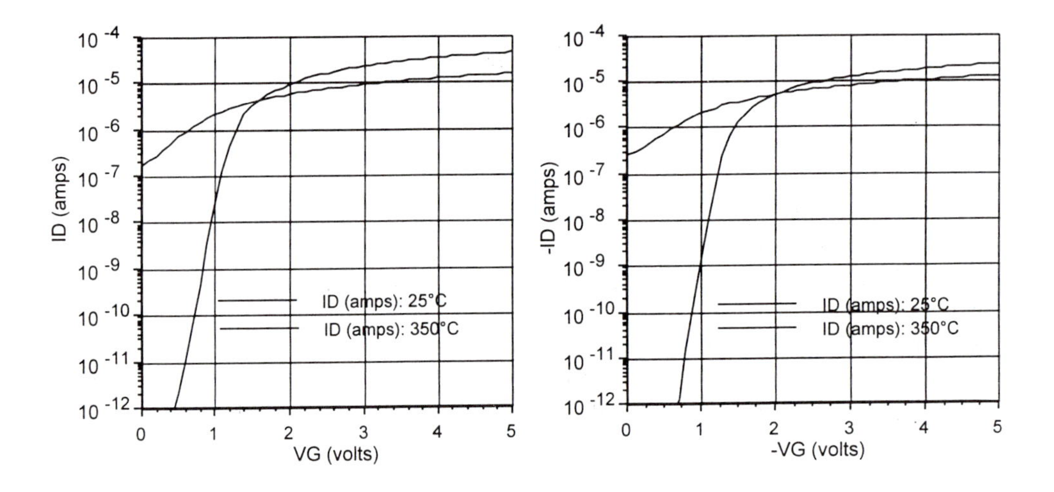

Figure 2.22 Transfer characteristics of NMOS and PMOS transistors at 25 and 350°C.

The temperature dependence of the standby current of bulk and SOI CMOS 64KSRAMs is illustrated in Figure 2.23. At low temperatures the defects tend to give the SOI device higher standby current. At higher temperatures, the standby current for the bulk part is more than two orders of magnitude higher than that of the bulk CMOS 64K SRAM. The performance of these circuits, as measured by the timing parameters, was found to be predictable up to 200°C [Brusius et al. 1994].

2.4.3 SOI device reliability

SOI devices can be reliable, even at high temperatures. The first concern of most reliability engineers is metal electromigration and packaging, subjects addressed elsewhere in this book. The oxide integrity, hot electron lifetime, and device lifetime are discussed here.

The integrity of the gate oxide is often a concern with SOI devices. While the quality of the oxide film may be similar to that of oxide grown on bulk devices, the film over the dislocations (principally in SIMOX) may be defective, especially for thin gate oxides [Colinge 1991]. One study [Brown et al. 1994] found a strong dependence on material preparation and similar breakdown voltages for bulk and SOI oxides; levels of gross defects on SOI devices were acceptably low. In another instance, SIMOX gate oxide capacitors had better Qbd performance than BESOI devices [Yallup 1993]. Of particular interest for high-temperature processes is the fact that the physical mechanism of oxide breakdown does not change

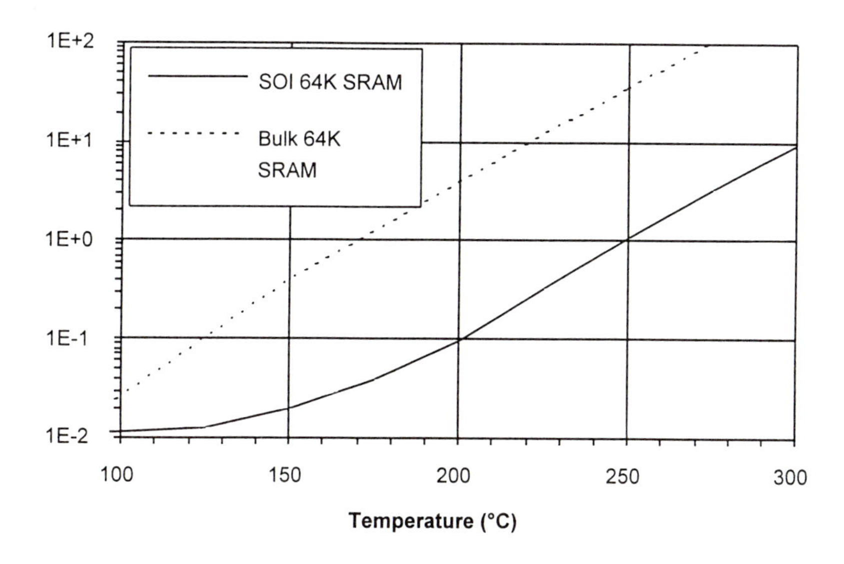

Figure 2.23 Temperature dependence of standby currents of bulk and SOI 256K SRAMs

significantly up to temperatures of 400°C [Suehle 1983]. Thus, properly made gate oxides can be reliable at high temperatures.

The hot electron lifetime is usually evaluated by DC bias stressing [Takeda and Suzuki 1983] of the devices as a function of time at the peak substrate-current bias condition [Simox 1994]. Minimum change in performance is important for analog operation, and more difficult to attain for high-voltage devices. One of the test results using reverse current as a measure of performance is shown in Figure 2.24. The lifetime is defined for the change of the reverse current to decrease to 90% of the original value. This testing was performed with 12 volts across the drain to source. The projection of the lifetime at 10 years is given by the dotted line in Figure 2.24. Clearly, the hot electron lifetime is longer than 10 years. Some details of the evaluation of the hot electron lifetime involving thinner gate oxide have been published [Jenkins and Liu 1994].

Indeed, SOI devices can be made reliable for high-temperature operation. Operational amplifiers, using transistors of the type shown in Figure 2.24, have successfully passed life tests of 1000 hours at 250 and 300°C, all with a 14-volt power supply.

2.5 Metallization for High Temperature Electronics

The major factors limiting the life of IC conductors and contacts operating at elevated temperatures are electromigration and corrosion. The material properties controlling these two factors and other important effects must be considered when choosing a conductor. The interrelations between several of the properties complicates this process.

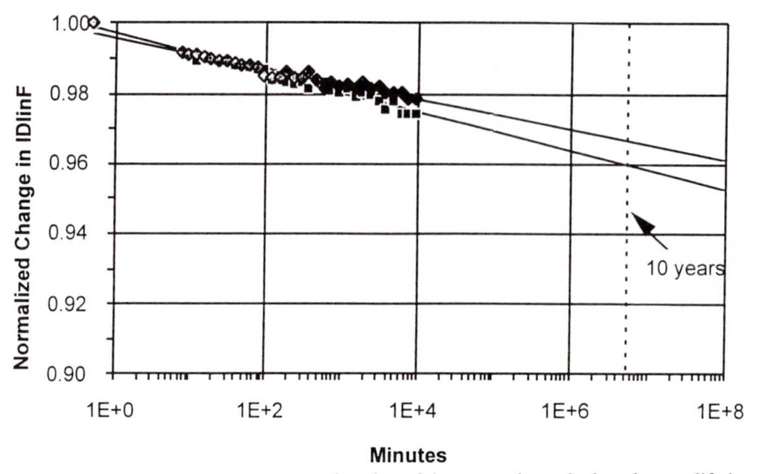

Figure 2.24 DC stressing of the reverse current as a function of time to evaluate the hot electron lifetime.

2.5.1 Electromigration factors

Current density (J): The most important factor in electromigration is the electrical current density. While there are other significant causes of atomic migration, such as temperature gradients, stress gradients, and so on, there can be no electromigration without electrical current. The most cited expression for the median time to failure of a thin-film conductor due to electromigration is the empirically derived "Black's equation." As originally formulated, the median time to failure varied as J^{-2}. Other publications presented test results with median time to failure as J^{-n}, with n running from 1 to over 3. The only published model of electromigration controlled median time to failure derived from physical considerations for real circuits yields a value of 2 for n^2. This assumes a boundary that permits electrical current flow while restricting vacancy flow. Such a condition is realistic for ICs because there is at least a partial blockage of vacancy flow at all contacts between different materials, and presumably total blockage at all contacts with effective diffusion barriers. Experimental values of n less than 2 can be obtained when the test structures have limited barriers to vacancy (or atomic) flow. Values of n greater than 2 can be obtained when the test structures exhibit localized heating due to the high current densities used in electromigration tests (almost all tests use J 10^6 A/cm^2). Consequently, it is reasonable to expect a median time to failure as J^{-2} for circuits in which electromigration is the major failure mechanism. Because of this strong dependence on current density, at a given temperature, applications requiring appreciable power will require more electromigration resistant metal than they would for typical power levels. This is important, since high-temperature electronics is frequently required to control electrical power in a hot environment.

Grain structure: As might be expected, given the importance of vacancies to atomic flow, electromigration occurs primarily along grain boundaries and non-passivated film surfaces. This suggests that uniformly large grain size will suppress electromigration, which has been experimentally verified [Thompson and Kahn1993]. The grain structure of thin films can be influenced by several factors — substrate surface condition, substrate deposition temperature, deposition rate, and so forth. The geometry of the grain boundaries also influences conductor life. In general, any grain-boundary triple point will produce a divergence in the flow of vacancies (or atoms). This produces either an accumulation of vacancies (causing a void) or an accumulation of atoms (causing a hillock). The eventual failures are either opens or shorts. The importance of grain-boundary geometry is clearly demonstrated by the electromigration resistance of aluminum films with a "bamboo" structure [Valdya et al. 1980]. In this structure, nearly all grain boundaries are perpendicular to the conductor line. Increasing grain size to about twice the width of a conductor by high-temperature annealing markedly increases the median time to failure (for the same J). This is related to the decrease in grain-boundary triple points as the grain structure becomes bamboo-like.

Electromigration activation energy: In Black's equation [Black 1969] and other analytical expressions for the electromigration rate, there is always an exponential dependence upon an activation energy (Q_{cm}) divided by kT. Unfortunately, published information on these activation energies is limited, with little data available for several metals of interest. Furthermore, almost all published activation energies are determined by median time to failure tests, rather than by actual measurements of metal migration. Because median time to failure is influenced by several factors in addition to the basic electromigration activation energy (such as grain structure, conductor geometry, and corrosion), there is significant spread among these determinations of the electromigration activation energy. Fortunately, Mogro-Campero, [1982] has shown the linear relation between the electromigration activation energy of several metals and their melting temperatures (T_m). From this, he estimates that Q_{em} (eV) = 7.4 X 10^{-4} X T_m (^{0}K). Such a direct connection between Q_{em} and T_m is physically reasonable, since both are related to the energy an atom must have to break its lattice bonds. Since aluminum melts at only 660^{0}C, it can be used only for the lower portion of the elevated temperature range, while only refractory metals should be considered for use at higher temperatures.

Surface migration: The surface of a thin film contains the largest defect in its structure by orders of magnitude. Unless the surface atoms of a metal film are immobilized, their migration will dominate the electromigration of the film. Striking evidence of this was observed during electromigration tests at 400^{0}C of gold films coated with a surface sealing layer. The current was supplied through bare gold wires with a cross-section sixty times larger than that of the gold films. After 370 h, there was no electrically or microscopically detectable evidence of electromigration damage to the films; but the electromigration damage to the wires was extensive [Goetz 1995]. Surface electromigration is not observed with conventional ICs (even at 125^{0}C) because standard aluminum-alloy conductor films are covered with surface oxides. They are also frequently buried under deposited layers of oxides or nitrides, which prevent movement of the metal surface atoms.

An adherent coating on a conductor produces both a direct and an indirect reduction of electromigration damage. The direct reduction of electromigration as a result of reducing the motion of surface atoms is obvious. However, strong reaction bonding between the coating and the surface atoms of the conductor is required. The indirect reduction of damage results because a large number of new vacancies is necessary to form voids within a film. Since the only source of new vacancies is injection from the surface, a coating that immobilizes surface atoms will suppress vacancy injection and the formation of electromigration voids. This has been demonstrated in a study of two sets of aluminum conductors; one set with a mm thick anodic oxide and the other set bare. Under the same electromigration test conditions, the anodized conductors had an average median time to failure over ten times longer than did the bare conductors [Learn 1973].

Less quantitative tests with gold conductors also indicated the clear benefit of a coating that suppresses vacancy formation. In this case, two similar sets of gold conductors were tested, one set with a sealing coating to suppress vacancy formation and the other without. After 280 h of electromigration testing with J = 1 MA/cm^2 at 400 ^{0}C, the unprotected conductors had some opens and much microscopically observable electromigration damage in the form of voids and hillocks. Similar testing of the coated gold conductors (except with a larger J of 2.7 MA/cm^2) produced no sign of any electromigration damage, either microscopically or electrically [Goetz 1995]. Since gold is practically non-reactive, it is clear that with gold conductors the observed improvement is not related to the coating suppressing conductor corrosion.

2.5.2 Corrosion factors

Oxide characteristics: Currently, guaranteed hermetic packages are not available for the higher end of the temperature range. Thus, only metal schemes that can be exposed to the atmosphere at the intended operating temperature are considered here. Under these conditions, nearly all corrosion of the IC metalization is oxidation related. Several properties of a conductor's oxide are important. Obviously, an oxide that self-limits its growth is highly desirable. Mechanical strength and adherence are also important. The poor chalk-like characteristics of WO_3 is an important reason why a simple tungsten conductor is a poor choice for high-temperature metallization, even though tungsten has a low resistivity and the highest electromigration activation energy of any metal. Providing protection by depositing a passivating coating is a possibility for all conductors, but this increases the process complexity. Also, as shown in the following section, electromigration can cause rupture of a coating, thereby reducing its corrosion protection.

Passivation coatings: Several factors must be considered in using a deposited passivating coating. One that is often overlooked is stress. The stress transferred from a coating to the conductor will be opposite in sign. Normally, some compressive stress is desirable in a coating to increase its fracture strength. However, this causes a tensile stress in the conductor that enhances the formation of voids. Large stresses of either type promote the formation of either voids or hillocks. Passivating overcoats have been shown to produce failures in aluminum conductors [Hinoda et al. 1987].

Depositing flaw-free, strongly adherent coatings over the complicated morphology of a top-level metal is a problem. Moreover, electromigration-produced hillocks will eventually fracture a passivating coating and expose unprotected material to corrosion. Although a passivating coating can significantly increase the life of any reactive metal, metal with lower reactivity with the atmosphere will have a longer life, even with a passivating coating.

Compatibility with silicon: Several metals can seriously reduce carrier lifetimes when present in silicon devices. Although their effects are not normally defined as corrosion, their reaction with the silicon lattice has a detrimental influence on silicon's electrical properties. Since this resembles the effects of corrosion, it is discussed here.

The best-know problem metals in silicon are copper and gold. Since gold is normally used in certain phases of silicon IC manufacturing, it can obviously be contained with the proper procedures. The same can be said for copper which has been undergoing active development as a metallization for ICs [Edelstein 1995]; however, it clearly requires the use of barrier layers and encapsulating structures. Similar precautions may be needed for other materials, particularly at high temperatures. Knowledge and caution are needed in working with these materials, but they need not be completely avoided. However, the cost of using them must be factored into a balanced decision on whether the benefits are sufficiently worth the necessary effort and precautions.

2.5.3 Other factors

Resistivity: Obviously, the resistivity of a conductor metal is important. Power loss, local heating effects, and speed all favor the use of the lowest resistivity conductor possible. The value placed upon low resistivity is so dominant that only the third and fourth lowest resistivity metal has ever been used as thin film conductors in large-scale production of semiconductor products (gold and aluminum). Currently, significant efforts are in progress to develop a conductor process using copper (second lowest) because its resistivity is 42% lower than standard aluminum conductor alloys. (The possible importance of copper for elevated-temperature electronics will be discussed in more detail later.)

Much work has been done on decreasing the electromigration of aluminum by alloying it with small amounts of other metals. It is likely that the improvements achieved are related to precipitates of the added metal forming in the aluminum grain boundaries. This is a general mechanism for retarding atomic movement along grain boundaries. However, these metal additions increase the resistivity of the aluminum and the actual values are rarely given with the reports of improved median time to failure. There have been many electromigration studies of aluminum alloys for over two and a half decades. Since dilute copper alloys with aluminum have been in heavy use for some time [Ames et al. 1970], it is unlikely that any other major improvement in the electromigration-vs.-resistivity compromise has been or will be found. Of course, there can always be applications (such as elevated-temperature electronics), where the sacrifice of an increased resistivity is accepted to obtain the needed median time to failure.

Step coverage: Since all metallization on ICs has to traverse steps (in contacts, if nowhere else), the "step coverage" property of a metal system is an important consideration. In general, vacuum-deposited metal films grow with a columnar grain structure — the higher the melting point of the metal, the more pronounced the columnar structure. The angle of the column growth depends on the angle at which the arriving atomic stream meets the nucleating surface. A fracture cross-section of the grain structure of a tungsten film deposited over a surface with a series of etched groves clearly shows the dependence of the column angles upon the angle of the surface. It also shows that there could be a problem with columns from the side walls shadowing the concave corners and crating voids. Continuity of the tungsten film is only obtained by applying a RF-induced bias to the growing film to resputter and increase the surface mobility of the deposited tungsten atoms [Goetz and Johnson 1994].

Because step coverage can be a problem, it must be considered in the development of a metallization process. Physical vacuum deposition methods generally require some special technique to obtain reliable metal continuity in concave corners. While metal reflow has been successful for aluminum in contacts, some type of etch-and-reposition procedure is needed for more refractory metals. Sputter deposition of the standard titanium tungsten alloy provides excellent sidewall coverage, presumably because of the high atomic surface mobility provided by the titanium addition, but it comes at the cost of higher resistivity. Commercial (CVD) systems for tungsten deposition have been developed to produce films with excellent step coverage; however, they are costly (>$\$10^6$).

2.5.4 Specific temperature ranges

From 100 to 200^{0}C: It is evident from existing technology that aluminum will satisfy almost all applications at the low end of this temperature range. Aluminum is unique in having the combined properties of low resistivity, excellent adhesion to oxides and nitrides, a high-quality self-passivating oxide, and the ability to be reactive-ion etched with submicron resolution. Other conductors need some type of adhesion layer unless they incorporate a highly adherent component, such as the 30 atomic percentage titanium in titanium tungsten. Since many types of adhesion schemes have been used, the following discussion of candidate conductors will not address them in detail, but it does assume that an appropriate one has been incorporated into the metal system.

Aluminum has a relatively low electromigration activation energy (low melting temperature). Consequently, at the upper limit of this temperature range, the addition of small percentages of one or more other elements is needed to suppress the electromigration if sizable currents are involved. As indicated above, any additive will increase the resistivity of pure metal films and resistivity information is frequently missing in reports of improvements to aluminum median time to failure. Consequently, the use of any additive other than copper to improve aluminum median time to failure at higher temperatures probably will require some experimental investigation of the life-resistivity tradeoff.

For resistivity reasons, significant efforts are underway to develop a copper conductor process compatible with IC manufacturing [Edelstein 1995]. The copper is patterned with the damascene technique, which involves depositing an oxide film several mm thick, etching trenches, depositing a copper film that overfills the trenches, and chemical-mechanical polishing away all the copper above the oxide surface. If a sealed passivating layer is deposited over the top surface of the copper conductor, it will function as an excellent conductor over the entire lower temperature range.

From 200 to 300^{0}C: Present indications are that this range will have the most applications for high-temperature electronics for some time to come. It is also the range for which there are the most candidates for conductor materials.

In the lower end of the range, aluminum with an additive is a prime candidate. Even if the additive increases the resistivity significantly, it still may be less than for other likely candidates. Furthermore, additives usually decrease the temperature coefficient of resistance of a metal, so the resistance ratio of an alloy to a pure metal is less at elevated temperatures than at room temperature. Finally, in some applications, the use of more electromigration-resistant aluminum alloys at higher temperatures may be worthwhile, even at the cost of a higher resistivity, because of the large available knowledge base for aluminum conductors.

If a satisfactory damascene technique with a top-surface seal is developed for copper on ICs, copper could also be an attractive conductor in the 200 to 300^{0}C temperature range. This speculation is based on the results for electromigration tests of gold conductors covered with sealed layers (reported above). In tests at 400^{0}C with current densities of up to 4 X 10^6 A/cm^2, the gold conductors exhibited no microscopically or electrically detectable evidence of electromigration damage in samples tested for 370 h. Extrapolation to a current density of 4 X 10^5 A/cm at 300^{0}C indicates a median time to failure of many years [Goetz 1995]. Similar performance would be expected for copper with a sealed passivating coating, since its melting temperature is 19^{0}C above that of gold.

It is clear that the above results are due to the sealing coating and not an inherent property of gold. We believe the apparent reduction of gold electromigration with the sealing coating is related to its suppression of the nucleation of vacancies at the metal surface, thereby suppressing electromigration. Gold films with a proper coating also are candidates for IC conductors in the 200 to 300^{0}C range. Even without a damascene technique, gold can be used as a top metal shunting layer. But with a proper damascene technique, copper would be preferable both in cost and performance.

If somewhat higher resistivities are acceptable, several more refractory metals could be candidates, including two platinum metals (iridium and rhodium) and two transition elements (molybdenum and tungsten). Because of problems with etching the platinum metals, (particularly by reactive ion etching), they have been virtually ignored as possible IC conductors. Since there is no technology or experience base for ICs with these metals, it is doubtful that at this time they are good candidates for development as elevated temperature conductors. In contrast, both molybdenum and tungsten have been evaluated as IC conductors, and systems and processes have been developed for CVD of blanket tungsten. Fine-line reactive ion etching of both molybdenum and tungsten has been demonstrated [Chow and Stecki 1984]. Except for applications requiring the lowest possible resistivity, these metals would be good candidates for conductors over the entire elevated temperature range. However, the high cost of commercial CVD systems for tungsten may be prohibitive in many cases. Also, it should be remembered that a passivating coating would be needed to prevent corrosion for both molybdenum and tungsten over much of the high temperature range.

From 300 to 400°C: These temperatures are in such a physically challenging range that some compromises in the resistivity and the life of conductors must be expected. The only established candidates that will have reasonable electromigration-limited median time to failure are molybdenum and tungsten. Both can be expected to be useful at the low end of the range, with tungsten having an advantage at the upper end because of its 800°C higher melting temperature (higher electromigration activation energy). However, for applications involving large temperature cycles, molybdenum may have an advantage over the entire temperature range because of its much greater ductility. In this range, reliable passivation coatings are critical to the life of both molybdenum and tungsten films. Real ICs using these metals without virtually flaw-free passivation will have their life limited by corrosion rather than electromigration. Since corrosion tends to cause a gradual increase in the conductor resistance, the median time to failure of a circuit will also depend upon how much tolerance to this change is designed into the circuit.

2.6 Availability and Use of Commercial-Grade Silicon Devices in Power Converters at High-Temperatures

Potential markets, such as the automotive industry, have led semiconductor manufacturers to begin developing silicon devices and integrated circuits designed for high temperatures. The utilization of silicon devices in an elevated-temperature environment is motivated by two factors: (1) some applications, such as the automotive industry, may not have temperature requirements that exceed 200°C and (2) an immense, global infrastructure already exists for the processing of silicon. Some discrete transistor families have now been rated to operate at temperatures up to 200°C [Marshall 1992]. Silicon-integrated circuits can be fabricated that can operate at temperatures up to 250°C [Erskine et al. 1994]. Recently, Texas Instruments studied an automotive fuel-injector power IC for operation to 200°C. Other evaluations of the performance of digital and low-power analog devices at high temperatures may be found in the literature [Prince et al. 1980, Palkuta et al. 1979, Draper and Palmer 1979, Sponger et al. 1991]. The manufacturer's databooks reveal a few commercially available silicon-integrated circuits rated for operation up to 200°C. For example, National Semiconductor Corporation has a 1.0 Amp power operational amplifier (part no. LH0021-200) with an operating temperature range of -65°C to 200°C; this component is packaged in a 3-pin T0-3 metal package.

Nevertheless, most silicon power semiconductors are rated for operation below 150°C. While manufacturers are beginning to provide devices that are rated for operation up to 175 °C, the devices are derated to zero operating power at this temperature. Researchers have been investigating the effects of temperature on power semiconductor device parameters such as on-resistance or on-state voltage, breakdown voltage, leakage current, and switching speed [Godbold et al. 1992, Bromstead et al. 1991, Wayne 1993, Menhart 1992, Hudgins 1991]. The influence of temperature on these parameters is essential information for the circuit designer. For instance, the effect of temperature on the on-state voltage is important because power losses in the device are related to it.

All the devices discussed here were tested in various power converter prototypes, placed in an oven where the temperature was varied up to 200°C. The elevated temperatures dictated that the power dissipation in the devices had to be kept low because any dissipation raises the junction temperature. Device derating can be utilized to limit the rise in junction temperature. For example, the power dissipation in a MOSFET rated for 20 A, utilized at a current level of 1 A, is greatly reduced.

In power converter prototypes, the power losses and switching stresses for the power semiconductor devices result in an increase in the junction temperature. The losses may be divided into two categories: conduction and switching. Conduction losses are ohmic losses and may be controlled either by using a device with a low on-resistance or by operating the device at a reduced current level. Since the on-resistance or on-state voltage typically increases with temperature, the conduction losses increase with temperature. Leakage currents also increase with temperature as does their contribution to the conduction losses.

Switching losses result from the simultaneous presence of voltage and current during the turn-on and turn-off of the power device [Mohan et al. 1989]. In contrast to conduction losses, switching losses are reduced through the selection of an appropriate circuit topology. For example, buck, bost, and similar converters are often controlled by pulse-width modulation (PWM) and are said to be hard-switched because the semiconductor switches are turned on and off under voltage and current. Soft-switched converters, such as resonant and resonant-switch converters, turn on and off at zero voltage/current. The voltage and current waveforms for the switch in a converter, such as the boost, can be shaped by the addition of components such as inductors, capacitors, and diodes to permit zero-voltage/zero-current switching. This soft-switching is achieved at the expense of increased peak voltages and currents. In designing a resonant-switch converter for high temperatures, the variations in device characteristics with temperature must be considered.

Resonant converters also allow zero-voltage/zero-current switching [Mohan et al. 1989]. In contrast to the resonant-switch concept, which employs a waveshaping circuit for each switch, soft-switching in resonant converters is achieved with the aid of a filter network of capacitors and inductors connected between the source and load. These converters may be controlled by varying the ratio of the switching frequency to the resonant frequency, which is related to the network inductances and capacitances. If the inductance or capacitance varies with temperature, the resonant frequency will vary with temperature, making control more difficult. Inductors can be constructed whose inductance value is insensitive to temperature, particularly if the magnetic properties of the core material are unaffected by temperatures over the relevant operating-range. Capacitors with large capacitance values (greater than 1 μF) are sensitive to variations in temperature; the capacitance decreases with increasing temperature. On the other hand, ceramic NPO capacitors with values less than 0.1 μF are relatively insensitive to variations in temperature. Resonant converters typically use the larger capacitor values, with the smaller values being utilized for snubber applications.

2.6.1 Component characterization

Designing power converter prototypes for elevated temperatures requires knowledge of how the component varies with temperature. Power converters typically are composed of such components as semiconductor devices, inductors, transformers, capacitors, and resistors. Commercial-grade silicon semiconductor devices are seldom characterized at temperatures exceeding 150°C. Magnetic materials used in transformer and inductor design should have curie temperatures in excess of the maximum core operating temperature. Materials are available for power converters operating at ambient temperatures up to 200°C. A limited number of commercial capacitors have a capacitance relatively insensitive to temperature variations; most of these have values less than 0.1 μF.

Typical silicon power semiconductor devices used in power converters are diodes,

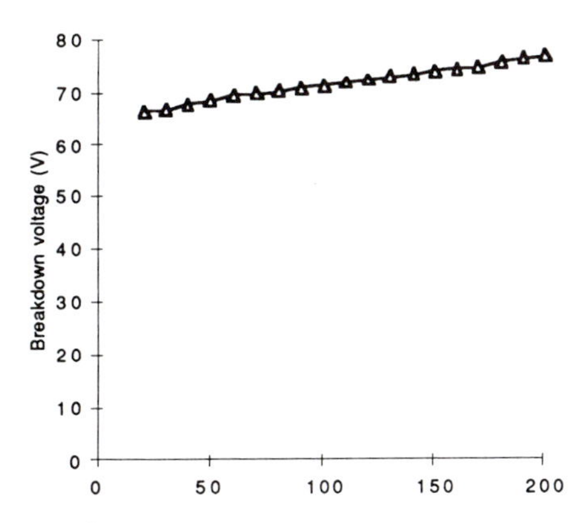

Figure 2.25 MOSFET breakdown voltage vs. temperature.

BJTs, MOSFETs, insulated-gate bipolar transistors (IGBTs), and MOS-controlled thyristors (MCTs). The device manufacturers normally don't provide data above 150°C; however, this data is needed by the circuit designer. Variables of interest to the circuit designer would include on-resistance or on-state voltage, leakage current, and breakdown voltage. The power dissipated in the device while it is conducting is directly proportional to the on-resistance or on-state voltage. When the device is in the off state, power is dissipated due to leakage current. This power loss is normally negligible at normal operating temperatures, but may become significant if the leakage current becomes large enough at elevated temperatures. Increases with temperature in either the on-resistance or the leakage current result in increased power losses, causing the device junction temperature to increase.

One of the MOSFETs utilized for power converter prototypes is International Rectifier's IRF044. This device is rated for 60 V and 30 A at a case temperature of 25°C. The on-resistance is listed as 21 mΩ typical and 28 mΩ maximum at Tj = 25 °C; these values are measured with V_{GS} = 10 V and I_D = 33 A. It is derated to zero power at a case temperature of 175°C. This device has been characterized over a temperature range of 20 to 200 ° C; the breakdown voltage, leakage current, on-resistance, and gate-to-source threshold voltage were measured over this range. Figure 2.25 shows a plot of the breakdown voltage vs. temperature. This data was measured using a Tektronix 371 curve tracer with V_{GS} = 0. The breakdown voltage at 200°C was 77.5 V. The on-resistance vs. temperature is plotted in Figure 2.26 This data was obtained at a drain current of 10 A with V_{GS} = 15 V. The resistance increases by a factor of approximately 2 over the temperature range. Figure 2.27 is a plot of the leakage current as a function of temperature. It was measured at the rated breakdown voltage of 60 V and varies from 0.5 nA at 20°C to approximately 1.1 mA at 200°C. The gate-to-source threshold voltage was measured as a function of temperature and is plotted in Figure 2.28. Using V_{DS} = 5 V and the Tektronix 371 curve tracer, the transfer characteristics of the MOSFET were obtained as a function of temperature. This data was then plotted as the square root of I_D vs. V_{GS}. Using a straight-line approximation to this curve, the threshold voltage was determined from the intersection of this approximation with the I_D = 0 axis [Mohan 1989]. The threshold voltage varied from 2.92 V at 20°C to 1.75 V at 200°C. A more detailed discussion of the physics behind the relationships plotted in Figures 2.25 through 2.28 may be found in reference [Wayne 1993].

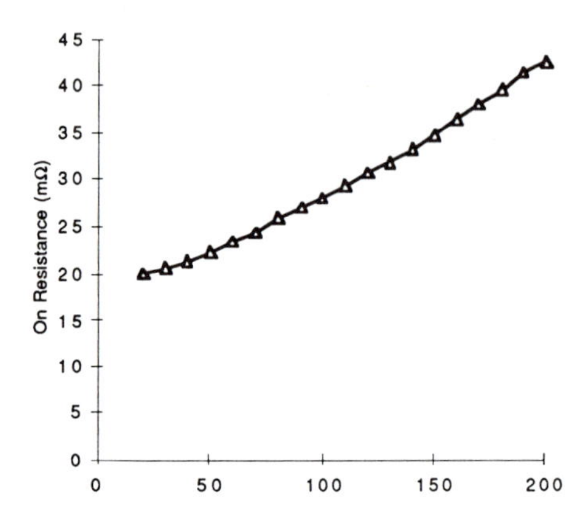

Figure 2.26 MOSFET on-resistance vs. temperature.

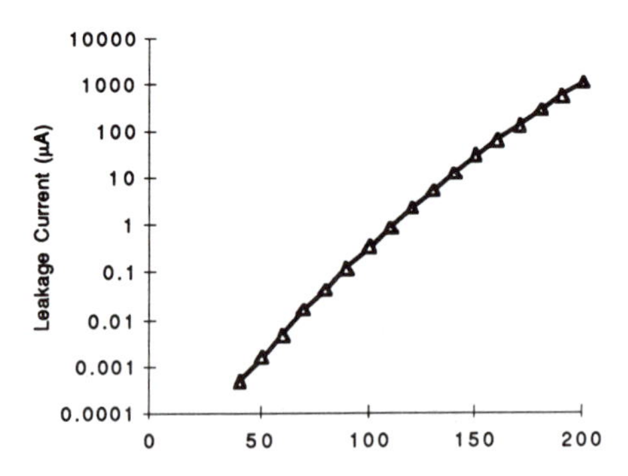

Figure 2.27 MOSFET leakage current vs. temperature.

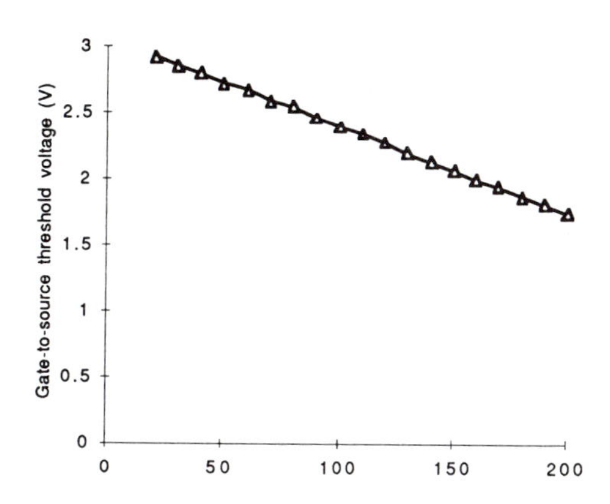

Figure 2.28 MOSFET gate-to-source threshold voltage vs. temperature.

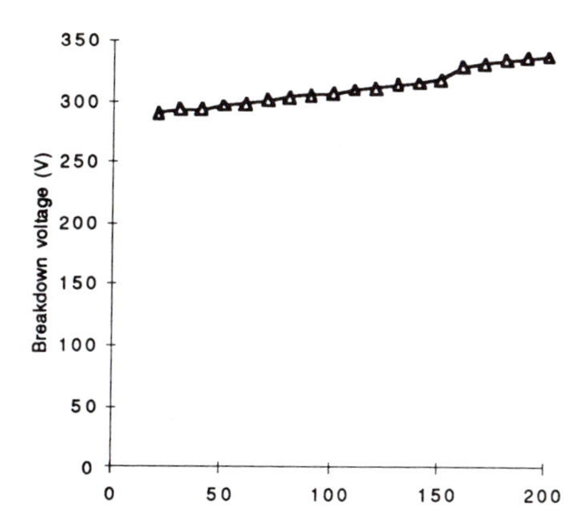

Figure 2.29 Diode breakdown voltage vs. temperature.

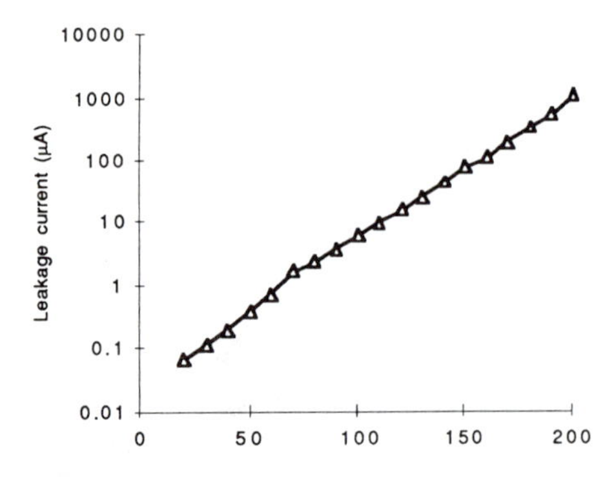

Figure 2.30 Diode leakage current vs. temperature.

A Motorola MUR5015 ultrafast diode, manufactured in a DO-5 package, has also been characterized over a temperature range of 20 to 200°C. The voltage and current ratings of this device are 50 A and 150 V, respectively, and the maximum junction temperature is 175°C. Similar to a MOSFET, it is derated to zero power at a case temperature of 175° C. Figure 2.29 shows a plot of the breakdown voltage over this temperature range. The breakdown voltage increased from 290.5 V at 20°C to 338 V at 200°C. The breakdown voltage of 150 V specified by the manufacturer is the minimum value of VRRM. Figure 2.30 shows the leakage current as a function of temperature at a reverse bias of 150 V. The leakage current varies from 6.8 nA at 20°C to slightly more than 1 mA at 200°C. The diode forward voltage drop is plotted as a function of temperature in Figure 2.31. This voltage decreases from 0.71 V to 0.42 V.

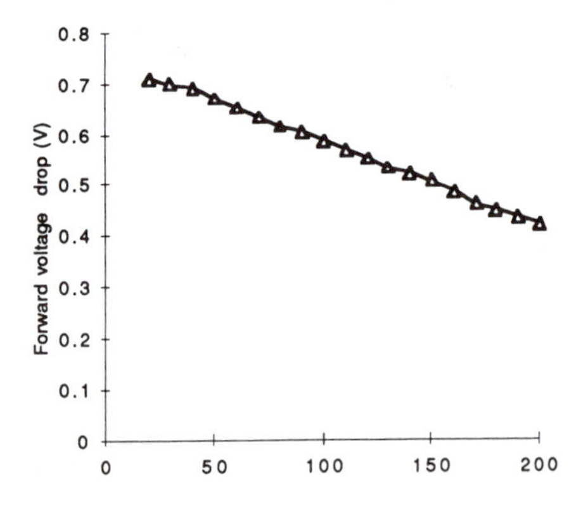

Figure 2.31 Diode forward voltage drop vs. temperature.

2.6.2 Power converter test results

Four different power converters were tested at temperatures up to 200°C. All of the converters utilize soft-switching techniques to reduce switching losses and to avoid the stress of both high current and high voltage for the semiconductor switches.

Parallel-loaded resonant inverter: The parallel-loaded resonant inverter [Mapham 1967], shown in Figure 2.32, was the first power converter tested for operation at elevated temperatures. This converter was designed for a switching frequency of approximately 20 kHz and a peak device current of approximately 20 A. As seen in this figure, one leg of the inverter utilized power MOSFETs, while the other was constructed using IGBTs. Both the MOSFET and IGBT are manufactured by Harris Semiconductor. The MOSFET has a rated breakdown voltage of 50 V and a rated current of 75 A. A Motorola MBR3535 diode is utilized to block the intrinsic body diode of the MOSFET. The IGBT is rated at 34 A, with a breakdown voltage of 1000 V. The drive signals for these devices are generated using a Unitrode UC3860 control chip [Product & Applications Handbook 1993-1994] and an International Rectifier IR2110 [International Rectifier] driver. The signals from the IR2110 are supplied to bipolar drivers through optocouplers. The inverter is operated in an open-loop fashion. A split inductor design for the inverter produces half-wave resonant current in the switches. Fast recovery diodes, part no. 1N3891, were used as anti-parallel diodes to commute the second half of the resonant current. Only the power devices and diodes were placed in the oven mounted on an aluminum heat sink. The driver circuitry, which remained outside the chamber, was connected to the power devices via twisted pairs of 12-gauge silver-plated teflon-coated wire.

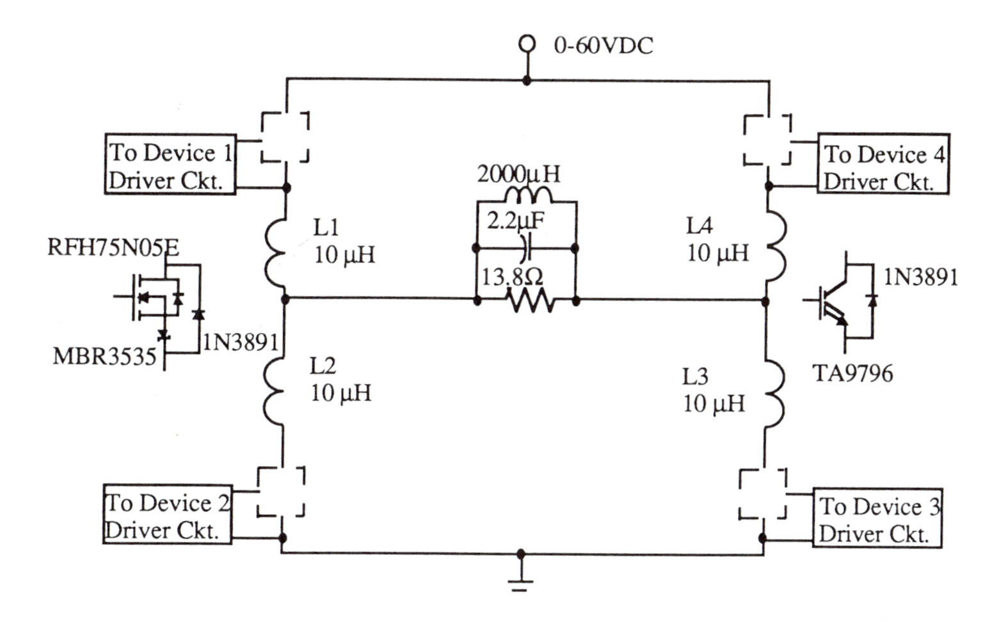

Figure 2.32 Parallel-loaded resonant inverter

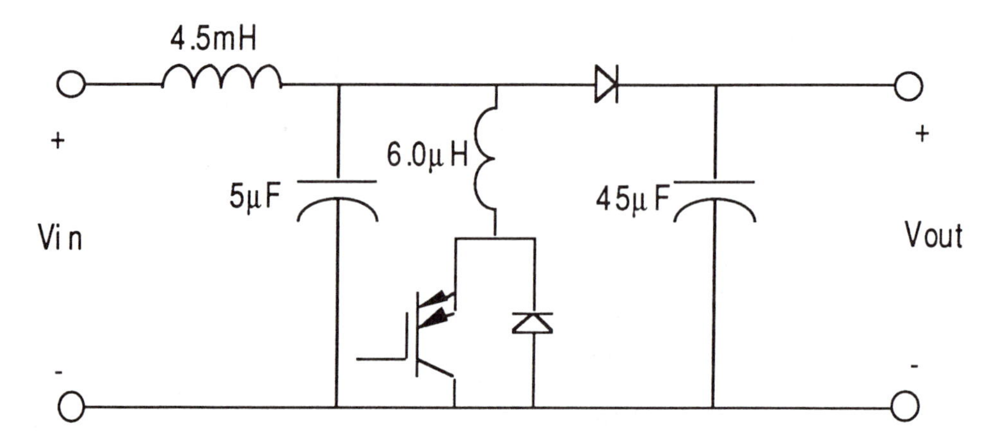

Figure 2.33 Zero-current-switched boost converter.

When this inverter was operated at 200°C, the case temperatures for the IGBT and MOSFET were measured at 214 and 208°C, respectively. No degradation in performance was observed at elevated temperature. The inverter constructed using only IGBTs was successfully operated at 40A peak current for 72 h at 200°C.

Zero-current-switched boost converter: The next power converter tested at elevated temperatures was a zero-current-switched boost converter [Liu et al. 1987], seen in Figure 2.33. The power switch in the basic boost-converter topology is transformed to a resonant switch by the addition of the diode, the 6 μH inductor and the 5 μF capacitor. These components shape the current through the IGBT/diode pair to permit softswitching.

This converter was designed to supply a 42 V, 100 W load from a 28 V source. The Harris TA9796 IGBT utilized in this converter is rated at 1000 V and 34 A. Even though this is obviously poor device utilization, the device was tested in an actual circuit since it had been characterized over the temperature range of interest. This poor device utilization was apparent in the efficiency measurements for this converter because this IGBT had an on-state voltage of 3 to 4 volts. The 1N3893 fast-recovery diode used to complete the resonant switch is rated to carry 12 A and block 400 V. A fast-recovery diode was selected to reduce switching losses and to minimize the voltage oscillations at turn-off.

Zero-current switching is realized through a constant on-time scheme using a Unitrode UC1860 control chip. The on-time is related to the 5 μF capacitor and 6 μH inductor and is selected to provide zero-current turn-off of the IGBT at all temperatures. If either the capacitance or inductance varies with temperature, then the soft switching of the IGBT may be lost. The 6 μH inductor was constructed using a Magnetics, Inc. 55717-A2 powdered permalloy torroid core. Testing verified that the inductor value remained constant with temperature. The 5μF capacitor must also be insensitive to temperature variations. Since suitable high-temperature capacitors could not be located, this capacitor was placed outside the test oven.

Two other elements in the converter are the input inductor, which has a value of 4.5 mH, and the 45 μF output filter capacitor. A Magnetics Inc. 58867-A2 powdered ferrite torroid core was used for the input inductor. Testing verified that the inductor value was relatively insensitive to temperature. The output filter capacitor value was set by the allowable ripple voltage on the output. A value of 45 μF was selected for this converter. This capacitor was located outside the test oven because of the difficulty in locating suitable high-temperature capacitors.

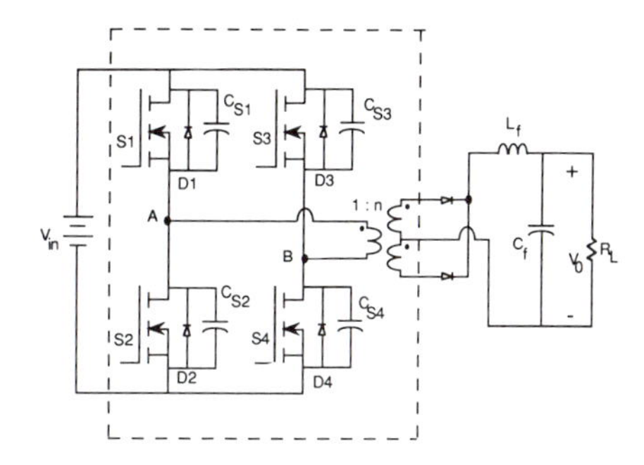

Figure 2.34 Phase-shifted PWM converter.

The control circuitry for this converter was also placed in the test oven. Again, the Unitrode UC1860 control chip was utilized to control the output voltage of the converter. This chip is contained in a ceramic package and is rated for operation to 125°C. Testing revealed that the 5 V precision reference and the output drive stage on this chip did not function properly at the elevated temperatures. Variations in the precision reference caused the converter output voltage to vary. The output drive stage was capable of sinking and sourcing enough current to switch the IGBT until the temperature reached 170°C. Operation above these temperatures required the addition of a current buffer consisting of an NPN and a PNP bipolar transistors in TO-18 metal packages.

This power converter, excluding the control circuit and the two capacitors, operated as designed at ambient temperatures between 30 and 200°C. Although no extensive life testing was done, the circuit was found to be generally reliable. It operated through repeated temperature cycling and at 200°C for periods up to seven hours without a single failure.

100 W Phase-shifted PWM converter: The next power converter tested at elevated temperatures employed a full-bridge, zero-voltage-switched (ZVS) PWM converter [Sable and Lee 1991, Sabate et al. 1991] as shown in Figure 2.34. With the exception of capacitors Cs1-Cs4, the dashed box in this figure indicates that part of the circuit that is tested in the oven. The capacitors Cs1-Cs4 are lossless snubber capacitors with small capacitance values. Difficulty in locating suitable snubber capacitors capable of operation up to 200°C prevented their inclusion in the circuit. Switches S1-S4 are implemented with power MOSFETS. Diodes D1-D4 could be the intrinsic diodes of the MOSFETs or discrete diodes. A slight improvement in efficiency was achieved with the use of discrete diodes in this case. Since this circuit is a buck-derived topology, a transformer is needed to step the voltage up to the desired level. This transformer is a center-tapped design to improve efficiency by reducing the number of diodes in the output rectifier. The inductor, L_f, and capacitor, C_f, form an output filter. This inductor helps to achieve zero-voltage switching of the MOSFETs. The large capacitance value of C_f prevented its inclusion in the oven.

This converter is designed to provide 100 W at 42 Vdc from a 28 Vdc input. The output voltage is controlled using a phase-shifted PWM control scheme, using a Unitrode UC 3875 control chip [Product & Applications Handbook 1994] that generates the switching signals for S1 - S4. This control chip and its auxiliary components were located outside the oven. The switching frequency is arbitrarily selected as 25 kHz. The components in the dashed box of Figure 2.34 are interconnected, using teflon-coated wire and nickel-plated solderless terminals. The MOSFETs and diodes are mounted on ananodized aluminum heatsink.

Switches S1-S4 in Figure 2.34 are International Rectifier's IRF044 MOSFETs. This

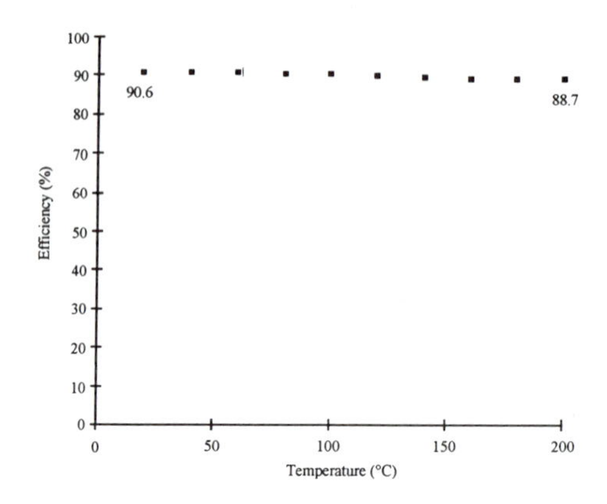

Figure 2.35 Efficiency vs. temperature for 100 W phase-shifted PWM converter.

device is rated for 60 V and 30 A at a case temperature of 25°C. It is not rated for operation above a case temperature of 175°C. Diodes D1-D4 are Motorola's MUR5015 ultrafast diodes with voltage and current ratings of 50 A and 150 V, respectively, and a maximum junction temperature of 175°C. The center-tapped transformer was designed using a Magnetics powdered iron core (part no. 58867-A2-4). It has 22 turns on the primary and 44 turns on each of the secondary windings. The magnetizing inductance of this transformer is 54 μH; this value was relatively insensitive to variations in temperature. The output diodes are also MUR5015 diodes. The inductor in the output filter is composed of 45 turns on a Magnetics core (part no. 58867-A2-4) and has an inductance of 136 μH. The filter capacitance is 330 μF.

With an output voltage of 42 Vdc and a load resistance of 15 Ω, the output power for this prototype is actually 117.5 W. Efficiency data was obtained through the following procedure. The oven temperature was adjusted to the desired value. After 30 min at this temperature setting, voltage and current measurements were taken with digital multimeters, and the temperature was moved to the next setting. The efficiency is plotted as a function of temperature in Figure 2.35; it starts at 90.6% for a temperature of 20°C and decreases to 88.7% at 200°C. The efficiency numbers may be improved slightly by adding capacitors Cs1 - Cs4. The power required for the control circuitry is also included in calculating the efficiency data. In life testing, this converter was operated at 117.5 W and at a temperature of 200°C for a period of 1000 hours without failure. The efficiency held constant at 88.7% during this test period. The output voltage remained at 42 Vdc throughout the test.

500 W phase-shifted PWM converter: A 500 W version of the phase-shifted PWM converter was also designed, constructed, and tested at elevated temperatures. The switching frequency remained at 25 kHz, and only minor modifications were made to the control circuitry to accommodate the increased power level. In an effort to keep conduction losses low, switches S1-S4 are actually two IRF044 MOSFETs connected in parallel.

Because of the increased power level, a new transformer was constructed using a Magnetics Square Permalloy Tape Wound core housed in an aluminum case. These cores are designed to withstand temperatures up to 200°C. The aluminum case prevents distortion of the core at elevated temperatures to maintain the magnetic properties of the core. The tape for this core has a thickness of 0.001 in. and a width of 1.0 in. The primary winding is composed of six turns of four 10 AWG teflon-coated conductors, while each secondary winding consists of 12 turns of one 12 AWG teflon-coated conductor. All transformer inductances were relatively insensitive to temperature variations; however, the winding resistances did increase with increasing temperature.

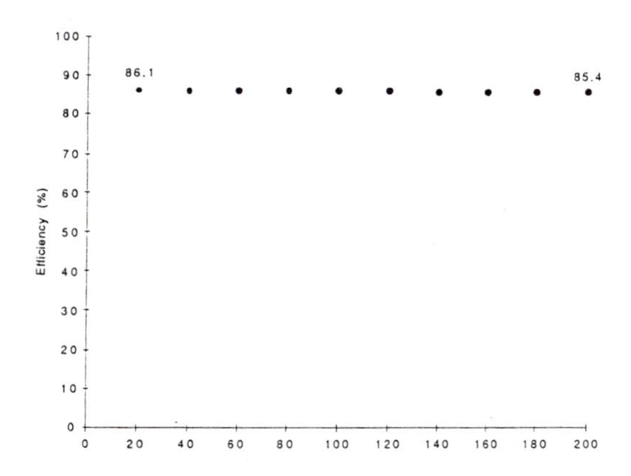

Figure 2.36 Efficiency vs. temperature for the 500 W phase-shifted PWM converter.

When this power supply was operated on the laboratory bench, the efficiency was determined to be 89%. The MOSFETs, diodes, and transformer were then placed in the oven. Because of the physical size of the oven, long feedthroughs were required to connect the input supply to the H-bridge and the output of the transformer to the output rectifier. The resistance of these long feedthroughs caused the efficiency to decrease to 86.1% when the power converter was operated in the oven at 20°C. The efficiency was measured over a temperature range of 20 to 200 °C. The output voltage and input voltage are measured with digital multimeters. Both of the currents are measured with 50 A current shunts. The power required by the control circuitry is included in the input power. The converter was operated at each temperature level for 30 minutes before any measurements were recorded. As can be seen in Figure 2.36, the efficiency decreased from 86.1% to 85.4% as the temperature increased from 20 to 200 °C. After this test, the power converter was subjected to life testing. This unit operated for approximately 3600 h at 200°C. The efficiency held constant at 85.4%, while the output voltage remained at 42 Vdc.

2.7 Summary

In summary, the performance of SOI devices is well enough understood and characterized to enable the design and fabrication of circuits that will function reliably at temperatures to 300°C. The greatest limitation is the cost, quality, and availability of the substrate material. Applications for high-temperature devices are being developed, and a number of organizations are evaluating the business aspects of becoming suppliers to meet these needs

Several major factors influence the life of conductors and contacts in IC electronics operating at elevated temperatures. These include electromigration, corrosion, resistivity, grain structure, step coverage, and operating temperature. Problems with some information in the literature on activation energies of electromigration and published insights on activation energies and current density influence have been identified. Candidates for conductors in each of these temperature ranges have been suggested, with some discussion of both their advantages and disadvantages.

Even though materials such as SiC and diamond are being developed for high-temperature electronic circuitry, silicon devices may continue to play a particular role in this area. Two reasons for this are the existence of an immense, global infrastructure for processing silicon and of some applications, such as automobiles, for which the ambient temperature does not exceed 200°C. This section has reported on the operation of commercial-grade silicon power devices in power converters operating at ambient temperatures up to 200°C. Four different power converters, constructed from commercial components, were tested under various operating conditions. Although relatively few, these tests have demonstrated that it is possible to operate commercial- grade silicon power devices at elevated temperatures. The devices should undergo more extensive life testing and be subjected to thermal cycling. This information could provide insight into the impact of the elevated temperatures on device reliability.

One conclusion derived from this testing is the importance of minimizing the power dissipation in silicon power devices. Power losses cause the device junction temperature to be near the ambient temperature. Conduction and switching losses are the two components of device power losses. One method to reduce conduction losses is to operate the device well below its ratings, in essence derating the device. Switching losses may be reduced through the use of soft-switching techniques in the power converter. An additional benefit of soft-switching is the reduction of switching stresses, which is more important at elevated temperatures. The change in efficiency with temperature, and not the absolute efficiency of the power converters, is a key issue.

Chapter 3

WIDE BANDGAP SEMICONDUCTORS

3.1 GaAs and Other III-V MESFET and Bipolar Devices

Gallium arsenide (GaAs) is the next most commonly used semiconductor material after silicon. GaAs is a member of the III-V compound semiconductor family. These semiconductor materials have properties that allow fabrication of devices whose functions cannot be performed with silicon devices. The most important of these properties are higher electron mobility and direct bandgaps. III-V processing is a fairly well-developed science [Williams 1990] and there are substantial industries that manufacture high-frequency (microwave and digital) discrete components and integrated circuits, and optoelectronic devices such as LEDs, solid-state lasers, and photodetectors. Metal semiconductor field effect transistors (MESFETs) and their derivatives are the most commonly used transistors in GaAs. This section reviews MESFET operation, degradation of GaAs MESFET performance at high temperatures, GaAs reliability, contact metallization, and work done with other III-V compounds to demonstrate operation at temperatures up to 550°C.

Semiconductors can be formed by alloys of the III-V compounds, commonly gallium, aluminum, and indium from column III of the periodic table, and arsenic and phosphorus from column V. A diagram of energy bandgap vs. lattice mismatch on GaAs is shown in Figure 3.1. These compounds have bandgaps which vary from 0.3 eV to nearly 2.5 eV. Because many of the bandgaps are larger than silicon's, they offer potentially better high-temperature performance than can be achieved with silicon devices. For example, the GaAs bandgap is 1.34 eV at room temperature. The crystal lattice spacings of these compounds are close enough to that of GaAs that they can be grown epitaxially on GaAs wafers by various techniques. Heterostructures formed by layering different compounds can take advantage of favorable properties of the compounds to address different aspects of device fabrication and performance. (Potential heterostructure approaches to developing high-temperature devices will be outlined in the next section.)

For device fabrication purposes, an essential difference between silicon and III-V compounds is that Si forms a tough, adherent native oxide, while III-V oxides are not adherent. For this reason, MOSFETs cannot be made in III-V's; MESFETs are the most common transistor type in GaAs ICs. Similarly, there are no dielectrically isolated GaAs technologies analogous to silicon-on-insulator (SOI). Device isolation is achieved using reverse-biased junctions and semi-insulating GaAs substrates. Semi-insulating GaAs has a resistivity of 10^7-10^8 ohm-cm, due to its crystal growth with a high density of traps [Ferry 1985]. At high temperatures this method fails because of increased conductivity in the semi-insulating material. Surface passivation is a more difficult issue, typically addressed with a deposited silicon nitride or oxide coating.

Figure 3.2 is a sketch of a MESFET cross-section. A depletion-mode MESFET is similar to a MOSFET, in that a negative gate voltage is used to deplete an n-type channel that

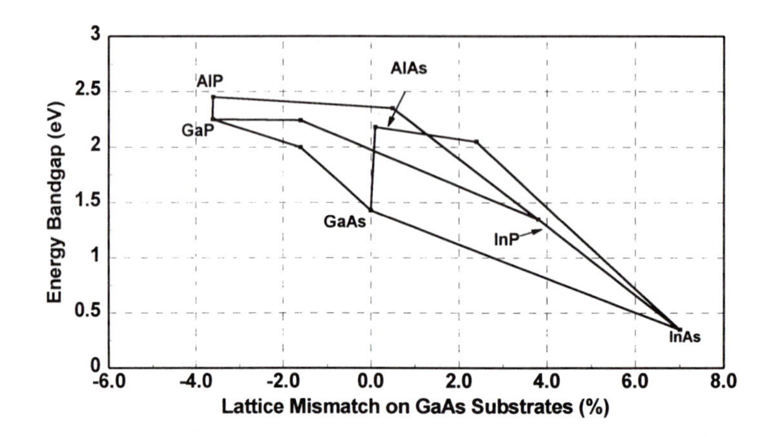

Figure 3.1 The bandgap-lattice mismatch diagram for elements commonly used in manufacture of III-V compound semiconductors. The labeled points are for binary compounds. The slope discontinuities, for example between AlAs and InAs, indicate that the properties do not vary smoothly with Al in In concentrations.

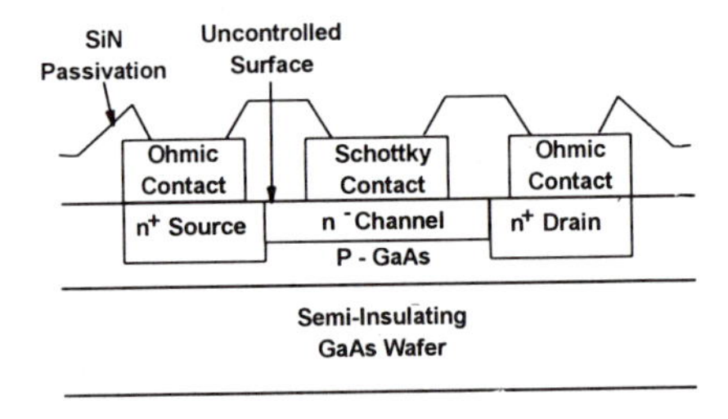

Figure 3.2 The basic MESFET structure showing source and drain ohmic contacts, the Schottky gate contact, and the channel stop.

connects n^+ source and drain. Rather than using an MOS capacitor to deplete the channel, a Schottky diode, discussed below, is used instead. The MESFET is typically made by first growing a thin layer of lightly doped p-type GaAs on a semi-insulating substrate. The n-type source, drain, and channel are formed by ion implantation. The Schottky contact and ohmic contact are formed by vapor deposition and heat treatments; the order of the processing steps varies depending on the materials used for the contacts. Finally, a passivation is applied. High-temperature performance of conventional MESFETs is limited by leakage currents at the gate, drain-to-substrate, in the substrate, and by the reliability of the ohmic contacts. Enhancement-mode MESFETs are also made, but have limited versatility because they require forward-biasing the gate.

P-channel devices are rarely used because the hole mobility in GaAs is very low compared with the electron mobility. This discussion will pertain almost entirely to n-channel MESFETs.

The electronic properties of metal-semiconductor interfaces are described in many textbooks [Streetman 1990]. However, a review of some of the phenomenology and fabrication methods for rectifying contacts and ohmic contacts is useful for understanding high-temperature performance limitations. In the simple theory, rectifying (Schottky) contacts, similar to pn junctions, are formed between a metal and a semiconductor if the work function of the metal is greater than that of the n-type semiconductor. A depletion region is formed in the semiconductor, whose thickness depends inversely upon the doping level of the semiconductor. Applying reverse bias to the junction increases the depth of the depletion region, modulating the drain-to-source conductance. The pinch-off voltage, V_p, depletes the entire depth of the channel and is given by:

$$V_p = \frac{q N_d a^2}{2\varepsilon} \tag{3.1}$$

where a is the channel thickness. Note that V_p is temperature-independent. The voltage-current characteristics of the Schottky junction (gate) are similar to those of a pn junction:

$$I_{GS} \propto e^{-q\varphi_B/kT} \left(e^{qV/kT} - 1\right) \tag{3.2}$$

where φ_B is a barrier height, which is a property of the metal semiconductor interface and depends upon the metal, semiconductor, and fabrication method. Unlike a MOSFET, there is gate leakage current that increases exponentially with temperature. Large barrier heights are beneficial for high-temperature operation. Schottky contacts are readily formed by metal deposition directly on the semiconductor. The contacts are usually annealed at high-temperature to stabilize or improve their properties. Older technologies used aluminum Schottky contacts. Today, Ti/Pt/Au, Ti/Pd/Au, and refractory metal nitrides such as TiN and TiWN are common in commercial devices.

It is important to distinguish between idealized and actual ohmic contacts. An idealized ohmic contact is formed if the work function of the metal is less than that of the n-type semiconductor, so that majority carriers accumulate in the semiconductor near the interface. In a practical sense, the ohmic contacts need only have resistances that are small compared with the parasitic resistance of the MESFET; they need not be truly ohmic. A low-resistance contact is formed if the semiconductor is very heavily doped. Even if the semiconductor does not go into accumulation, heavy doping reduces the potential barrier and narrows the depletion region, allowing majority carrier tunneling through the depletion region to the metal. Heavy doping is done in two or three ways. In an alloyed contact, several metals are typically deposited on the semiconductor, followed by a heat treatment. One of the metals will penetrate and dope the semiconductor. Numerous other chemical and metallurgical reactions may take place. Alloyed contacts are not likely to be stable near their formation temperatures. In an as-deposited contact, the semiconductor is doped by some means, such as ion-implantation (and annealing) or during its epitaxial growth. A relatively inert metal may then be deposited on the semiconductor to form a contact with a thermodynamically stable interface. Most commercial devices use an alloyed contact of gold, germanium, and nickel. Evidently germanium is driven into the GaAs where it is activated to heavily dope the GaAs n-type, and the remaining GeAu forms a near-eutectic mixture with a melting temperature of about 356°C.

High-temperature operation raises questions about both device functionality and lifetime or reliability. Both issues have been addressed in the literature for GaAs devices. Shoucair and Ojala [1992] evaluated the functional performance of MESFETs on a test chip produced using a commercial process (MOSIS design rules, with Vitesse as the foundry) over a temperature range from 25 to 400°C. They found that the MESFETs were functional over this range, although the characteristics degraded steadily with increasing temperature. Drain leakage currents were proportional to n_i, reaching 10% of the saturation current at between 350 and 400°C. The leakage was actually two orders of magnitude larger than for a silicon MOSFET with similar geometry. The gates leakage increased, and the small-signal gate input resistance decreased fairly logarithmically from about 10^{10} to 10^4 ohms between 25 and 400°C. Like silicon MOSFETs, the MESFETs exhibited a zero thermal coefficient point for the drain current. They concluded that considerable effort would be required to develop integrated circuits that operate above 250°C. Their work also suggests that conventional devices should be functional somewhat above 125°C.

Because of increased reaction rates for all types of chemical and metallurgical reactions at elevated temperature, reliability is an important concern. Data collected by Magistrali et al. [1990] from the literature on accelerated lifetime tests of gold-containing Schottky barriers typically show electrical changes after a few hundred hours in the 200 to 300°C temperature range. Refractory metal Schottky contacts have been developed [Papanicolaou 1994] using materials such as TiWN, WSi_x, $TiWSi_x$, and TiW. These contacts are typically annealed for short times at up to 800°C, so that they may be expected to have long lifetimes at lower temperatures such as 500°C, although extended life tests have not been reported. Au-Ge ohmic contacts can be expected to degrade near 300°C because of the eutectic temperature of 356 °C. Magistrali's tabulation of this data shows wide variation in ohmic contact lifetimes, but one can conclude that they do not have long life above 300 °C. Sokolich et al. [1991] described microwave FETs with an 800-h lifetime at 250°C, apparently limited by ohmic contact degradation. Triquint Semiconductor, one of the major GaAs manufacturers, gives median device lifetimes of about 1000 h in 260°C testing [Quality and Reliability Assurance Manual, 1992].

Progress has been made in developing high-temperature ohmic contacts and more appears to be possible if metallurgical and thermodynamic principles are applied to the problem. These principles rely on understanding the miscibility of different metals, interdiffusion, and barrier heights between the metal and semiconductor. These principles are applied with varying degrees of deliberateness so that three approaches are reported in the literature for forming high-temperature ohmic contacts.

One approach uses diffusion barriers to keep gold away from the GaAs. Fricke and co-workers [1989] have worked on developing high-temperature GaAs technology for over a decade. They describe MESFETs using ohmic contacts with a WSi diffusion barrier functioned to 400°C, with >1000 h lifetimes at 300°C. The lifetime at 400°C was <100 h. A simple operational amplifier built with these MESFETs operated at 300°C.

A second approach uses alloyed contacts without gold, such as GeInW and NiInW [Murakami et al. 1988], NiInGeMo [Swirhun et al. 1991] PdIn [Fricke et al. 1994] or GeInNi [Lee et al. 1994]. Vapor-deposited metals are subjected to heat treatments at temperatures as high as 1000°C in order to alloy indium with the GaAs.

A third approach forms an as-deposited ohmic contact on InGaAs. For an indium fraction >70%, InGaAs forms ohmic contacts directly with several metals for x>0.7 [Papanicolaou et al. 1994, Woodall et al. 1981]. In this method, a compositionally graded layer of n-type InGaAs is grown on a GaAs substrate, and the metal contact is vapor-deposited. This property of InGaAs is most likely responsible for the success of alloyed contacts containing In.

To summarize, the upper limit on GaAs IC operation is likely to be set by junction leakage below 400°C. To achieve this with 10^4 h lifetimes will require substantial development of ohmic contacts and metal interconnections.

GaAs BJTs have only received a small amount of attention for high-temperature operation. Doerbeck et al. [1982] found increased leakage current and contact degradation similar to that observed with MESFETs. This early work demonstrated operation (exposed to air) at up to 400°C.

Gallium phosphide has also been studied for high-temperature operation. Zipperian et al. [1982] reported on the fabrication and performance of a GaP pn junction and a BJT operated at temperatures up to 450°C. The unencapsulated pn junction remained stable for 1000 h of operation at 300°C. Weichold described a GaP MESFET operated at nearly 300°C [Weichold et al. 1982]. Photodiode operation at temperatures up to 400°C has also been described [Dreikeet al. 1994, Sims et al. 1994].

3.2 Heterojunction Bipolar Devices

Ternary and quaternary III-V material systems grown on GaAs, GaP, or InP substrates offer significant potential advantages for high-temperature component development. Materials such as AlGaAs, AlGaP, and AlInGaP are lattice-matched to GaAs and GaP, as shown in Figure 3.1. Heterostructures of these materials offer many potential solutions to the electrical problems associated with the use of bulk III-V compound materials; different materials may be chosen to selectively address individual problems of reverse leakage, device isolation, forward voltage drops, contacting, or chemical passivation. Heterojunction systems have been shown to aid contacting problems, reduce resistive parasitics, and reduce both forward voltage drops and reverse current leakage in pn junction structures used in diodes, bipolar junction transistors (BJTs), and semiconductor-controlled rectifiers (SCRs) [Zipperian et al. 1982, Sze 1981]. These advantages exist without compromising excellent high-temperature junction performance because of the wide bandgaps.

A properly designed p^+N heterojunction (an upper case "N" denotes the wider bandgap material in a heterojunction) can have a reverse leakage comparable to that of a pn homojunction made from the wider bandgap material, while having a forward voltage drop comparable to that of a pn homojunction made from the narrower bandgap material. The complete physics of a heterojunction is more complicated than that of a homojunction, but some important points can be drawn from the discussion of pn junctions in Section 2.1. The reverse leakage current is from minority carriers generated in or near the depletion region that drift across the junction due to the built-in potential. In an abrupt p^+n junction, the depletion region is wider on the n-side of the junction by the ratio N_d/N_a, based on a charge conservation argument, independent of the semiconductor bandgaps. So, particularly in direct semiconductors like GaAs, essentially all current is carried by minority carriers (holes) generated thermally in the n-side depletion region, and drifting across the junction. If, however, a p^+N heterojunction is constructed, the ratio of the widths of the depletion regions is unchanged, but the hole generation rate on the N side will be reduced exponentially by the larger bandgap. The reverse current may be further reduced if the N material is an indirect semiconductor. Then, in forward bias, current in the pn junction is due largely to majority carriers diffusing from the bulk to the junction, with sufficient energy to surmount the bias-reduced built-in potential. The heterojunction's built-in potential will lie between the built-in potentials for the wide and narrow bandgap homojunctions. A sketch of a p^+Nn heterojunction is shown in Figure 3.3a.

The high-temperature performance of heterojunctions was first studied by Zipperian and coworkers [Zipperian et al. 1982a, b]. Reverse current density in abrupt junctions made from Si, GaAs, GaAs/$Al_{0.3}Ga_{0.7}As$/GaAs, and GaP is shown in Figure 3.4. The optimum heterojunction performance was achieved with a 30% Al fraction. Though not shown by this data, the GaAs/$Al_{0.3}Ga_{0.7}As$/GaAs diode had a forward voltage drop about 35% smaller than the GaP diode under all conditions.

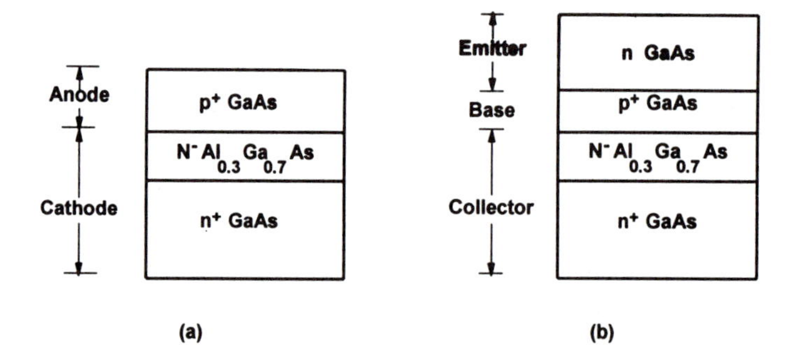

Figure 3.3 (a) A GaAs/$Al_{0.3}Ga_{0.7}As$ p^+N heterojunction. (b) A GaAs/AlGaAs/GaAs heterojunction bipolar transistor for high-temperature operation.

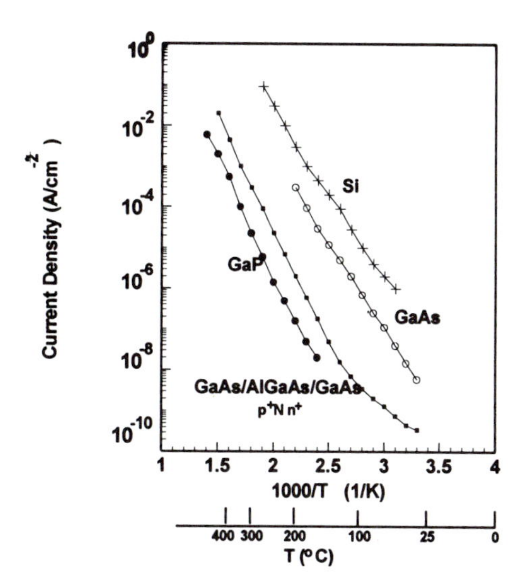

Figure 3.4 Reverse current density in pn junctions made from semiconductors with different bandgaps. Note the reduction in current density with increasing bandgap, and the effect of AlGaAs in the heterojunction [Zipperian et al. 1982b]

High-temperature operation of bipolar transistors is limited by reverse leakage current in the base-collector junction. This reverse current can be reduced in a heterojunction bipolar transistor (HBT) using a wide bandgap semiconductor in the collector. The structure of a high-temperature HBT is shown heuristically in Figure 3.3b.

It is worth noting that the junctions optimized for high-temperature in an HBT are different from those in a conventional HBT [Sze 1981]. A wide-bandgap emitter is used to improve both electron injection into the base and electron transport across the base.

Experimentally, 450°C operation of high-temperature diodes, HBTs, and SCRs, using AlGaAs/GaAs heterojunctions grown on GaAs substrates has been demonstrated. Hetero-structure AlGaP/GaP diodes, BJTs, and SCRs grown on GaP substrates have demonstrated useful electrical characteristics to temperatures in excess of 550°C [Zipperian et al. 1982, Sims 1994, Frost and DiNetta 1986, Fricke 1982]. Unencapsulated, unpassivated diodes in this chemical system have also survived 1000-h life tests at 300°C, with minimal changes in their electrical characteristics [Zipperian et al. 1982a].

The AlGaAs/GaAs heterojunction system has two other advantages over either GaAs, AlGaAs, or GaP alone, based on several physical and chemical effects. First, wide-bandgap materials tend to have more resistive ohmic contacts than narrower bandgap materials; heterojunction systems could be designed in which all ohmic contacts are made to low-bandgap GaAs. Second, a drawback of using AlGaAs is that it is very chemically reactive; heterojunction systems such as those described here alleviate the chemical problems by burying the AlGaAs between relatively inert layers of GaAs. Another application of heterostructures is to make an analog of a SOI wafer by growing semi-insulating, wide-bandgap AlAs on a GaAs wafer. As shown in Figure 3.1, AlAs and GaAs are very well lattice-matched. This approach has been adopted by Lee et al. [1994] who fabricated GaAs MESFETs on top of 0.15 and 0.25 mm thick AlAs buffer layers that were grown epitaxially on GaAs. These MESFETs also used indium-based ohmic contacts, as described in Section 3.1.1. A sketch of this MESFET layout is shown in Figure 3.5. MESFET drain leakage on the buffered substrate was reduced by a factor of thirty, compared with the same MESFET fabricated on a semi-insulating

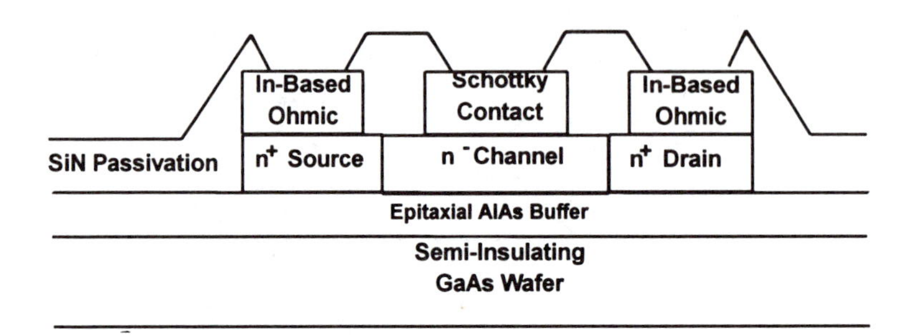

Figure 3.5 A sketch of a mesa-type MESFET constructed on an AlAs buffer layer deposited on a GaAs wafer, of the type studied by Lee et al. [1994] and Wilson and O'Neill [1994].

GaAs substrate. Wilson and O'Neill [1944] have used a commercial device simulator (Atlas2/BLAZE by Sylvaco) to evaluate the performance of AlGaAs/GaAs MESFET and HEMT designs at temperatures in the 300 to 400°C range. This methodology allows them to isolate different current leakage paths, and develop means to control them. Integrated circuits based on this approach appear feasible and promising.

In addition to the fairly well known AlGaP/GaP and AlGaAs/GaAs heterojunction systems, where significant component development has already occurred, a new chemical system, AlInGaAs/AlGaAs/GaAs holds much promise for high-temperature service. Electrically injected red vertical-cavity surface emitting lasers operating at room temperature have recently been demonstrated [Schneider et al. 1993]. This system offers the possibility of forming lattice-matched very wide bandgap/narrow bandgap heterojunctions. This would be useful in a variety of high-temperature component types. AlInGaP is also a direct-bandgap semiconductor. This offers the intriguing possibility of a high-temperature optoelectronic component technology. Much work would need to be done to make this a viable commercial technology.

3.3 Materials and Devices

3.3.1 Material properties for semiconductor devices

For semiconductor devices that operate at high temperatures or with high power at microwave frequencies, silicon carbide (SiC) offers a unique combination of electronic and physical properties. For example, the bandgap of single-crystal 6H polytype SiC is 3.02 eV, compared with 1.12 eV for silicon (Si) and 1.42 eV for gallium arsenide (GaAs) at 300 K. Compared with Si or GaAs, the larger bandgap of SiC results in an increased maximum operating temperature. SiC diodes and FETs have been operated at 650°C with little loss in electrical properties. In addition, to satisfactory operation at high temperatures, SiC exhibits a higher thermal conductivity, a higher critical electric field at which breakdown occurs, and a saturated carrier velocity equal to that of GaAs at the high fields desirable for high-power

Figure 3.6 Lely cavity (a) and grown platelets (b). Reprinted from D.L. Barett et al. J Crystal Growth, 109, 17, 1991 with permission

devices. Table 3.1 is a compilation of properties that affect device performance of silicon, GaAs, 3C-SiC (cubic), and the hexagonal 6H and 4H-SiC polytypes.

The Johnson figure of merit [Williams 1990] for high power and frequency, based on critical field and saturation velocity, gives 6H-SiC a figure of merit 700 times that of Si and 100 times that of GaAs. The Keyes figure of merit [Ferry 1985] for switching speed, based on thermal conductivity, saturated velocity, and dielectric constant, gives 6H-SiC a figure about 6 times Si and 12 times GaAs. In addition to these attractive electrical properties, SiC can be optically transparent and is physically rugged due to its high mechanical strength, which may be an advantage in device preparation.

3.3.2 Advances in crystal growth technology

The advantageous properties of SiC have been recognized for three decades [Streetman 1990]. Until relatively recently, single-crystal SiC was produced by the Lely growth method and existed only as small platelets of varying size (up to 10 mm). A Lely [Sze 1981] growth cavity and platelet crystals are illustrated in Figure 3.6. In this technique, platelets are grown within a cavity formed in a charge of SiC by heating to 2550°C under essentially isothermal conditions in an argon atmosphere. Under these conditions, platelets nucleate randomly along vapor transport flow paths into the cavity [Sze 1981]. The platelets shown in Figure 3.6 contain growth junctions formed by doping the charge with aluminum (p-type) and over-doping with nitrogen (n-type) during the latter portion of the growth run [Shoucair and Ojala 1992]. The junction area can be viewed as the clear transparent regions in the grown junction platelets.

Although grown junction and diffused junction devices were fabricated using Lely platelets, [Shoucair 1992, Magistrali 1990] they generally displayed non-uniform physical and electrical properties, and the individual platelets were difficult to handle. Consequently, work on SiC material and devices virtually stopped in the U.S. in the mid 1970s [Papanicolaou et al. 1994]. Recent advances in the growth of large single-crystal boules and substrates of SiC [Sokolich et al. 1991, Quality and Reliability Assurance Manual 1992, Fricke et al. 1989, Murakami et al. 1988, Swirhun et al. 1991, Fricke et al. 1994, Lee et al. 1994] and the preparation of SiC epitaxial layers [Woodall 1981] have enabled the fabrication of practical SiC devices, including blue-light-emitting diodes, [Quality and Reliability Assurance Manual 1992] HBTs, MESFETs, and MOSFETs [Woodall et al. 1981].

Table 3.1 Comparison of Fundamental Properties for Devices (300 K)

	E_g (eV)	E_b (MV/cm)	V_{sat} (10^7cm/s)	σ_T (W/cmK)	ε
Si	1.12	0.6	1	1.5	11.8
GaAs	1.42	0.6	2	0.5	12.8
3C-SiC	2.39	2.0	2.5	~4.9	9.7
6H-SiC	3.02	3.2	2.0	~4.9	10.0
4H-SiC	3.26	3.0	2.0	~4.9	—

Large-diameter SiC boule growth owes much of its initial development to the modified sublimation technique of Tairov and Tsvetkov [Sokolich et al. 1991]. In this technique, SiC vapor species are transported under near-vacuum conditions from a subliming SiC source to a seed crystal held at a lower temperature.

3.3.3 Crystal growth system design

The development of SiC boule growth has established the potential for obtaining SiC crystals and wafers suitable for microelectronic device fabrication. Boule growth of SiC is being developed with the goal of fabricating large-diameter (2 in. and greater) crystals with high purity and improved perfection. For this purpose, a sublimation vapor transport system has been designed, consisting of a growth cavity with a quartz-wall induction-heated system that uses high-purity graphite and carbon parts within the hot zone (Figure 3.7). The system is vacuum-tight and capable of high-temperature bakeout to reduce nitrogen, a common source of shallow donors in SiC.

Physical vapor transport (PVT) growth proceeds by sublimation of a SiC source and deposition of the vapor species upon a high-quality SiC monocrystalline seed wafer. The growth rate is controlled by ambient partial pressure, seed-to-source *T*, and temperature gradient at the solid-vapor interface. For experiments described here, the growth rate was varied between 0.25 and 1 mm per hour. Undoped crystals were grown in a 20 Torr high-purity argon ambient provided by the boil-off of ultra-high pure (UHP) liquid argon. N^+-doped crystals were prepared by adding controlled amounts of high purity nitrogen to the argon ambient.

The SiC used for the sublimation source was prepared commercially by a modified Acheson technique, in which carbon- and silicon-bearing components, consisting principally of coke and SiO_2, are first reacted to form a metallurgical grade SiC, which is then purified further [Zipperian et al. 1982].

As-grown crystals were precisely oriented and machined to the required substrate dimensions. Wafers were sliced using conventional Si wafer technology modified to accommodate the increased hardness of SiC. Substrates were prepared using a diamond-based abrasive polish technique. 6H and 4H-SiC substrates fabricated from undoped high-resistivity crystals are colorless and transparent, in contrast to N^+ nitrogen-doped substrates, which transmit mainly in the green part of the visible spectrum for 6H, due to the well-known Biedermann mechanism, [Weichold et al. 1982] and are yellow to amber in color for the 4H-polytype. A typical 2-in. diameter SiC boule grown by physical vapor transport is shown in Figure 3.8 and undoped, high-resistivity, sliced wafers are shown in Figure 3.9.

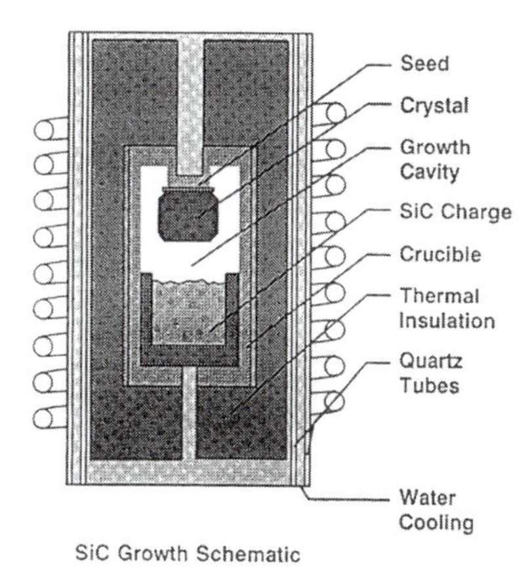

Figure 3.7 Schematic of sublimation vapor transport growth cavity.

Figure 3.8 SiC boule grown by seeded sublimation vapor transport. Reprinted from H.M. Hobgood et al., J. Crystal Growth, 137, 181, 1994 with permission

Figure 3.9 50-mm diameter SiC boule, along with 37-mm diameter high resistivity SiC wafers.

3.3.4 Materials characterization

The distribution and density of crystalline defects in as-grown crystals have been determined by chemical etching, X-ray topography, and transmission electron microscopy (TEM). Dislocation etch pit densities ranging from 10^4 to 10^5 cm^{-2} are revealed by etching polished substrates in molten KOH at 450 to 500°C for 20 to 30 min. Etching experiments and X-ray topographs can also reveal the presence of subgrain boundaries and micropipes or pores within the substrate. Micropipes appear as dark, faceted pits corresponding to tube-like voids of micron-size or less, which can extend the full substrate thickness along the c-axis growth direction.

Micropipes in SiC substrates represent a particularly serious defect from the point of view of device fabrication and yield of working devices, since such defects can render the device nonfunctioning. Although the mechanism of micropipe formation is at present not fully understood in all details, some empirically determined characteristics are now clear. Koga et al. [1992] have related micropipe formation to the purity of the starting SiC source material [Dreike et al. 1994].

Based on recent work, micropipe generation and propagation also appears to be sensitive to the specific seeding techniques employed and to such growth parameters as pressure and temperature. In c-axis crystals, micropipes originate in the vicinity of the seed wafer, lie generally along the growth direction, and are often associated with subgrain boundaries, suggesting an interaction with the dislocations comprising the boundary. These observations are supported by TEM studies confirming a strong tendency for micropipes to align along dislocation cores, where they are often bounded by periodic arrays of stacking faults (Figure 3.10). At present, micropipe densities can range from 100 cm^{-2} to 10^3 cm^{-2}, depending upon the crystal growth conditions employed.

Figure 3.10 TEM micrograph showing micropipe defect surrounded by stacking fault structures in 6H-SiC crystal. Reprinted from H.M. Hobgood et al., J. Crystal Growth, 137, 181, 1994 with permission

Glow discharge mass-spectroscopy (GDMS) and secondary ion mass spectroscopy (SIMS) have been used to monitor relative dopant concentrations in sources and crystals. Under certain growth conditions, boron is the dominant residual acceptor in unintentionally doped crystals (aluminum is the next most abundant). To produce semi-insulating behavior [Sims and DiNetta1994], boules have been intentionally doped with deep-level donor impurities, vanadium in particular. Resistivity values for these crystals were monitored using Lehighton contactless RF eddy-current measurement techniques supplemented by temperature-dependent Hall-effect measurements. N^+ crystals were produced by nitrogen doping and temperature-dependent Hall-effect measurements were used to obtain the carrier concentration and the activation energy of the nitrogen level. Transmission electron microscopy (TEM) and energy dispersive X-ray spectroscopy (EDX) were used for defect analysis.

3.3.5 Results and discussion

N^+ crystals with resistivities <0.02 -cm have been produced by controlled nitrogen doping. Figure 3.11 compares the temperature dependence of carrier concentration in lightly nitrogen-doped (T70-1) and more heavily doped (T110-8) 6H-SiC crystals. The lightly doped crystal exhibits a room temperature carrier density of 2×10^{18} cm^{-3} and a well defined linear region corresponding to an activation energy of approximately 83 meV, in good agreement with that commonly observed for nitrogen occupying the hexagonal lattice position in 6H-SiC. The deviation from linearity at temperatures below about 100 °K is attributed to the formation of an impurity band, leading to roughly constant conductivity as the temperature is reduced. For

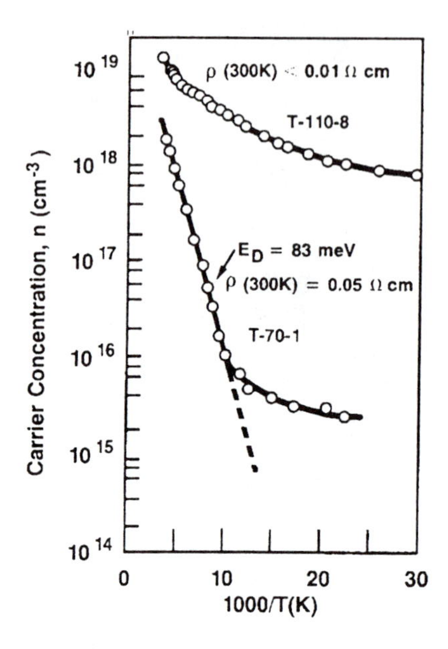

Figure 3.11 Carrier concentration vs. 1000/T for N^+ nitrogen-dopped 6H-SiC at two doping levels. Reprinted from H.M. Hobgood et al., J. Crystal Growth, 137, 181, 1994 with permission

the heavily doped sample, the n vs. 1/T data reveal no well defined linear region. The onset of impurity banding occurs at temperatures approaching 200 °K. These observations are consistent with those of Alekseenko et al., [1987] indicating that an increase in nitrogen doping density in 6H-SiC crystals results in a gradual reduction in the doping vs. 1/T activation energy and a transition to a weak, "metallic-like" dependence of doping on temperature.

Under certain growth conditions, undoped crystals are p-type and can exhibit a range of resistivities, as shown in Figure 3.12, where the room-temperature resistivity of undoped crystals is plotted as a function of crystal length (expressed as increasing wafer number). The resistivity varies from approximately 100 to 5000 cm (the upper limit of the Lehighton technique). Hall-effect measurements of carrier concentration as a function of temperature display activation energies in the 0.350 eV range, consistent with the ionization energy of the lowest boron acceptor level in 6H-SiC. The spread in the observed resistivity is attributed to variability in the contamination level of boron, as well as variability in compensation effects induced by residual shallow donors and deep-level impurities.

Much higher resistivities can be obtained with intentional compensation of shallow impurities by deep-level impurities. Figure 3.13 shows resistivity determined by Hall-effect measurements plotted as a function of reciprocal temperature for intentionally vanadium-doped 6H-SiC samples. For these measurements, samples in the van der Pauw configuration were used with contacts consisting of sputtered titanium annealed at 900°C for 15 minutes. Since direct measurements of the resistivity near room temperature are complicated by the exceedingly high sample resistance and spurious surface leakage effects, temperatures in the 160 to 470°C range were employed. A linear fit of the data over this temperature range yields an activation energy of approximately 1.48 eV, consistent with the expected energy of the vanadium deep-donor level in 6H-SiC. The extrapolated resistivity at 300°K is in the range

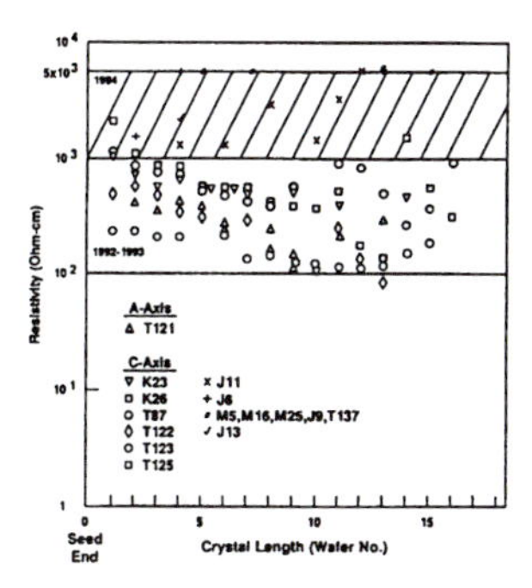

Figure 3.12 Room temperature resistivity of undoped 6H-SiC crystals as a function of crystal length, expressed as increasing wafer number. Reprinted from R.C. Glass et al. with permission from Institute of Physics Publishing, London

of 10^{15} cm. The semi-insulating behavior of the vanadium-doped 6H-SiC crystals is attributable to compensation of residual acceptors by the deep-level vanadium $V^{4+}(3d^1)$ donor center located near the middle of the bandgap [Hobgood 1994]. Doping with different concentrations of vanadium has allowed the determination that there is an upper limit in the mid 10^{17} cm^{-3} range above which vanadium silicide precipitates form (Figure 3.14), thereby degrading crystal quality.

To optimize the compensation of the SiC, the background impurity levels in grown crystals were reduced by reducing source-related impurities (Table 3.2). The effective drop in the concentration of the principal impurities during transport depends on the specific impurity. The reduction of metallic impurities is significant in crystals grown from once-transported source material (1 to 2 orders of magnitude per transport step). Reduction of nitrogen and boron, however, is not as efficient as reducing metallics. The segregation coefficient ([concentration after transport]/[concentration before transport]) of B and N is 0.2 to 0.5, while that of metallics is 0.01 to 0.08.

Therefore, to prepare crystals with controlled high resistivity, it is of fundamental importance to reduce the N and B levels in the starting source, while metallics are more easily reduced *insitu*. When a source with reduced B and N is used (last two lines in Table 3.2), the overall impurity levels reduce, permitting a concomitant reduction in the concentrations of compensating dopants.

In summary, N^+ SiC crystals with resistivities of 0.02 cm have been produced by controlled nitrogen doping. After determining the origin and segregation behavior of the principal residual impurities in PVT-grown SiC, high-resistivity crystals with reduced background impurity concentrations have also been produced. Resistivity levels were extended into the semi-insulating range (10^8 cm at 200°C) by controlled addition of vanadium deep-level impurities. In 6H-SiC, the semi-insulating behavior is attributed to compensation of residual acceptors (mainly boron) by the vanadium deep-donor.

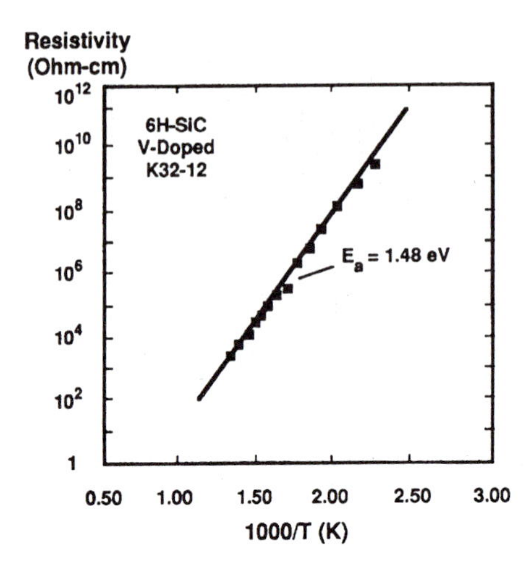

Figure 3.13 Resistivity as a function of reciprocical temperature for a vanadium-doped 6H-SiC grown by the physical vapor transported method. Reprinted from H.M. Hobgood et al. Appl. Phys. Lett., 66(11), 1364, 1995 with permission

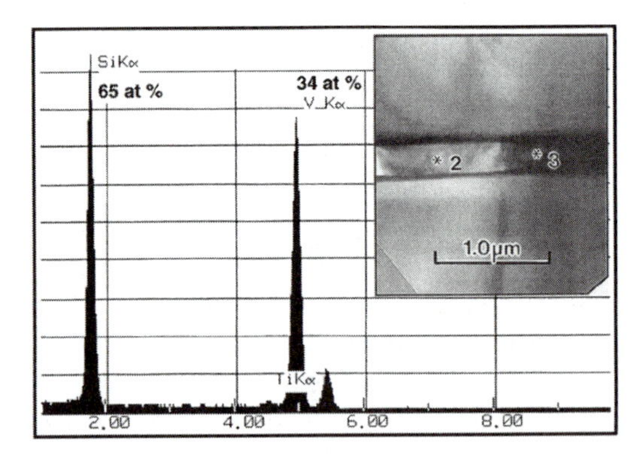

Figure 3.14 EDX scan and TEM micrograph of vanadium silicide precipitates (2 and 3) in SiC with high vanadium concentration added to source. Reprinted from R.C. Glass et al. Sillicon Carbide and Related Materials, 1995 with permission from Institute of Physics Publishing, London.

Table 3.2 Chemical Analysis of SiC Sublimation Source Material and SiC Crystals Grown From Both Original and Once-Transported Sources

	N	B	Al	V	Ti
Source 1	-	8-30	5-48	30-140	35-49
Crystal from - Source 1	9	6	4	0.3	3.0
Crystal from - Transport of 1	2	5	0.8	<0.03	1
Source 2	--	4	8	210	41
Crystal from - Source 2	1	2	2	3	2

Note: ppm wt from SIMS and GDMS data: '--' indicates that the concentration of N was undetermined with these techniques
Reprinted from R.C.Glass et al., Silicon Carbide and Related Materials, 1995 with permission from the Institute of Physics Publishing London

3.3.6 SiC epitaxial growth

SiC microelectronic devices are being actively developed due to their potential for high-power and high-temperature operation, from DC to microwave frequencies. While vapor-phase epitaxial (VPE) growth is currently the most practical means of SiC active layer formation, SiC materials technology, including active layer formation, is still immature [Burk 1996]. In large part, this is due to the high temperatures required to grow bulk and epitaxial SiC. VPE growth of SiC is typically accomplished by passing hydrogen carrier gas containing several hundred parts per million of silane and propane reagents over a bulk SiC substrate at 1450 to 1600^{0}C [Powell et al. 1990]. Because of the high growth temperature, water-cooled quartz growth chambers with inductively heated SiC-coated susceptors are typically employed.

3.3.7 Vapor-phase homoepitaxial growth of 6H- and 4H-SiC

Introduction: Significant breakthroughs in the quality of SiC epitaxial layers have recently been accomplished, enabling the development of SiC-based microwave devices, such as metal-semiconductor field effect transistors (MESFETs) and static induction transistors (SITs) [Sriram et al. 1994, Clarke et al. 1995]. In this section, the VPE growth of device-quality SiC is described. The growth of SiC layers with background doping less than 10^{14} cm^{-3} and intentional n- and p-type doping from 10^{15} to 10^{19} cm^{-3} is used to demonstrate the recent advances in SiC epitaxy. Device profiles with abrupt doping transitions are given. The types and origins of some of the remaining morphological defects in SiC VPE layers are also discussed.

Growth technique: A water-cooled, quartz horizontal VPE reactor was employed for the data presented in this section. All growths occurred at atmospheric pressure using resin-purified silane and propane in Pd-cell purified hydrogen carrier gas at 1450 to 1520^{0}C. The temperature of the inductively heated SiC-coated susceptor was determined using a sapphire light pipe calibrated by melting Si samples. Both n-type (0.01 Wcm) 6H and 4H-SiC c-axis (0001) substrates oriented 3 to 4^0 off towards the a-axis direction, < 1210 >, were used. Intentional n-type doping was accomplished using N_2. Growth rates were from 1.5 to 2.5 μm/hr.

Results and discussion: A comparison of the morphology of two specular, 5 μm-thick 6H and 4H-SiC epitaxial layers produced in the same growth run at 1520^{0}C is shown in Figure 3.15. Growth of SiC on these off-axis substrates occurs by the propagation of steps along the downstep a-axis direction [Ueda et al. 1990, Burk et al. 1994]. Relatively shallow defects, which appear as rounded triangles (hereafter denoted as "amphitheaters"), as well as polishing scratches are observable in the figure. While not shown, 4H-SiC can also exhibit larger, faceted pits with a tetrahedral shape. These were minimized by growing at a temperature above 1500^{0}C and selecting only a high quality 4H-SiC substrate. Both amphitheaters and tetrahedral pits occur because of interruptions in the propagation of steps across the growth surface, either by the substrate surface or by some *in situ* deposit or etch pit developing prior to growth.

Si droplets are a common precursor for epitaxial layer morphological defects. Si droplets occur due to the etching of SiC by hydrogen and also during growth at Si/C ratios higher than one. Figure 3.16 shows photos of a 6H-SiC sample which was removed from the reactor immediately after growth nucleation and then again after a subsequent nucleation and growth of 5 μm of epitaxial SiC. The start-growth conditions were intentionally selected to minimize the occurrence of Si droplets. Despite this, a few submicron Si droplets are still observable after the first nucleation. Upon superposition of the images, every epitaxial layer morphological defect in the final layer can be attributed to a detectable pore, Si droplet, or impression (left behind from a transient droplet) that was observed after the first nucleation. Before the first nucleation, only the micropores and a few faint scratches were visible. The fact that no new population of defects occurs after the second nucleation and growth indicates that the VPE growth conditions have effectively eliminated spontaneous morphological defect formation and that the remaining Si droplets are decorating pre-existing flaws in the substrate or its surface.

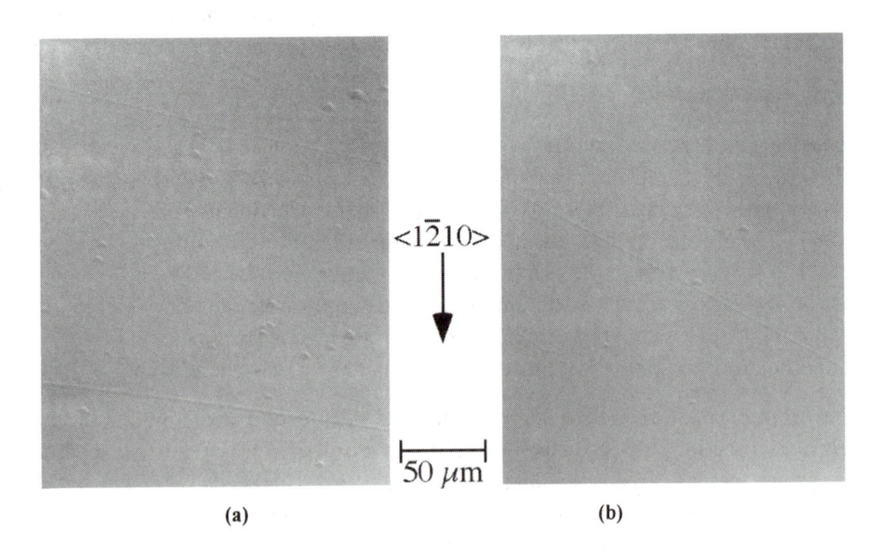

Figure 3.15 Surface morphology of 5 μm-thick VPE layers of (a) 6H-SiC and (b) 4H-SiC. Note the presence of small, shallow rounded triangles ("amphitheathers") and polishing scratches. Reprinted from A.A. Burk, Jr. et al. Silicon Carbide and Related Materials, 1995 with permission from Institute of Physics Publishing London.

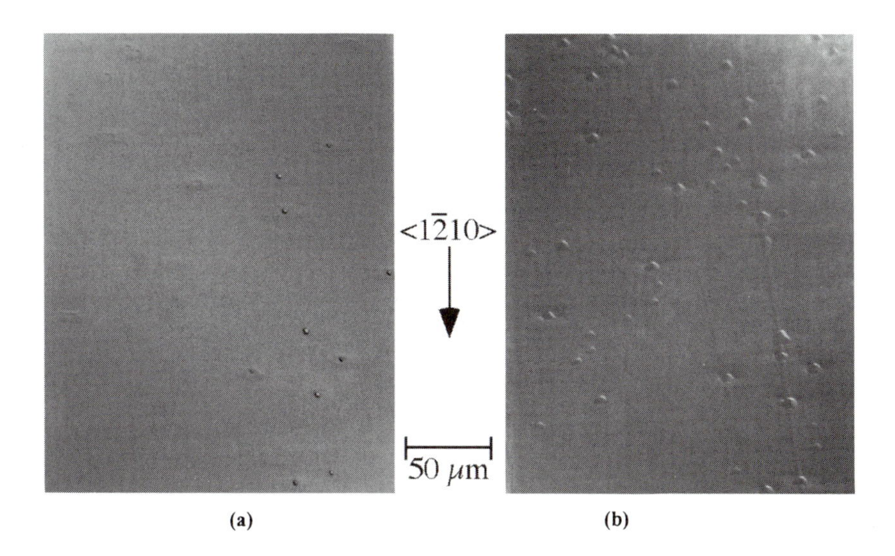

Figure 3.16 Surface morphology of (a) 6H-SiC layer after SiC growth nucleation showing Si droplets and impressions from evaporated Si droplets, and impressions from evaporated Si droplets, and (b) 5 μm-thick SiC VPE layer grown on area shown in (a) without ex situ cleaning. Reprinted from A.A. Burk, Jr. et al. Silicon Carbide and Related Materials, 1995. With permission from Institute of Physics Publishing London.

The ability to produce epitaxial layers with a wide range of doping is accomplished by introducing dopant atoms and varying the input Si/C ratio. Figure 3.17 shows the carrier concentration of a large number of SiC epitaxial layers as a function of Si/C ratio. The data includes both unintentionally doped films and films intentionally doped n-type, using an N_2 mole fraction of 1.2×10^{-4}. The effect of Si/C ratio on doping was first reported for SiC by Larkin et al. [1994] who referred to it as "site-competition epitaxy". Both the 6H and 4H polytypes give approximately the same carrier concentration for a given Si/C ratio. Note that carrier concentrations of less than $1 \times 10^{14}\,cm^{-3}$ are possible at a Si/C ratio of about 0.20. The high purity of undoped SiC VPE layers has been confirmed by low-temperature photoluminescence, as well as by secondary ion mass spectrometry (SIMS) [Clemen et al. 1994].

By precise control of the Si/C ratio and intentional dopant incorporation, films of both n- and p-type SiC can be grown with carrier concentrations from less than 1×10^{15} cm^{-3} to greater than 1×10^{19} cm^{-3}. The carrier concentration can be changed abruptly throughout this wide doping range, allowing growth of active layers for microwave devices. Figure 3.18 shows a SIMS nitrogen profile for a MESFET structure. MESFET devices fabricated on these layers indicate breakdown voltages in excess of 70 V and channel currents of 230 mA/mm. Small-signal RF measurements (taken at a drain bias voltage of 40 V) yield 8.5 dB gain at 10 GHz and an f_{max} of 25 GHz. Power measurements on 2 mm periphery SiC-MESFETs yielded 3.5 W at 6 Ghz [Sriram et al. 1994].

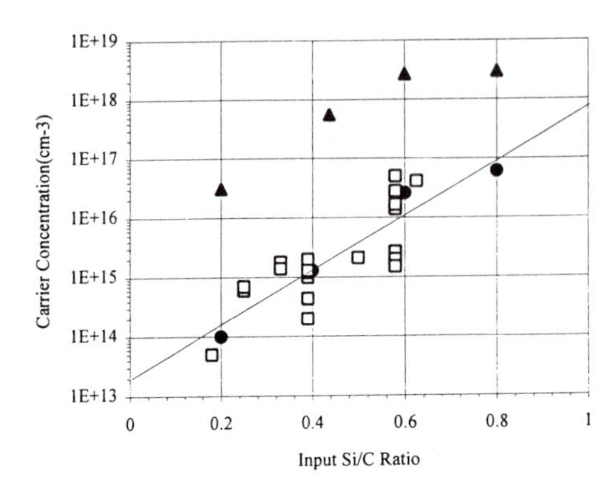

Figure 3.17 Effect of Si/C ratio on n-type doping level for VPE SiC, from CV measurements for undoped layers at 1 μm/hr (circles) and doped layers using $[N_2]=1.2\times10^{-4}$ at 2 μm/hr (triangles). Reprinted from L.B. Rowland et al. Proceedings IEEE/ Cornell Conf. Adv. Concepts High Speed Semicond. Dev. Circ. 1995, ©IEEE.

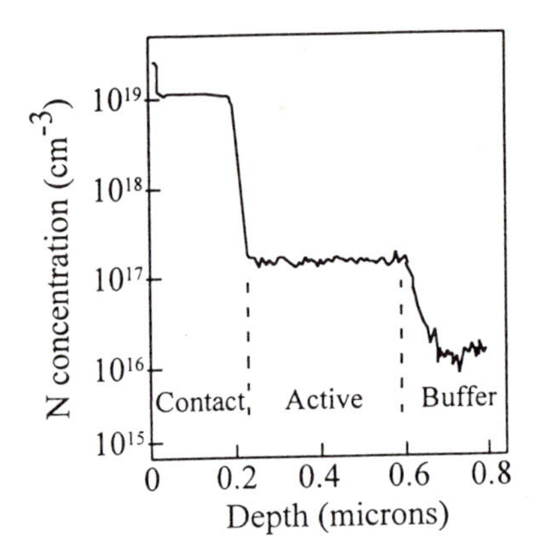

Figure 3.18 Nitrogen concentration as a function of depth as determined by SIMS for a MESFET layer grown using VPE. Reprinted from L.B. Rowland et al. Proc. IEEE/Cornell Conf. Adv. Concepts High Speed Semicond. Dev. Circ. 1995, ©IEEE.

IC devices: Silicon carbide has been heralded as the semiconductor of choice for high-temperature, high-frequency power devices due to its high saturated electron velocity (2×10^7 cm/sec), high thermal conductivity (4.9 W/cm K), and high breakdown field strength (2.5×10^6 V/cm). Devices made with silicon carbide have the potential to fulfill the needs of certain high-temperature aerospace applications, such as propulsion controls, space power systems, and satellites, in addition to earth-based applications such as automotive electronics, power electronics, and nuclear reactor instrumentation.

6H-SiC MOSFET: Recently there has been a push to demonstrate high frequency, high power SiC MESFET transistors [Sriram et al. 1994]; however, it is generally accepted that the instability of a metal-semiconductor interface at high temperatures may preclude its use in devices destined for high-temperature operation. Many researchers have examined metal-oxide-semiconductor interface, devices including the vertical MOSFET [Palmour et al. 1994], the inversion-mode lateral MOSFET [Palmour et al. 1991], and the depletion mode lateral MOSFET [Krishnamurthy et al. 1994]. Most n-channel inversion-mode MOSFET devices to date suffer from a poor oxide-semiconductor interface, due to incorporation of aluminum or carbon in the oxide and at the interface from SiC during oxidation. It is likely that impurities incorporated at the interface or in the oxide will lead to gate instability and long-term reliability problems. On the other hand, very good oxidation characteristics have been demonstrated on n-type SiC [Brown et al. 1991], making an n-type depletion mode MOSFET an attractive high-temperature device in SiC.

A depletion mode 6H-SiC lateral MOSFET for high temperature applications is described in this section. The device fabrication closely follows SiC MESFET fabrication, [Sriram et al. 1994] with the exception that before a gate is applied to the device, a thermal oxide is grown on the active channel area. Due to the use of submicron e-beam lithography, channel currents of 225 mA/mm and a maximum transconductance of 12 mS/mm have been obtained from these 6H-SiC MOSFETs. Additionally, simple digital and analog demonstration circuits have been fashioned with these devices. Details regarding device fabrication, resulting data, and circuit operation are described in this section.

Device fabrication: One-inch-diameter, 6H silicon carbide wafers were used as the starting material. [Barrett et al. 1993]. Appropriately doped n- and n+ epitaxial layers were grown, using vapor phase epitaxy. The devices were mesa isolated, using a fluorinated reactive ion etch. Channel recessing was performed and a thermal, wet oxidation at 1150°C was employed to grow approximately 500 Å of silicon dioxide.

Contacts to the drain and source were made with nickel, and sintered using a rapid thermal annealing process. Electron-beam direct-write lithography was used to define the gates to obtain precise alignment and dimensional control. Finally, peripheries as high as 2 mm were obtained using a gold-plated air-bridge technique similar to that used in certain GaAs technologies.

Measured results: The resulting MOSFETs are of the depletion mode type (normally on) and are turned off by applying negative gate voltage. The DC characteristics from a 2 mm-periphery, 0.5-mm gate-length device, shown in Figure 3.19, reveal a channel current of 225 mA/mm and a maximum transconductance of 12 mS/mm, the highest reported transconductance for a 6H-SiC MOSFET. These devices also perform well when driven by positive bias, further increasing the available source current by 100 mA. Drain-source breakdown voltage is greater than 80 V. The MOSFETs were examined for high-temperature DC performance by probing with a temperature-controlled chuck to 300°C. The DC characteristic of a 160-μm wide device operating at 300°C is shown in Figure 3.20. The transconductance at 300°C is reduced to approximately half the room temperature value, and the devices still exhibit excellent MOSFET performance at the elevated temperature.

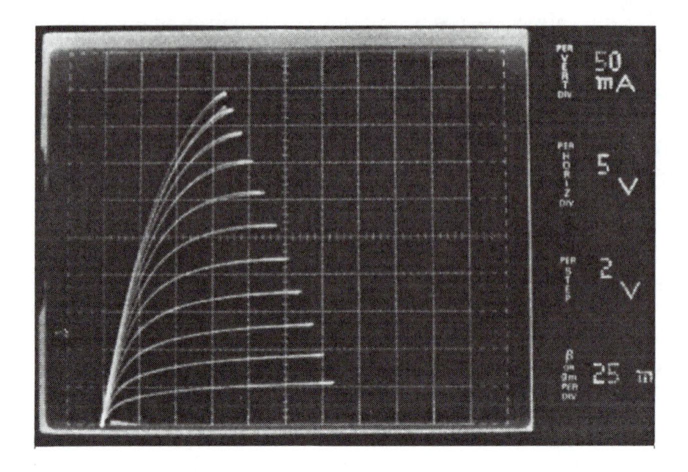

Figure 3.19 Output characteristics for 2 mm device, revealing a 225 mA/mm channel current and a transconductance of 12mS/mm.

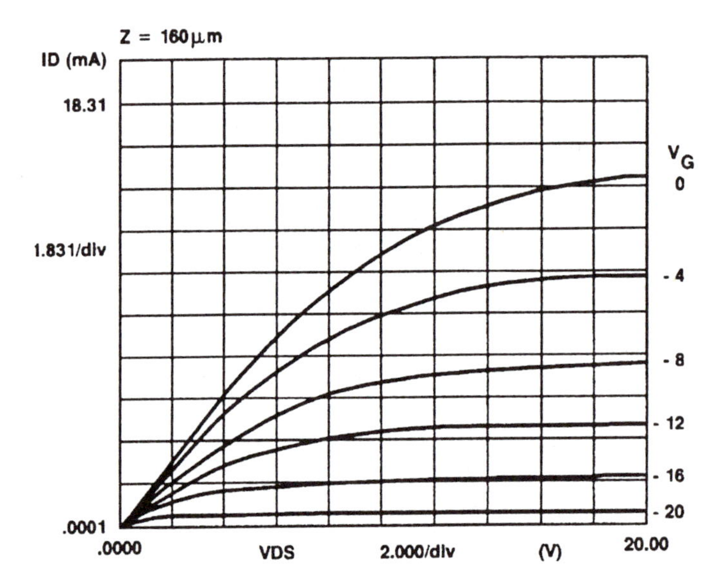

Figure 3.20 High-temperature (573K) DC operation of the 6H-SiC MOSFET.

MOSFET chips were laser scribed and mounted into 40-pin ceramic packages, shown in Figure 3.21, to demonstrate preliminary monolithic SiC circuits. Circuits were fashioned by wirebond connections directly on the die and by off-chip connections via the lead package pins. SiC analog circuits, such as single-stage transistor amplifiers, depletion-loaded differential pairs, and resistive-loaded differential pairs, were tested. Functional SiC digital building blocks, such as inverters and NOR gates, were also tested.

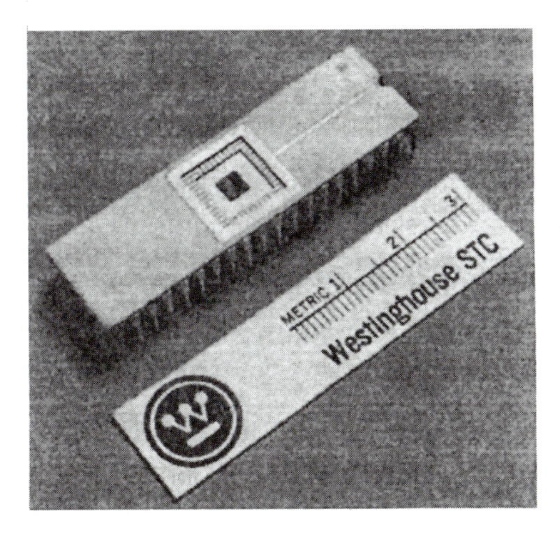

Figure 3.21 Finished 6H-SiC MOSFET die mounted in a 40-pin ceramic package for testing.

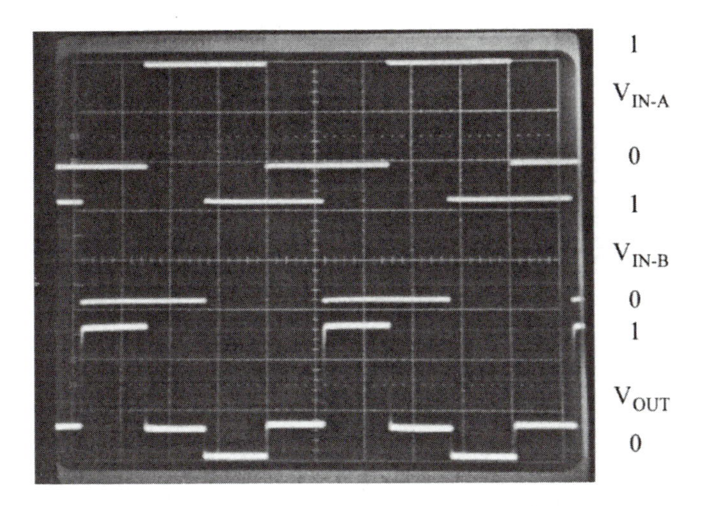

Figure 3.22 Two-input digital NOR circuit is constructed with three SiC MOSFETs.

The input and output wave forms from a two-input digital NOR circuit are shown in Figure 3.22. The input was a 15 V-magnitude 10-KHz square wave. Proper operation of the digital circuits was achieved from DC to frequencies greater than 1 MHZ; however, some distortion in the output characteristics was observed at 1 MHZ due to parasitic wiring capacitance.

4H-SiC UMOSFET: SiC power devices offer several potential performance advantages over silicon power devices in terms of higher reliability, while operating at higher junction temperatures, higher CW power ratings, and reduced switching losses. When combined with the higher thermal conductivity of SiC, significant savings in system cooling requirements, weight, reliability, and cost can be expected in power electronic applications. The potential candidates for SiC power devices include the vertical metal oxide semiconductor field effect transistor (VMOSFET or UMOSFET due to the U shaped active channel region), insulated gate bipolar transistor (IGBT), and gate turn-off thyristor (GTO). This section focuses on the potential performance advantages of the 4H-SiC-based UMOSFET.

In order to illustrate the performance advantages of the 4H-SiC UMOSFET, consider a first-order design of 600 V devices in silicon and 4H-SiC materials. A schematic cross-section of the device structure is shown in Figure 3.23. In the forward conduction mode, a positive gate bias turns on an inversion layer of electrons along the sidewalls of the U-shaped grooves, providing a conduction path for electrons from source to drain. For a small drain bias (0-5 V), the resistance of the inverted channel is given by

$$R_{ch} = \left(\frac{L}{NW}\right)\frac{d_i}{\mu_{n,ch}\varepsilon_0 K_i(V_G - V_T)} \tag{3.3}$$

where L represents the length of the inversion channel; NW is the total width of N sidewalls, each with a width of W; d_i is the thickness of the gate insulator; $\mu_{n,ch}$ is the mobility of electrons in the inversion layer; ε_0 is the permittivity of the free space; K_i is the dielectric constant of the gate insulator; V_G is the gate bias with respect to the grounded source terminal; and V_T is the threshold voltage of the MOS device.

The lightly doped drift layer serves to block the drain voltage under off-conditions. The thickness of this layer, X_{epi}, and the n-type doping density, N_d, are designed to provide the required blocking voltage, V_B:

$$V_B = 0.8\left(\frac{E_s X_{epi}}{2}\right) = 0.8\left(\frac{qN_d X_{epi}^2}{2\varepsilon_0 K_s}\right) \tag{3.4}$$

where E_s is the maximum field at the junction consistent with the maximum negative electric field, E_i., acceptable for the dielectric and the breakdown field, E_c , of SiC, and K_s is the dielectric constant of the semiconductor. It has been assumed that only 80% of the ideal parallel plate breakdown voltage is available with existing edge-termination techniques. As the drain bias is increased with gate at zero volt (the device is off), the maximum electric field occurs at the junction of the p-type doped channel region and the n-type drift layer. In a simplified one-dimensional case, the entire drift layer gets depleted when the maximum electric field at the junction reaches the breakdown field, E_s. At this point, an avalanche breakdown is initiated and the device cannot block any more drain voltage than V_B. However, under the on-condition (V_G-V_T > 0 V) and low drain bias (V_D = 0 to 5 V), the n-type drift layer presents a resistance, R_{epi}, to the current flow from source to drain:

$$R_{epi} = \left(\frac{1}{q\mu_n N_d}\right)\frac{X_{epi}}{A} \tag{3.5}$$

where μ_n is the bulk mobility of electrons in the drift layer, and A represents the total cross-sectional area of the device perpendicular to the current flow, or the footprint of the active area of the device. The area, A, is determined by the cell pitch, S, and the total number of sidewalls, N:

$$A = \left(\frac{N}{2}\right) WS \tag{3.6}$$

The specific on-resistance of the device is given by

$$\rho_{on} = (R_{ch} + R_{epi})A \tag{3.7}$$

Equations 3.3 to 3.7 represent a simplified, first-order, one-dimensional design of the UMOSFET and can be effectively used to compare Si and 4H-SiC materials. Given a required value of the breakdown voltage, N_d and x_{epi} are calculated from Equation 3.4. This requires the knowledge of E_s, which is determined by the breakdown field of Si in the case of silicon devices. In the case of SiC, the breakdown field, E_c, of the semiconductor is about 2 MV/cm, which results in an unacceptable field of about 5 MV/cm in the gate oxide. Typically, gate oxides are not characterized under negative gate bias because the oxide field in silicon MOSFETs rarely exceeds about 0.9 MV/cm, due to the low breakdown field strength of silicon (E_c = 0.3 MV/cm). Here, we assume a value for the field in the oxide under negative gate bias to be about E_{i-} = 4 MV/cm, which translates into an E_s = 1.56 MV/cm for SiC. Then, for a given area, A, for example, 1 cm^2 and a design pitch S of 10 m, the product WN is calculated from Equation 3.6 and used in Equation 3.3 to compute R_{ch}. The positive gate voltage, V_G, depends on the maximum total electric field, $E_{i+} = V_G/d_i$, (in the insulator over the N^+ region) that the gate insulator may be stressed while avoiding long-term reliability problems like time-dependent dielectric breakdown (TDDB). Finally, R_{epi} is calculated from Equation 3.5 and on from Equation 3.7. Table 3.3 summarizes such calculations for 600 V designs in Si and

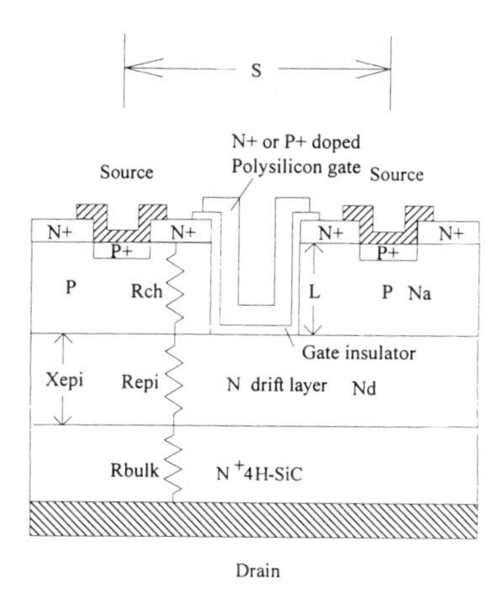

Figure 3.23 A schematic cross-section of a unit cell of the 4H-SiC UMOS, showing all the active layers. In a device, there are N cells of each of width W and pitch S. Reprinted from A.K. Agarwal et al. Proc. 8th Int. Symp. Power Semicond. Dev. Ics, 1996 with permission, ©IEEE

4H-SiC.

Although the calculations in Table 3.3 are simplified, the results are interesting. The high breakdown field of 4H-SiC helps in two ways. First, it reduces the required thickness of the drift layer, x_{epi}, down to 9.6 μ m from 50 μm for the silicon case. In fact, it can be reduced further by designing the drift layer for punch-through. Second, the doping in the drift layer increases by about a factor of 23. The combination of these two effects results in a much lower value for the drift-layer resistance, R_{epi}, for the 4H-SiC UMOSFET. Due to a lower assumed value of the inversion channel mobility of electrons, the channel resistance is somewhat larger for 4H-SiC.

It is interesting to note that, whereas the silicon design is completely dominated by the drift-layer resistance, the 4H-SiC-based design has the potential to be dominated by the channel resistance if surface electron mobility is much less than 100 cm^2/Vs. Considering the bulk electron mobility of 600 cm^2/Vs along the c-axis (direction of the current flow in vertical devices) in 4H-SiC, an inversion-layer mobility of 100 cm^2/Vs is quite reasonable. However, the measured values of the inversion-layer electron mobility are much lower at present due to the poor gate insulator/SiC interface characteristics. The above calculation brings out another critical problem with 4H-SiC UMOS devices. Note from Table 3.3 that E_{i-} = 4 MV/cm is used as the maximum allowable electric field in the insulator under negative bias, and E_{i+} = 2 MV/cm as the maximum field in the insulator under positive bias.

Thus, the gate insulator is required to withstand high electric field stress of either polarity. This may present a long-term reliability problem like TDDB. During the on-condition, the gate drive, (V_G-V_T), can be reduced, which will result in somewhat higher channel resistance. Generally, in the silicon industry, the electric field in the high-quality thermal oxide gate insulator is limited to 2 MV/cm to avoid TDDB problems, even though the breakdown field of the thermal oxide tends to be in excess of 10 MV/cm. However, during the off-condition, the high field stress in the insulator is unavoidable if we are to take advantage of the high breakdown field of 4H-SiC. This is not a problem in the silicon case, due to the much lower breakdown field of silicon.

In light of these considerations, the following technological goals need to be reached for achieving a specific on-resistance in the range of 4 to 8 m Ωcm^2:

(1) high-quality gate dielectric that can withstand a stress of at least 2 MV/cm under positive gate bias and at least 4 MV/cm under negative gate bias on a long-term basis;

(2) high-quality gate dielectric/p-SiC interface along the sidewalls, leading to an inversion-layer mobility of about 100 cm^2/Vs. This implies low density of fast and slow interface traps, $D_{it} < 5x10^{10}\ cm^{-2}eV^{-1}$ and low fixed charge in the gate insulator, $Q_i < 1x10^{11}\ cm^{-2}$.

Accomplishing the above goals will result in a 600 V 4H-SiC UMOS power transistor with ρ_{on} of 4 to 8 mΩ -cm^2, and a cell pitch of 10 μm. It means that the device can support a current density in the on-state, $J_{on} \approx 375$ to 750 A/cm^2 at a forward voltage drop of $V_f = 3$ V, resulting in on-state conduction losses of 975 to 1950 W/cm^2. This is likely to raise the junction temperature to above 200°C, depending on the cooling method employed. Fortunately, at higher junction temperatures, the channel mobility will reduce, thus lowering the current density and preventing thermal runaway. This is a very useful attribute of a MOSFET, in contrast to bipolar transistors, which tend to thermally run away. The high thermal conductivity of SiC substrates, along with higher junction temperatures, tends to simplify cooling requirements.

Table 3.3 Device design and performance calculations for V_B=600V UMOSFET in Si and 4H-SiC materials

Device & Material Parameters	4H-SiC	Si
E_{i+} (MV/cm)	2.0	2.0
E_{i-} (MV/cm)	-4.0	-4.0
E_c (MV/cm)	2.0	0.3
E_s (MV/cm)	E_{i-}(3.9/10)=1.56	0.3
x_{epi} (m)	9.6	50.0
N_d (cm^{-3})	9.0×10^{15}	3.9×10^{14}
A (cm^2)	1.0	1.0
μ_n (cm^2/Vs)	600	1400
R_{epi} (m)	1.1	57.2
L (m)	2.0	2.0
d_i (Å)	500	500
S (m)	10.0	10.0
WN (cm)	2000	2000
μ_{nch} (cm^2/Vs)	100	500
Na (cm^{-3})	1.0×10^{17}	2×10^{16}
n_i (cm^{-3})	1.0×10^{-7}	1.45×10^{10}
$2\varphi_F = 2(kT/q)\ln\{N_a/n_i\}$ (V)	2.87	0.73
φ_{ms} (V) [P^+ poly gate]	-1.52	+0.2
Q_i+Q_{it} (Coulomb/cm^2)	q(5×10^{11})	q(1×10^{11})
V_T (V)	4.33	1.7
V_G (V) {E_{i+}=2 MV/cm }	10+1.4=11.4	10+1.1=11.1
R_{ch} (m)	2.05	0.31
ρ_{on} (m -cm^2)	3.15	57.5

Preliminary results, shown in Figure 3.24, are very encouraging. The device made on 4H-SiC had a periphery of WN = 0.3 cm, a thermal gate oxide thickness of about 100 nm, a cell pitch of 22 μm, a drift layer of 5 μm doped at about 1×10^{16} cm^{-3}, a p-type channel layer doped at 1×10^{17} cm^{-3}, and a channel length of 2 μm. The channel current showed activation, with temperature reaching a maximum at 250°C, thus indicating a large number of interface traps at the thermal oxide/p-SiC sidewall interface. The current density was 18.5 A/cm^2 at 3 V forward voltage, corresponding to a specific on-resistance of about 162 mΩ -cm^2 at 250°C. The maximum positive gate bias was 18 V to keep the positive field in the gate oxide overlapping the N^+ source below 2 MV/cm.

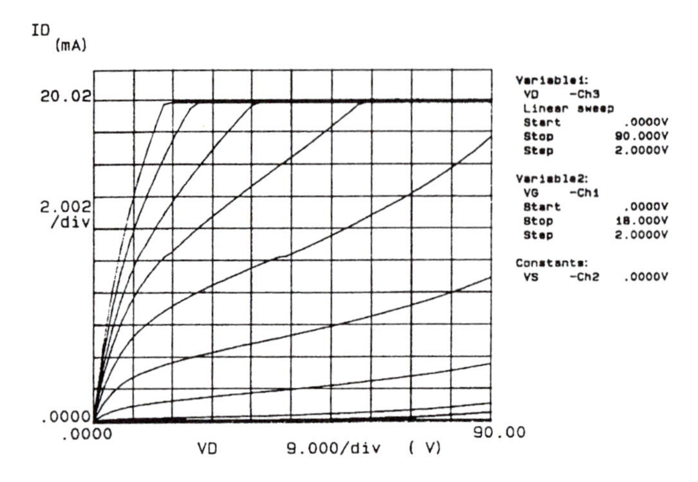

Figure 3.24 I-V characteristics of a 4H-SiC UMOSFET. V_G is stepped from 0 to 18 V in 2V steps. Reprinted from A. K. Agarwal et al. Proc. 8th Int. Symp. Power, Semicond. Dev. Ics, 1996 with permission, ©IEEE

It is clear that the on-resistance is dominated by the inversion-layer resistance due to the poor quality of the gate oxide/p-SiC interface along the vertical sidewall. The breakdown voltage was about 100 V for this device. No special edge termination technique was employed. The results are encouraging, and point towards several problems that need to be solved, such as good-quality gate oxides or alternative gate insulators, appropriate edge-termination, smaller cell-pitch down to 6 to 10 μm, and ways to round the sharp corners at the bottom of the vertical U-shaped grooves that tend to concentrate the electric field, with resulting low breakdown voltages. Moreover, the reliability of gate oxides or alternative insulators, such as oxide/nitride/oxide (ONO) dielectrics, needs to be established at high temperature under appropriate electric field polarity and magnitude.

3.3.8 SiC MESFET

It is well known that SiC is ideally suited to high power generation at high frequencies owing to its remarkable transport properties (v_{sat} = 2.7 x 10^7 cm/sec), very high breakdown field strength (E_{br} = 3 MV/cm), and thermal conductivity (α_T = 5.0 W/cm K). However, the shortcomings of single-crystal wafers and process technology have precluded serious power device investigations until recently.

4H and 6H polytype SiC single-crystal wafers are now available in research grades [Palmour et al. 1994, Hobgood et al. 1994] as 1" and 2" diameter wafers with n-type, p-type, and high resistivity [Hobgood et al. 1995] properties. These wafers, together with important advances in the controlled deposition of doped, high quality VPE films, has allowed an investigation of semiconducting SiC as the starting material in high frequency power transistors. This section describes the use of high resistivity 6H- and 4H-SiC wafers for microwave power MESFETs [Sriram et al. 1994].

Device fabrication: Both 6H and 4H polytypes of SiC have been used for the fabrication of microwave power MESFETs. A critical element in these devices is the high-resistivity substrate, which is essential to minimize capacitive and resistive parasitics that would otherwise degrade high frequency power performance.

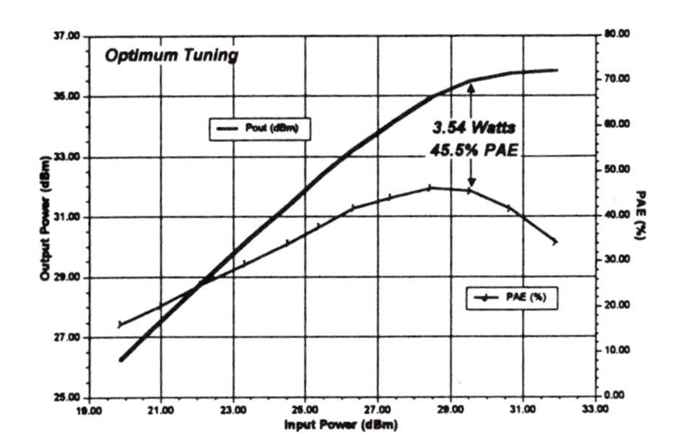

Figure 3.25 6H-SiC MESFET power performance at 6 Ghz (V_d=40V). Reprinted from S. Sriram et al. 53rd Annual. Dev. Res. Conf. 1996 with permission ©IEEE)

Single crystals of SiC grown in the c-<0001> axial direction and with diameters up to 50 mm were used for MESFET fabrication. The crystals were grown by a physical vapor transport technique [Hobgood et al. 1995]. Polished substrates with surface normals oriented approximately 3 degrees off the <0001> direction were fabricated from the crystals. Epitaxial layers were grown on high-resistivity substrates at 1520°C, using an atmospheric pressure vapor-phase epitaxial (VPE) reactor employing silane and propane as the reagents and nitrogen as the n-type dopant. The material structure consisted of an undoped buffer layer, a 0.4 μm channel layer doped to 2.5 to $5x10^{17}/cm^3$, and a 0.2 μm n+ layer doped in excess of $10^{19}/cm^3$.

The device fabrication process is similar to that reported previously [Nishizawa et al. 1975]. In this process, reactive ion etching (RIE) was used to fabricate the mesa isolation and channel recess. Nickel, sintered using rapid thermal annealing (RTA), was used for the ohmic contacts. With this procedure, the contact resistance is typically less than $5x10^{-6}$ ohm-cm^2 for either 6H or 4H SiC. A multi-layer metal system was used for the gates, with gold as the top layer to reduce the gate resistance. The gate length was 0.5 μm long and the gate-source and gate-drain spacings were 0.5 μm and 1 μm, respectively. Air-bridges were used to interconnect

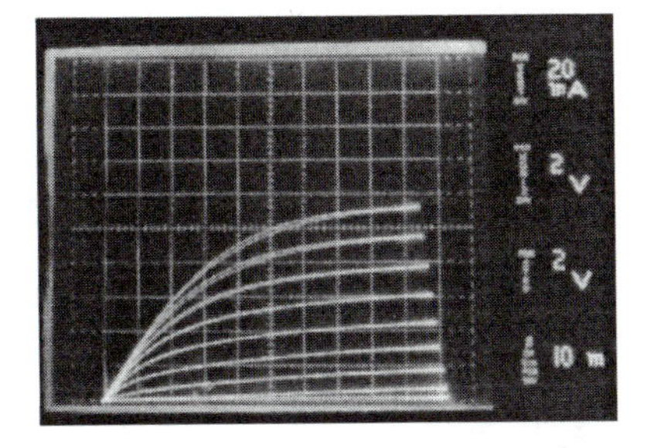

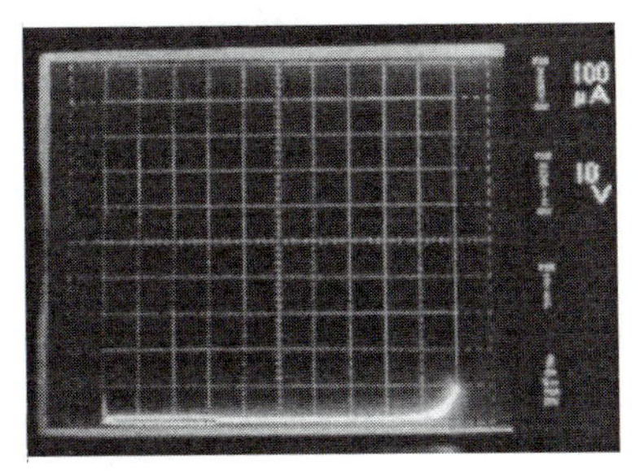

Figure 3.26 200 μm source-periphery 4H-SiC MESFET DC data, (a) I-V, (b) gate-drain breakdown. Reprinted from S. Sriram et al. IEEE Electron Dev. Lett. 17(7), 1996 with permission, ©IEEE.

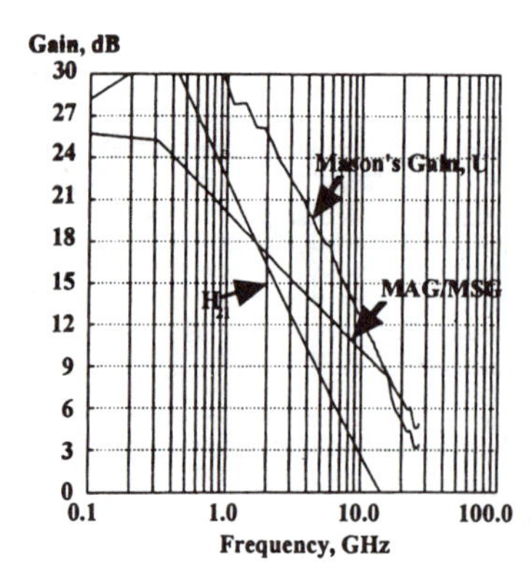

Figure 3.27 Small-signal RF characteristics for a 0.5 μm gate-length 4H-SiC MESFET. Reprinted from S. Sriram et al. IEEE Electron Dev. Lett, 17(7), 1996 with permission, ©IEEE.

the source fingers of multi-finger devices, and devices up to 6.4 mm periphery have been fabricated using this process. To reduce parasitic capacitance, the pads and interconnections were formed on a 2000Å thick SiO_2 layer.

Measured results: Figure 3.25 shows the measured power performance of a 2 mm-periphery 6H-SiC MESFET at 6 GHz. It can be seen that this device delivers 3.5 W with 45.5% power-added efficiency and 6dB of associated gain. The power output corresponds to 1.75 W/mm and is nearly three times larger than that normally obtained with GaAs MESFETs. Small signal measurements revealed an f_{max} of 25 GHz, an f_t of 10 GHz, and 8.5dB of gain at 10 GHz. To our knowledge, the power performance and the operating frequency reported here are the highest yet reported for 6H-SiC devices.

4H-SiC offers higher performance potential than 6H devices, due to the higher mobility and the shallower donor ionization energies in 4H-SiC. Figure 3.26(a) shows the DC characteristics of a two-finger MESFET with a gate periphery of 200 μm. The maximum drain current can be seen to be greater than 500 mA/mm and is nearly a factor of two higher than that obtained with 6H-SiC. The DC transconductance, g_m , is also higher and shows a maximum value of 40 mS/mm. These devices also exhibit high gate-drain breakdown voltages of about 100 V (at 0.5 mA/mm), as shown in Figure 3.26(b). These DC results clearly indicate the possibility of obtaining more than double the RF power density than can be obtained from similar devices on 6H-SiC.

Small-signal RF performance of 4H-SiC MESFETs is shown in Figure 3.27. These devices indicate a maximum available gain (MAG) of 5.1 dB at 20 GHz. Extrapolation of Mason's unilateral gain at 6 dB/octave yields an f_{max} value of 42 GHz, which is the highest reported to date for SiC devices. The unity current gain frequency, f_t, of 13.2 GHz was calculated by extrapolation of the short-circuit current-gain parameter, h_{21}, at 6 dB/octave. It is important to note the RF results were obtained at a drain bias voltage of 40 V. These results clearly demonstrate the simultaneous high voltage and high frequency capabilities of SiC devices.

SiC static induction transistor: The static induction transistor (SIT), invented by Watanabe and Nishizawa in 1950, consists of a multi-channel, vertical structure [Watanabe and Nishizawa 1950]. It controls current flow by modulating the internal potential of a single channel using a surrounding gate structure. In 1975 experimental SITs were fabricated, and the source-drain current of this device was shown to follow a space-charge injection model [Nishizawa et al.

1975]. More recently, very high power broad-band performance has been obtained from silicon SIT devices [Baliga 1982, Cogan et al. 1983, Regan et al. 1984, Kane and Frey 1984, Bencuya et al. 1985, Regan et al. 1985, Butler et al. 1986, Regan et al. 1986].

Device operation: SITs are a class of transistors with a short channel FET structure in which a current flowing vertically between source and drain is controlled by the height of an electrostatically induced potential energy barrier under the source. This electrostatic barrier develops at pinch-off when negatively charged opposing gate depletion layers converge to deplete the source drain channel of mobile charge carriers. Analogously to the vacuum triode, both the gate (grid) voltage and the drain (anode) voltage affect the source drain current.

Referring to the current voltage characteristic of Figure 3.28, the various regions of operation can be defined as (A) ohmic, (B) thermionic emission, © space-charge limited current (SCLC), and (D) velocity-saturated space-charge limited current. Examination of the DC SIT characteristic of Figure 3.28 under zero gate bias (point A), reveals conventional ohmic conduction of the doped channel dominated by the electron drift mobility, μ_n. With the application of a gate bias, the drain current moves toward pinch-off (point B) and the channel is depleted of carriers, as in a conventional MESFET. When the device is pinched off, the application of further drain bias causes the induced barrier between the opposed gates to be lowered (point C), and space-charge limited current begins to flow. Space-charge limited current, a signature of classic silicon-based SITs, is a function of device geometry, electron mobility, and the dielectric constant. A further increase in drain bias raises the electron velocity in the drift region until velocity saturation occurs, and the drain current as a function of drain bias changes slope (point D).

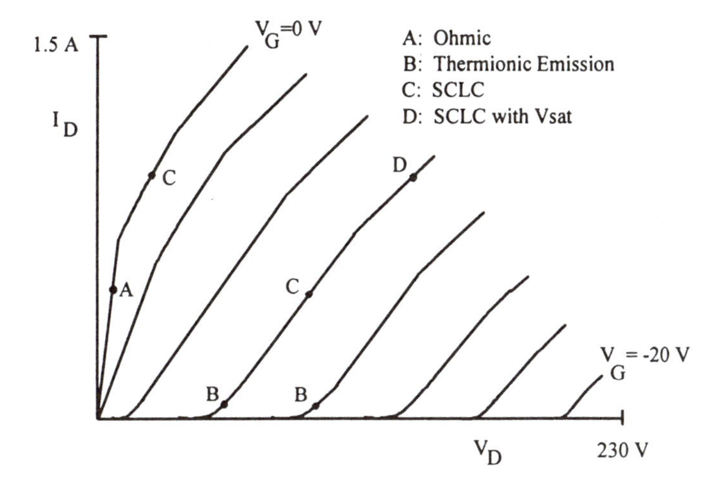

Figure 3.28 Current vs. voltage characteristic of a mixed-mode SIT. Reprinted from R.C. Clarke et al. Proc. IEEE/Cornell Conf. Adv. Concepts High Speed Semicond. Dev. Circ. 1995 with permission

In SIT devices, the gate-to-drain spacing, or drift region distance under the gate, comprises an undoped film several microns thick that can withstand the application of large gate-to-drain voltages and permit useful high blocking voltage in the transistors (200 V in this example). The maximum blocking voltage is a function of geometry and E_c, the electric field strength of the material. In classic SIT operation, the channel between the two gates is depleted of carriers when the gate is at zero bias, and application of drain bias is needed to turn the device on and produce source drain current. However, in order to maximize channel current and minimize the effect of knee voltage on device efficiency, mixed-mode SIT operation is usually employed. In mixed-mode operation, enough doping is added to the channel to provide ohmic currents at medium gate bias, while higher gate bias reverts to classic SIT operation, with its attendant large blocking voltage. Transconductance, transit time, unity gain frequency, and cut-off frequency are all strong functions of the electron mobility and saturated electron velocity.

This first-order discussion of SIT operation clearly identifies the material parameters most significant for high-frequency power performance: electron mobility, μ_n; saturated electron velocity, v_{sat}; dielectric constant, μ_r; and critical field strength, E_c. Relevant material parameters are compared in Table 3.4 for various semiconductor families. Based on first-order equations describing the SIT IV relationship and the data of Table 3.4, it may be deduced that SiC SITs, particularly the 4H-SiC polytype, will outperform SITs made from other materials with regard to power output, frequency response, and DC-to-RF conversion efficiency.

Device fabrication: SITs require two extremes of doping — a thick, ~ 5μm, unintentionally doped drift layer of $1\times10^{16}cm^{-3}$ carrier concentration and a very highly doped (1×10^{19} cm^{-3}), 0.2 μm thick contact layer. The growth chemistry for SiC employs propane as a source of carbon and silane as a source of silicon, transported via hydrogen carrier gas to an inductively heated graphite susceptor (1450°C). SiC is grown homo-epitaxially on c-axis-oriented n-type SiC substrates placed on the susceptor. A SIT carrier concentration vs. depth profile is shown in Figure 3.29. (The 0.2 nm-thick surface n+ is added subsequently:)

Table 3.4 Comparison of Material Parameters for High Frequency Power Performance

	Electron mobility (cm^2/Vs)	Saturated electron velocity (10^7cm/s)	Dielectric constant	Critical field strength (MV/cm)
GaAs	8500			
Si	1400			
6H-SiC	60(‖ to c-axis)	2.0	9.7	3.0
4H-SiC	800(‖ to c-axis)	2.0	9.7	4.0

Reprinted from R.C. Clarke et al. Proc. IEEE/Cornell Conf. Adv. Concepts High Speed Semicond. Dev. Circ. 1995 with permission.

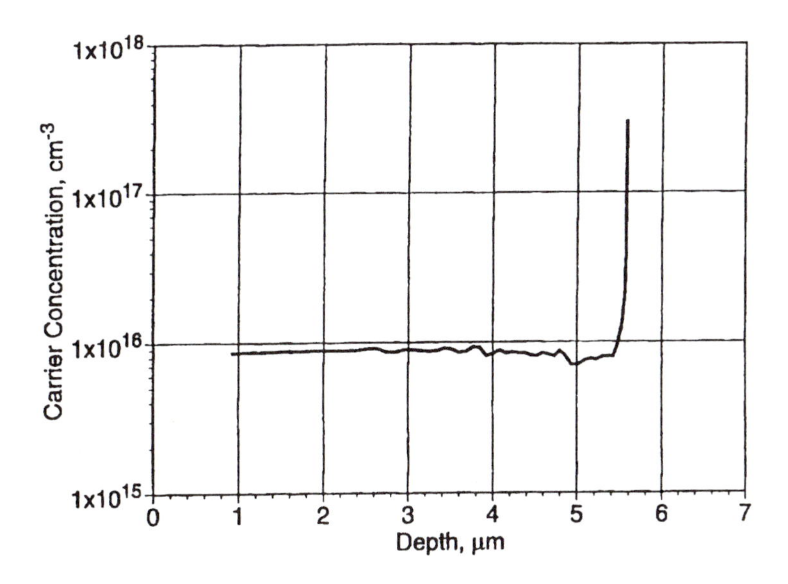

Figure 3.29 A carrier concentration depth profile of a SIT drift region formed by VPE. Reprinted from R.C. Clarke et al. Proc. IEEE/Cornell Conf. Adv. Concepts High School Semicond. Dev. Circ. 1995 with permission

Electron-beam lithography was used to define RIE masks and metal patterns to provide source ohmic contacts and Schottky metal gate electrodes. Dielectric passivation was applied, and thick metal was patterned to provide low-resistance source and gate interconnects to the surface of the device, with the drain electrode being the n+ substrate. An SEM photograph of a completed SIT is shown in Figure 3.30.

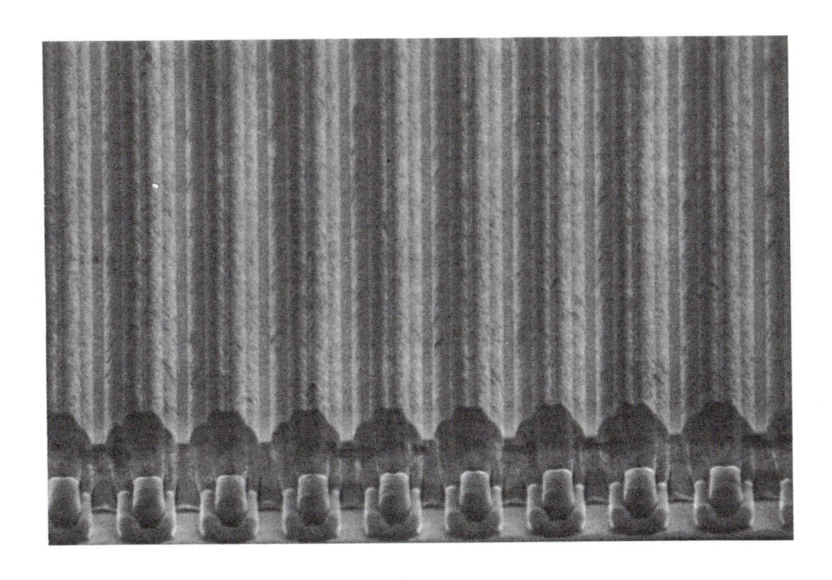

Figure 3.30 Structure detail of a SIC SIT. Seen are the individual fingers and the top source electrode.

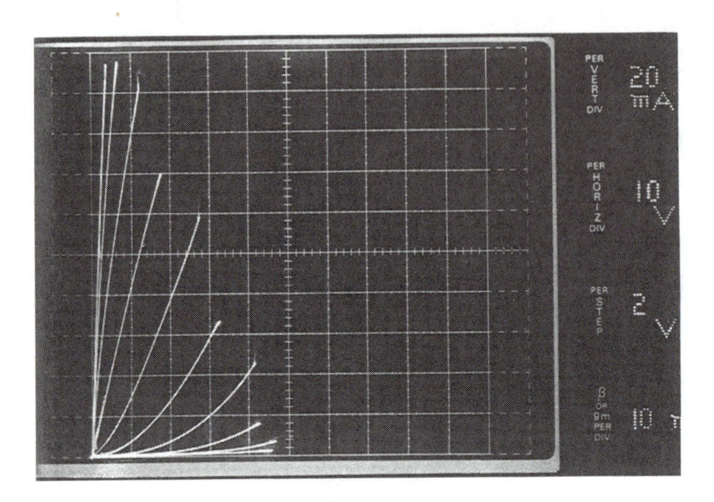

Figure 3.31 DC Characteristics of a 3-cm periphery SIT. Reprinted from R.C. Clarke et al. Proc. IEEE/Cornell Conf. Adv. Concepts High Speed Semicond. Dev. Circ. 1995, with permission.

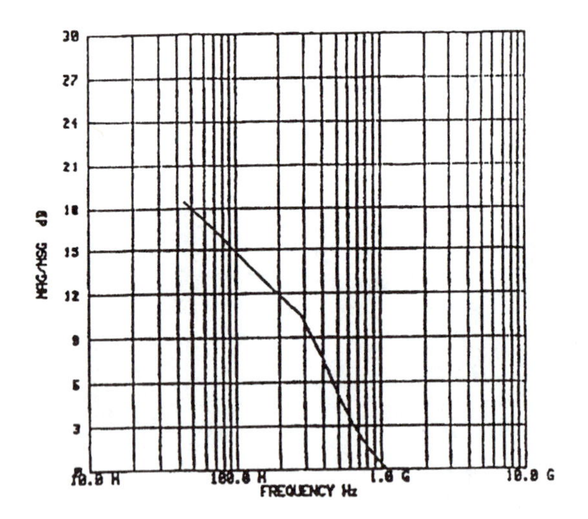

Figure 3.32 Small signal gain of a 1-cm periphery SIT. Reprinted from R.C. Clarke et al. Proc. IEEE/Cornell Conf. Adv. Concepts High Speed Semicond. Dev. Circ., 1995 with permission

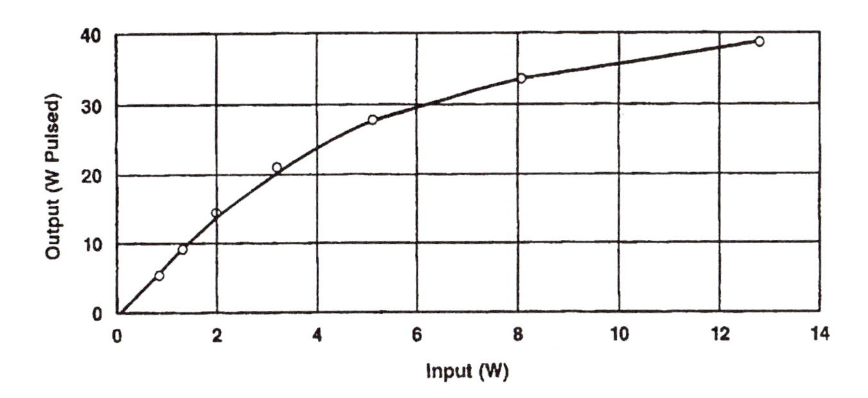

Figure 3.33 Pulsed power performance of a monolithic 11-cm periphery SIT. Reprinted from R.C. Clarke et al. Proc. IEEE/Cornell Conf. Adv. Concepts High Speed Semicond. Dev. Circ. 1995 with permission.

Device performance: Low-wattage DC characteristic of a 3-cm source periphery 6H-SiC SIT is shown in Figure 3.31. Measured data revealed a maximum channel current of 300 mA/cm, a transconductance of 30 mS/cm, and a voltage gain of 8. On-wafer small signal measurements were conducted using cascade probes and a common drain configuration. A small signal gain of 12 dB at 200 MHZ was measured, as shown in the gain frequency curve of Figure 3.32. Monolithic SIT chips were screened for DC performance, and wirebonded in parallel into a Kyocera package. Of pulsed output power, 37W was developed from an 11-cm periphery SIT at 175 MHZ, with a power-added efficiency of 60%, as shown in Figure 3.33. The associated gain was 10 dB.

Observed maximum currents in 6H-SiC SITs were nearly five times less than predicted by mathematical models. Reduced channel currents are a problem for power devices because both power gain and power added efficiency will be adversely impacted. The poor currents in 6H-SiC were found to be associated with anisotropic electron transport behavior. Currents traveling in a direction parallel to the c-axis (into the surface) of a 6H wafer were observed to be five times smaller than currents normal to the c-axis (along the surface) for the same applied voltage.

Experimental evaluation of current transport in SIT-like structures was conducted by measuring pulsed currents (to minimize heating effects). Figure 3.34 shows the extracted electron mobility and saturated drift velocity for 6H- and 4H-SiC parallel to the c-axis (in the SIT current direction) and reveals a much larger electron mobility (6X) in the 4H polytype than seen in 6H. The results of the extraction are lower than estimates given in Table 3.4 due to the material quality of the 4H material used.

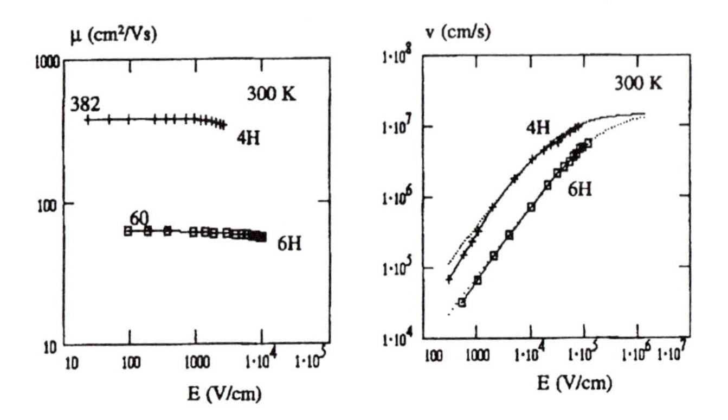

Figure 3.34 The measured electron mobility and saturated electron velocity for 6H and 4H SIT structures. Reprinted from R.C. Clarke et al. Proc. IEEE/Cornell Conf. Adv. Concepts High Speed Semicond. Dev. Circ. 1995 with permission

4H-SiC SITs showed a current density of 1 A/cm of source periphery, a voltage gain of 15, and a maximum transconductance of 75 mS/cm as seen in Figure 3.35. The device blocking voltage with a gate bias of -20 V was 200 V. Small-signal common drain measurements were performed on wafers using cascade probes. As can be seen in the computer-derived, common-source, gain-frequency curve of Figure 3.36, a cut-off frequency, f_{MAX}, of 4 GHz was obtained. Automated wafer mapping was employed to identify candidate 1.5 cm SITs. Transistors of combined 16.5 cm periphery were cut from the wafer, soldered, and wirebonded into a Kyocera package. This package developed 225 W output power with an associated gain of 8.7 dB at 600 MHZ (Figure 3.37). The maximum power-added efficiency was 47% and the power density was 13.5 W/cm. These results from 4H-SiC SIT devices are significantly better than silicon devices, in terms of power density, breakdown voltage, and frequency response.

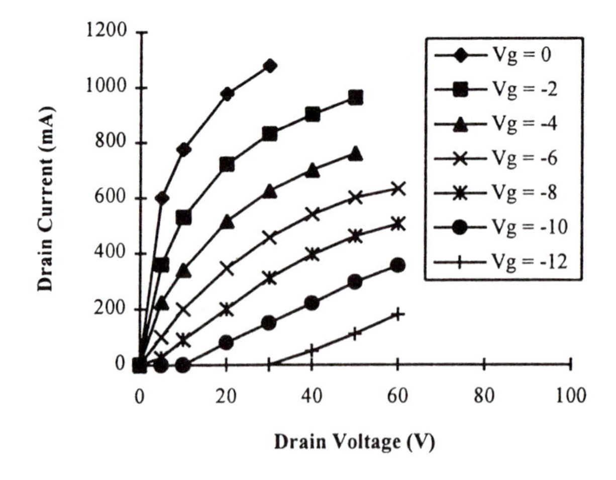

Figure 3.35 5 μs pulsed DC SiC SIT characteristic of a 1-cm source periphery device.

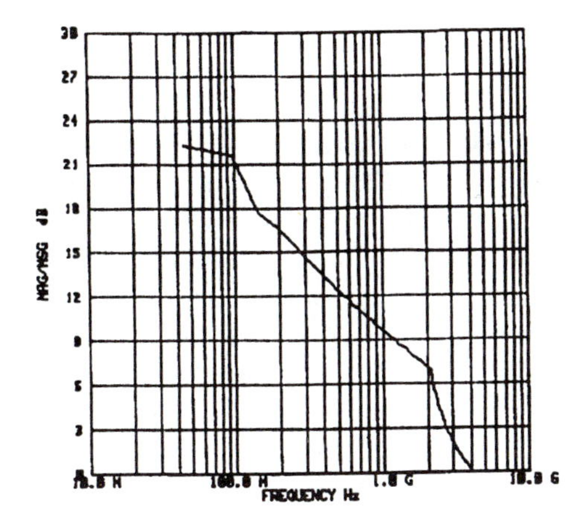

Figure 3.36 Small-signal data of a 1.5-cm 4H-SiC SIT (Vd = 20V, Vg = -4V)

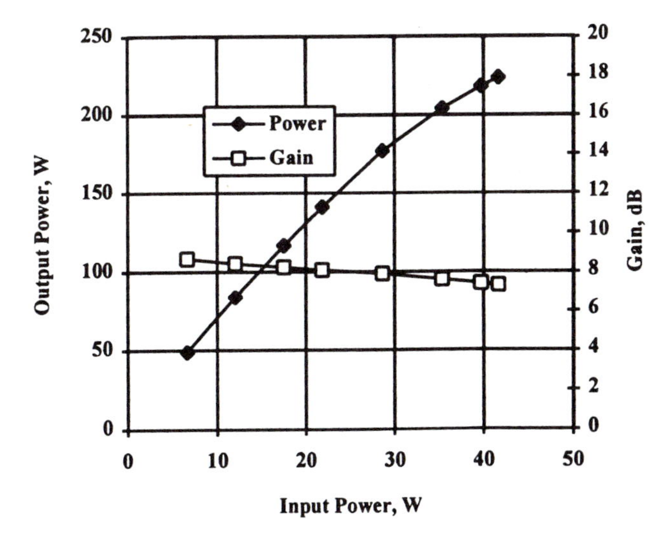

Figure 3.37 Pulsed power and gain curves for a 16.5-cm SIT.

3.4 Metal Contacts on SiC

3.4.1 Introduction

Metal-semiconductor (MS) contacts play a vital role in all electronic and opto-electronic semiconductor devices. They are required to transfer electrical power and signals to and from the semiconductor and external circuitry, and their quality and reliability critically affects device and circuit performance. There currently exists a myriad of projected military and commercial applications for high-temperature semiconductors that can operate at temperatures from ~ 200 to 600°C. At these temperatures, the stability and reliability of both the semiconductor and the metal contacts become important. In the case of silicon-based electronics, device operation is limited to temperatures of less than ~125°C due to semiconductor substrate and p-n junction current leakage, and the reliability of device contacts and packaging. However, while at higher temperatures the use of wide-bandgap semiconductors can ensure acceptably low leakage currents, the stability of metal contacts and packaging becomes of greater concern.

Currently, the most technologically mature wide-bandgap semiconductors for electronic device applications are the 6H- and 4H-silicon carbide polytypes. In the last five years, substantial progress has been made in the use of bulk and epitaxial-layer SiC. Consequently, a wide variety of discrete prototype SiC devices have been reported in the literature; these include p-n diodes, MOSFETs, MESFETs, JFETs, BJTs, and thyristors operating above 300°C. When these unpackaged devices are operated near 600°C in atmospheric environments, degradation of the metal contacts limits device lifetimes to less than a few hours. The responsible degradation mechanism is believed to be oxidation, as evidenced by reported contact stability data at 600°C for 55 h in vacuum. In addition to oxidation, at these temperatures metal contact interdiffusion and electromigration become of concern. These problems include metal-metal and metal-semiconductor reactions and interdiffusion. A large amount of data has been collected showing how such mechanisms limit the reliability of Si and III-V semiconductor devices. However, this is not the case for SiC, for which comparatively few experimental results have been reported. Very little has been reported on the reliability of metal systems on SiC, and even less on elevated-temperature failure mechanisms.

3.4.2 Metal-semiconductor contacts

MS contacts can be divided into two classifications based on their current-voltage behavior. Contacts with rectifying characteristics are called Schottky contacts; those with linear nonrectifying characteristics are called ohmic contacts. The contrast in their behavior stems from differences in the nature of space charges at the MS interface. Important semiconductor devices utilizing Schottky contacts include MESFETs, solar cells, and diodes. Schottky diodes are used as clamps, microwave mixers, and diagnostic tools for determining doping profiles and deep-level defects. The most important parameter describing a Schottky barrier is the energy barrier height, φ_B, where values of $\varphi_B > 0.5$ -1.0 eV are generally desired in many applications to minimize device leakage current, especially at elevated temperatures. Ohmic contacts are required by all semiconductor devices, as well as integrated circuits, to transfer current from the semiconductor to other circuitry. An ohmic contact must contribute a negligible contact resistance, R_c, compared to the total device on-resistance, R_{on}, and a negligible voltage drop appearing across the device. A significant degradation in device efficiency can result with a large R_c. For example, the efficiency of light-emitting diodes and lasers, and the gain and current-driving capability of FETs are strongly influenced by contact resistance. A small value for R_c is especially important in high-current power device applications, where power dissipated as joule heating is proportional to R_{on}.

$$q\varphi_{Bn} = q(\varphi_m - \chi) \quad (1)$$

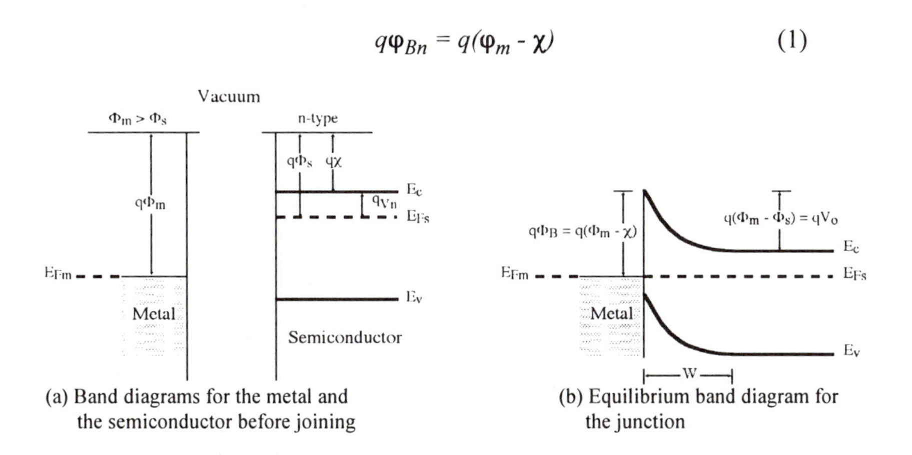

(a) Band diagrams for the metal and the semiconductor before joining

(b) Equilibrium band diagram for the junction

Figure 3.38 Formation of a Schottky barrier of an n-type semiconductor.

Ohmic contacts are best described using a parameter called specific contact resistance, R_c, or contact resistivity, ρ_c, where $\rho_c = R_c A_{eff.}$ and A_{eff} is the effective contact area.

3.4.3 Energy band diagrams and Schottky barrier height

When a metal makes intimate contact with a semiconductor, a transfer of charge occurs between the materials until an equilibrium is established in which the Fermi levels in the two materials are equal [Sze 1981, Mönch 1994]. As shown in Figure 38(a), a material work function is defined as the energy difference between the vacuum level and the material Fermi level, $E_{F(M,S)}$, as denoted by $q\varphi_m$ in the metal and $q\varphi_s$ in the semiconductor. Furthermore, $q\varphi_s = q(\chi + V_n)$, where $q\chi$ is the electron affinity, defined as the potential difference between the bottom of the conduction band, E_c ,and the vacuum level, and the quantity qV_n is the difference in energy between E_c and the Fermi level. In the ideal case, a Schottky contact is formed when the metal makes contact with an n-type semiconductor under the condition $\varphi_m > \varphi_s$, as shown in Figure 38(b). Figure 38(a) shows how, initially, the Fermi level of the semiconductor is higher than that of the metal. When contact is made, a transfer of electron charge from the semiconductor into the metal occurs (electrons always flow in the direction from a high- to low-Fermi level), lowering the semiconductor Fermi level by an amount equal to the difference between the two work functions. The transfer of electrons leaves positive "space charges" behind that give rise to band bending and the formation of a potential energy barrier, $q\varphi_{Bn}$, called a Schottky barrier. The Schottky barrier height, $q\,\varphi_{Bn}$, for injection of electrons from the metal into the semiconductor conduction band is given by

$$q\varphi_{Bn} = q\,(\varphi_m - \chi) \quad (3.8)$$

For the case of a contact formed between metal and a p-type semiconductor under the condition $\varphi_m < \varphi_s$, as shown in Figure 3.39(a), the barrier height is given as

$$q\varphi_{Bp} = E_g - q\,(\varphi_m - \chi) \tag{3.9}$$

where E_g is the semiconductor bandgap. As mentioned previously, Schottky contacts display rectifying current-voltage behavior. An ohmic contact, on the other hand, results when $\varphi_m < \varphi_s$ for n-type and $\varphi_m > \varphi_s$ for p-type semiconductors. An ohmic contact has no inherent potential barrier to prevent charge flow and results in linear current-voltage behavior.

Ideally the metal-semiconductor barrier height depends only on the metal work function, the semiconductor bandgap, and the electron affinity, as described by Equations 3.1 and 3.2. However, in practice it is difficult to control the barrier height by varying the metal work function. It has been experimentally observed that the barrier height for common semiconductor materials such as Ge, Si, GaAs, and other III-V materials is relatively independent of the work function of the metal. Schottky contacts are generally formed on both n-type and p- type semiconductors with $\varphi_B \approx E_g/3$ for both cases. The relative constancy of the barrier height with work function of metals is called Fermi level pinning, because the Fermi level in the semiconductor is pinned at some energy in the band gap to create a Schottky contact [Sze 1981, Schroder 1990]. In the case of SiC, Pelletier et al. [1984] attributed Fermi level pinning in 6H-SiC to intrinsic surface states, suggesting little dependence of barrier height on the work function of the metal. Porter et al. [1993] found that the barrier height differences of titanium, platinum, and hafnium contacts on n-type (0001) 6H-SiC were all within a few tenths of 1 eV, which is also characteristic of Fermi level pinning.

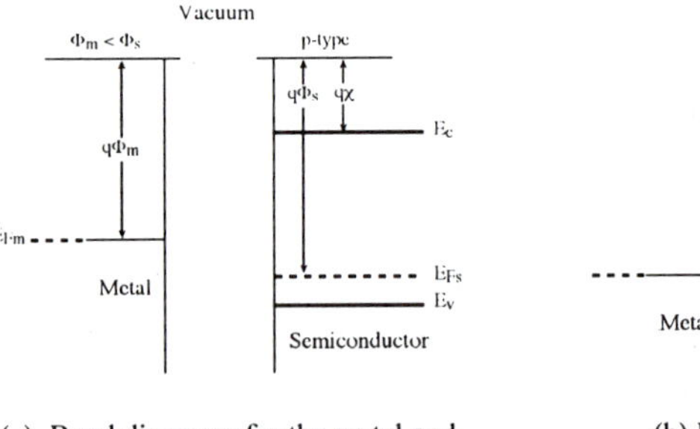

(a) Band diagrams for the metal and the semiconductor before joining

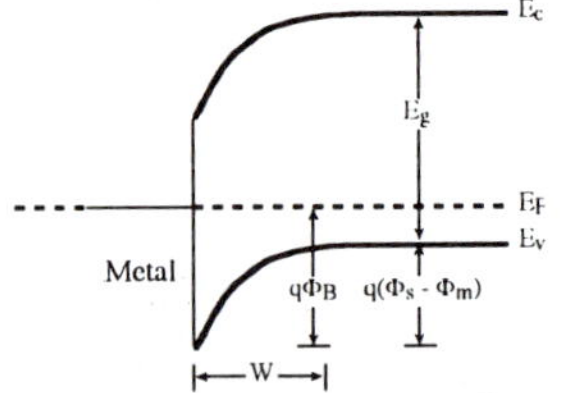

(b) Equilibrium band diagram for the junction

Figure 3.39 Formation of Schottky barrier of metal to p-type semiconductor.

On the other hand, Waldrop and Grant [1990, 1992] reported a strong dependence of barrier height on work function for metal contacts to 3C- and 6H-SiC. In Figures 3.40 (a) and (b), Schottky barrier height is plotted against metal work function, φ_m, for 3C- and 6H-SiC, respectively. The details of Schottky barrier formation are not yet fully understood. However, it appears that interfacial electronic states due to defects, metal-induced inner-gap states, and interfacial chemistry play important roles during contact formation.

3.4.4 Carrier transport processes for Schottky and ohmic contacts

The carrier transport mechanisms for the metal-semiconductor interface are strongly influenced by the doping concentration and temperature. Three typical cases are shown schematically in Figure 3.41 for an n-type semiconductor. Figure 3.41(a) depicts a lightly doped semiconductor ($N_d < 10^{17}$ /cm^3). In this case, the depletion width, *W*, is wide and the electrons cannot tunnel through the interface. The only way for the electrons to flow between the metal and the semiconductor is by thermionic emission (TE) over the potential barrier, φ_{Bn}. Figure 3.41(b) shows the band diagram of a metal contacting a semiconductor doped at an intermediate level ($N_d = 10^{17}$ to 10^{18}/cm^3). In this case, the electrons can partially tunnel through the interface and both thermionic and tunneling process are important, which is called thermionic-field-emission (TFE). When the semiconductor is extremely heavily doped ($N_d > 10_{18}$/cm$_3$), electrons can tunnel from the Fermi level within the metal into the semiconductor. This process is called field-emission (FE), and is shown in Figure 3.41(c).

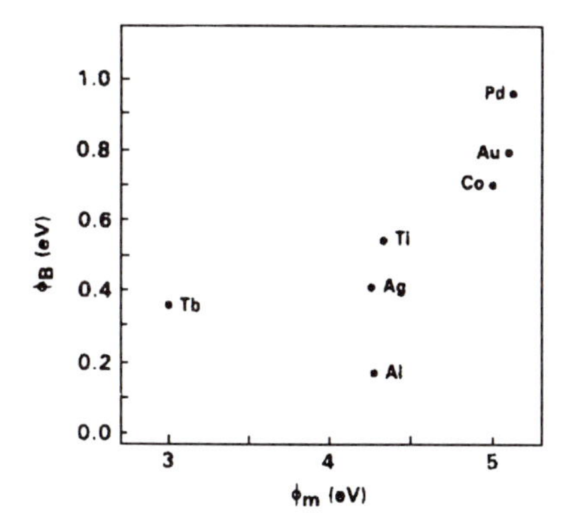

(a) Barrier height vs. work function for 3C-SiC

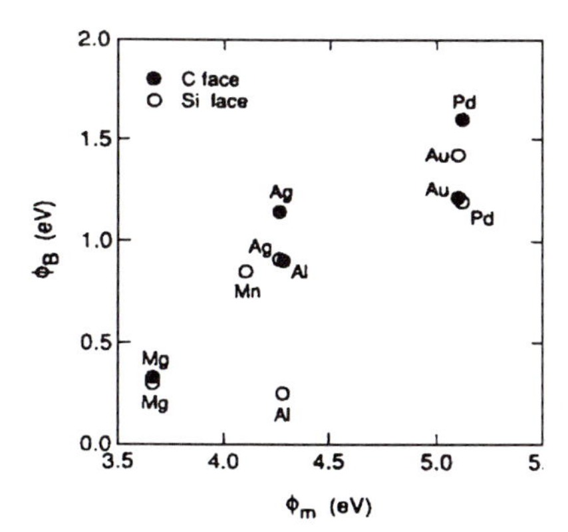

Figure 3.40 Schottky barrier height φ_B of various metal contacts to SiC versus metal work function φ_m [Waldrop and Grant 1990, 1992]

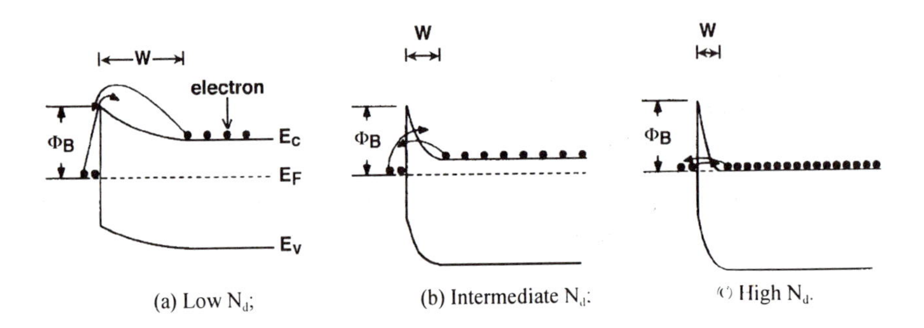

Figure 3.41 Conduction mechanisms through metal/n-type semiconductor interface with different doping levels.

A useful parameter describing the electron tunneling probability is given as kT/E_{00}, where E_{00} is a characteristic energy defined by

$$E_{00} = \frac{q\,h}{2}\sqrt{\frac{N_d}{m^{*}\epsilon}} \tag{3.10}$$

where h is Planck's constant divided by 2π, m^* is the effective mass of the tunneling electron, and ϵ is the dielectric constant of the semiconductor. The depletion width decreases with increasing doping density, N_d, making it easier for carriers to tunnel through. This indicates that when E_{00} is high relative to the thermal energy, kT, the probability of electron transport by tunneling increases. Therefore, the ratio kT/E_{00} is a useful measure of the relative importance of the thermionic process to the tunneling process. For lightly doped semiconductors, $kT/E_{00} >> 1$, and the dominant current mechanism is due to thermionic emission. For $kT/E_{00} \approx 1$, both thermionic and tunneling mechanisms dominate, and for $kT/E_{00} << 1$, carrier transport is dominated by carrier tunneling.

In the case of metal-semiconductor contacts with low semiconductor doping levels, thermionic-emission will dominate current transport and the resulting specific contact resistance can be expressed as [Sze 1985]

$$R_c = k\,/\,(q\,A*T)\,\exp\,[q\,\varphi_{Bn}/kT] \tag{3.11}$$

where k is Boltzmann constant, A^* is the effective Richardson constant, and T is the absolute temperature. Equation 3.4 shows that metals yielding low barrier heights should form ohmic contacts. Conversely, contact metals that yield a relatively high barrier height should form rectifying Schottky contacts. For contacts with high doping levels, the barrier width becomes very narrow, and the tunneling current can become dominant. In this case, tunneling current is proportional to tunneling probability, and the specific contact resistance can be expressed as [Sze 1985]

$$R_c \sim \exp[C \, \varphi_{Bn} \, / \, (N_d)^{1/2}] \quad (3.12)$$

This relation shows that contact resistance depends strongly on the doping level and barrier height, where parameter C depends on the semiconductor dielectric constant and carrier effective mass. Hence, a high doping level, a low barrier height, or both are required to obtain low values of R_c. Equation 3.5 also indicates that it is difficult to obtain rectifying Schottky contacts on highly doped semiconductors. The general approach used to form an ohmic contact is to elevate the doping level at the metal-semiconductor interface via diffusion of dopant impurities into the semiconductor. The source of the dopant is generally the metal itself. A metal film containing the desired doping species is typically evaporated or sputtered on the semiconductor, followed by high-temperature sintering to achieve dopant diffusion.

3.4.5 SiC metallization process and characterization

The process of forming ohmic and Schottky contacts on SiC is essentially the same as that used in Si and GaAs electronics. A brief review of typical techniques used for SiC wafer preparation, metallization, and metal contact characterization follows.

Wafer surface preparation: Preparation of the SiC surface prior to metal deposition is critical to the quality of the contact. Surface contamination can reduce metal adhesion, increase contact resistance, and create electronically active defects that alter current conduction. Contaminants and native oxide layers on the wafer surface must be removed prior to metal deposition. Solvents, such as trichloroethylene (TCE), trichloroethane (TCA), acetone, methanol, and propanol, are used to degrease the SiC wafer. In order to obtain a fresh oxide-free SiC surface, Evwaraye at al. and Chaudhry et al. [1991] suggested the growth of thermal oxide, followed by subsequent oxide removal via acid etching. This procedure also removes defects on SiC surface caused by mechanical polishing. Various acids and bases such as HF, H_2SO_4, HCl, NH_4OH:H_2O_2:H_2O, HCl:H_2O_2:H_2O, H_2SO_4:H_2O_2, HNO_3:H_2SO_4:H_2O, K_2CO_3, and KOH melt, are used to etch SiC wafer surface. Prior to metal deposition, in-situ surface cleaning, using sputter etching or irradiation of a high-energy laser beam is commonly performed.

Metal deposition techniques: The most common techniques for metal deposition include sputtering and thermal or electron-beam evaporation. In sputtering a metal target is bombarded with energetic ions, which knock off metal surface atoms by energy transfer. These released atoms land on the wafers and become part of a metal film coating. In metal evaporation, the deposition metal material is vaporized in a vacuum from its liquid phase, and the vapor is transported and deposited onto the wafer. The vacuum used for evaporation is generally higher than 10^{-5} torr. At this low pressure, gas molecules have a mean free path greater than 1 m. Consequently, the evaporated metal atoms suffer no collisions from any residual gas and achieve straight-line travel from the target to the surface of the wafer. The condensation of the metal vapor on the wafer produces the desired metal contact film.

Contact sintering (annealing): With a few exceptions, as-deposited metal contacts on SiC exhibit rectifying characteristics. High-temperature sintering (heating) is often required to form ohmic contacts with low contact resistivity and good thermal stability. As mentioned, diffusion of dopants from the metal into the semiconductor during sintering is used to form ohmic contacts. Since the metal sintering time required for ohmic contact formation on SiC is generally longer than that for contacts formed on, for example, GaAs, traditional sintering in a tube furnace is more often used than rapid thermal annealing (RTA). Sintering is usually performed in an atmosphere of argon, in a vacuum, or forming gas (3%H_2 in N_2). Depending on the contact system used, sintering temperatures can range from 300° to 1200°C, and sintering times can vary from a few seconds to a few hours. During sintering of SiC contacts, silicides and/or carbides are usually formed that play a role in decreasing both the Schottky barrier height and, consequently, the contact resistance.

Characterization: Key parameters used to describe ohmic and Schottky contacts are the contact resistivity, ρ_c (or contact resistance, R_c), and Schottky barrier height, φ_B, respectively. Several methods are used to determine contact resistivity. The transmission line measurement (TLM) technique is the most popular one [Schroder 1990, Berger 1972]. In this technique, a set of 5 to 8 metal fingers (tabs) of width z are formed adjacent to each other. The spacing between adjacent fingers decreases, typically over an order of magnitude, from one end of the pattern to the other. The resistances are obtained from the pattern, and the contact resistivity is calculated from the relation

$$\rho_c = R_c^2 z^2 / \rho_s = \frac{R_O}{2} z L_t \tag{3.13}$$

where R_c is the contact resistance, R_0 is the resistance obtained by extrapolating to a zero distance between fingers, $R_c = R_0/2$, ρ_s is the semiconductor sheet resistance, and L_t is the transfer length. Other common ways to determine contact resistivity include a circular transmission line method [Reeves 1980] and the four-point probe method [Terry and Wilson 1969, Kuphal 1981]. Experimental measurement of the Schottky barrier height can also be determined by several different techniques, including current-voltage measurements, capacitance-voltage measurements, and x-ray photoelectron spectroscopy [Cohen and Gildenblat 1987, Schroder 1990].

In order to study the interfacial reaction between metal contacts and semiconductors, and to identify compounds formed during annealing, Auger electron spectroscopy (AES), transmission electron microscopy, Rutherford backscattering spectrometry (RBS), and X-ray diffraction (XRD) techniques are often employed. A concise description on these analyzing techniques can be found in Waldrop and Grant [1993].

3.4.6 Schottky contacts on SiC

Many metals form Schottky contacts on SiC in the as-deposited condition. In order to obtain good Schottky contacts, suitable metals must possess a large barrier height on SiC. Another important consideration is high-temperature contact stability. At elevated temperatures, metals react with SiC to form silicides and/or carbides, which may increase or decrease the Schottky barrier height.

The formation of Schottky barrier contacts on n-type 3C-SiC (100) for palladium, gold, cobalt, titanium, silver, terbium, and aluminum was investigated by Waldrop and Grant [1990] These contacts exhibited values for φ_B over the range 0.16 to 0.95 eV, and the values depended strongly on the metal work function, in general accordance with the Schottky limit as specified in Equations 3.1 and 3.2. Waldrop [1992, 1993] also systematically studied the formation of Schottky barrier contacts on n-type 6H-SiC, using palladium, gold, silver, terbium, erbium, manganese, aluminum, and magnesium. The values of φ_B for these metals were found to extend over a range of 1.3 eV. It was also discovered that as-deposited values of φ_B were dependent, to varying degrees, on the 6H-SiC crystal face and that the changes in φ_B following sintering were also dependent on the crystal face where the C-face was found to be more reactive than the Si-face. Schottky barrier heights for the above metals are listed in Table 3.5.

Gold forms a Schottky contact on SiC with a relatively high barrier height. Yoshida et al. [1985] produced good-quality Schottky barrier contacts by evaporating gold onto chemically etched n-type 3C-SiC epilayers grown by CVD, and obtained barrier heights of 1.15 ± 0.15 eV and 1.11 ± 0.03 eV using capacitance-voltage and photoresponse measurements, respectively. Gold Schottky contacts on 3C-SiC were unaltered following a 1-h heat treatment at 300°C in argon. The contacts were rectifying after further heating at 500°C for 90 min; however, after heating at 700°C the contacts degraded and showed ohmic behavior. A gradual out-diffusion of silicon was observed by AES, which became more prominent at high sintering temperatures [Ioannou et al. 1987]. Surface preparation is very important in the formation of gold Schottky barriers on SiC. Dmitriev et al. [1992] pointed out that for bulk crystals the damaged layer can exceed 10 μm and all damaged surface must be removed before metal deposition. After sufficient etching, the value of φ_B for gold on n-type 6H-SiC for both bulk material and epitaxial layers was approximately 2 eV.

Electron-beam-deposited platinum contacts on n-type 3C-SiC exhibited superior thermal stability when subjected to short sintering cycles at temperatures as high as 800°C. When thermally treated in the range of 450° to 800°C, a combination of silicide and carbide was believed to have formed at the Pt/SiC interface, while the rectifying characteristics also improved. The interfacial reaction was dominated by the diffusion of Pt into the SiC layer. As the sintering temperature increased, the barrier height increased from 0.95 to 1.35 eV. The lowest value for the ideality factor was 1.5 after annealing at 450°C. $PtSi_x$ is a promising metallization on 3C-SiC for high-temperature applications [Papanicolaou et al. 1989]. Platinum deposited on 6H-SiC yielded Schottky contacts having φ_B= 1.06 eV with low leakage currents and low ideality factors, as determined from current-voltage measurements. Only slight changes in platinum contact behavior occurred following 20 min of sintering from 450 to 750°C in increments of 100°C. Throughout the sintering, the ideality factor and leakage currents remained low, whereas the Schottky barrier height increased slightly with sintering temperatures to 1.26 eV [Porter et al. 1993]. A 400V 6H-SiC Schottky barrier diode was fabricated using electron-beam-deposited platinum. These high-voltage Schottky barrier diodes were reported to have a low forward voltage drop (1.1 V for a J_F of 100 A/cm^2), small reverse leakage, and excellent switching characteristics [Bhatnagar et al. 1992].

A metal contact to SiC can demonstrate either Schottky or ohmic characteristics, depending on annealing conditions. Daimon et al. [1986] reported aluminum on n-type 3C-SiC showing ohmic behavior stable up to 400°C, but with distinct rectifying characteristics following sintering at 900°C. On the contrary, aluminum contacts on p-type 3C-SiC clearly changed from non-ohmic to ohmic with sintering at 900°C. The nickel metal system has demonstrated similar behavior. Steckl and Su [1994] developed a nickel metallization process to fabricate both rectifying and ohmic contact to n-type 3C-SiC by controlling the sintering temperature. These nickel Schottky diodes yielded high breakdown voltages (170V) for 3C-SiC. The nickel Schottky junction remained stable at sintering temperature as high as 600°C.

Sintering of contacts at 800°C resulted in low-resistance ohmic contact behavior. The thermally induced mechanism responsible for the rectifying-to-ohmic transition was attributed to formation of nickel silicides during high-temperature sintering, which was confirmed by X-ray diffraction and Auger analysis.

Cobalt contacts on n-type 6H-SiC were studied by Lundberg and Östling [1993]. The Schottky barrier height was found to be 0.79 eV for the as-deposited contact. Excellent rectifying behavior was demonstrated up to 700°C. Consecutive annealing from 300° to 800°C increased the barrier height from 0.8 to 1.3 eV. Heat treatments at 900°C resulted in ohmic contact behavior. RBS, AES, and XRD studies showed that Co reacted with SiC and formed Co_2Si and CoSi at elevated temperatures.

Silicides of $TaSi_2$ and $MoSi_2$ were deposited on n-type 6H-SiC by RF sputtering. As-deposited $TaSi_2$ films were rectifying with Schottky barrier heights of 1.8 and 1.2 eV on the C- and Si-faces, respectively. The reverse leakage currents were about 10^{-5} A at -10V. The as-deposited $MoSi_2$ contacts were Schottky contacts with a barrier height of 1.0 eV for both C- and Si-face. After sintering at 925°C for 2 minutes, the rectifying characteristics for both $TaSi_2$ and $MoSi_2$ contacts were deteriorated [Petit and Zeller 1992]. Schottky contacts on SiC were also studied for various metals such as palladium [Waldrop and Grant 1990, 1992], titanium [Porter et al. 1993, Waldrop and Grant 1990, Spellman et al. 1992], silver [Waldrop and Grant 1990, 1992], terbium [Waldrop and Grant 1990], manganess [Waldrop and Grant 1992], magnesium [Waldrop and Grant 1992] and hafnium [Porter et al. 1993]. Table 3.5 summarized recent studies of Schottky contacts to SiC.

3.4.7 Ohmic contacts on SiC

Ohmic contacts require several properties for practical use in electronic device applications: low contact resistivity, high-temperature stability, good metal-semiconductor adhesion, smooth surface morphology, and a relatively simple fabrication process. Generally, it is difficult to satisfy all these requirements using only one metal. This has led to the development of multilayer metal systems. In order to obtain ohmic contacts with low contact resistivity, metals possessing a low work function for contact on n-type SiC and a high work function for contact on p-type SiC should be chosen. However, due to Fermi level pinning and the complexities of Schottky barrier formation on SiC, the contact resistance of metals cannot easily be predicted this way and must be determined experimentally. For the purpose of ensuring high temperature stability, transition metals with high melting points, such as tungsten, tantalum, molybdenum, titanium, and nickel, have been considered as ohmic contact metals for SiC.

Like the properties of the metal used, the doping level at the semiconductor surface will also strongly influence carrier transport across the metal-semiconductor interface that controls the contact resistance. Although increasing the doping level will not decrease the metal barrier height, it will significantly enhance carrier tunneling and increase current flow. Therefore, an effective approach to decrease contact resistance is to ensure heavy doping in the semiconductor via introduction of dopants from the metal or heavily doping

Table 3.5 Schottky Contacts to SiC

Contact metallization	SiC	Annealing [°C]	SBH [eV]	Year published	Ref.
Au	3C, n-type		1.1-1.15	1985	[Ioannou 1987]
Al, Al-Si	3C, n-type	900	Rectifying	1986	[Steckl 1993]
Au	3C	As-deposited	1.2	1987	[Dmitriev 1992]
Pt	3C, n-type	As-deposited	0.95	1989	[Bhatnagar 1992]
$PtSi_x$	"	800	1.35	"	"
Pd	3C, n-type	As-deposited	0.95	1990	[Waldrop 1992]
Au	"	"	0.78	"	"
Co	"	"	0.69	"	"
Ti	"	"	0.53	"	"
Ag	"	"	0.4	"	"
Tb	"	"	0.35	"	"
Al	"	"	0.16	"	"
$TaSi_2$	6H, n-type, C-face	As-deposited	1.8	1992	[Spellman 1992]
"	6H, n-type, Si-	"	1.2	"	"
$MoSi_2$	6H, n-type, C-	"	1	"	"
"	6H, n-type, Si-face	"	1	"	"
Ni-Mo	6H, n-type, C-face	"	1.8	"	"
"	6H, n-type, Si-face	"	0.9	"	"
"	6H, n-type, C-face	825°C-2 min	1.2	"	"
"	6H, n-type, Si-face	825°C-2 min	0.9	"	"
Ni	6H, n-type, C-face	as-deposited	2.2	"	"
"	6H, n-type, Si-face	"	1.5	"	"
Au	6H, n-type	As-deposited	2	1992	[Papanicolaou 1989]
Ti	6H, n-type, Si-face	As-deposited	0.88	1992	[Crofton 1993]
"	6H, n-type, Si-face	700°C-60	1.04	"	"
Pd	6H, n-type, C-face	As-deposited	1.6	1992	[Murakami 1990]
"	6H, n-type, Si-face	"	1.11	"	"
Au	6H, n-type, C-face	"	1.14	"	"
"	6H, n-type, Si-face	"	1.4	"	"
Ag	6H, n-type, C-face	"	1.1	"	"
"	6H, n-type, Si-face	"	0.92	"	"
Mn	6H, n-type, Si-face	"	0.81	"	"
Al	6H, n-type, C-face	"	0.84	"	"
"	6H, n-type, Si-face	"	0.3	"	"
Mg	6H, n-type, C-face	"	0.33	"	"
"	6H, n-type, Si-face	"	0.3	"	"
Pt	6H, n-type	As-deposited	Rectifying	1992	[Rastegaeva 1992]
Ti	6H, n-type, C-face	As-deposited	1.0	1993	[Yoshida 1985]
"	"	400°C	0.98	"	"
"	6H, n-type, Si-face	As-deposited	0.73	"	"
"	"	400°C	0.97	"	"
Ni	6H, n-type, C-face	As-deposited	1.59	"	"
"	"	400°C	1.66	"	"
"	6H, n-type, Si-	As-deposited	1.24	"	"

Table 3.5 contd.

Ni	6H, n-type, Si-face	400°C	1.25	1993	[Yoshida 1985]
"	"	600°	1.39	"	"
Al	6H, n-type, C-face	As-deposited	0.84	"	"
"	"	600°C	1.66	"	"
"	6H, n-type, Si-face	As-deposited	0.3	"	"
"	"	600°	1.12	"	"
Ni	3C, n-type	As-deposited	Rectifying	1993	[Ioannou 1987]
Ti	6H, n-type, Si-face	As-deposited	0.85	1993	[Waldrop 1990]
Pt	"	"	1.02	"	"
Hf	"	"	0.97	"	"
Co	6H, n-type, Si-face	As-deposited	0.79	1993	[Petit 1992]
"	"	400°C	0.90	"	"
"	"	600°C	1.08	"	"
"	"	800°C	1.30	"	"
Ni	3C, n-type	As-deposited	Rectifying	1994	[Steckl 1994]

the starting semiconductor wafer. When the doping level at the interface is extremely high, the depletion region becomes very narrow, which enhances carrier tunneling. Figure 3.42 shows the dependence of contact resistivity on doping level for aluminum-titanium ohmic contacts on p-type 6H-SiC [Crofton et al. 1993].

Following high-temperature sintering of the metal/SiC interface, silicides and/or carbides may be formed. It has been found that a very simple linear correlation exists between Schottky

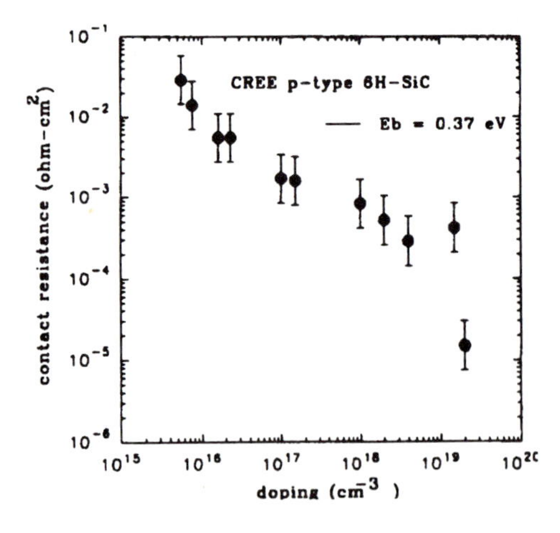

Figure 3.42 Contact resistivity vs. doping level for Al-Ti ohmic contacts on p-type 6H-SiC [Crofton et al. 1993].

barrier heights and heats of formation for transition metal silicides on n-type silicon [Andrews and Phillips 1975]. Unfortunately, sufficient information is not available for silicide and carbide contacts to SiC. It might be expected that a similar trend would exist. In other words, silicides (or carbides) possessing a high negative heat of formation would probably have a low Schottky barrier height. With this in mind, metals can be chosen that form stable silicides and/or carbides with SiC that possess a high negative heat of formation, in the hope of decreasing contact resistance.

Ideal Schottky theory predicts that when the work function of the metal is less than that for an n-type semiconductor, or the work function of the metal exceeds that of a p-type semiconductor, the contacts will be ohmic. In practice, a number of metals are reported to be ohmic on SiC in the as-deposited conditions. Daimon et al. [1986] found that an aluminum contact on n-type 3C-SiC was ohmic after deposition. Glass et al. [1991] used low-energy ion-assisted reactive evaporation to deposit TiN (work function ≈ 3.74 eV) onto n-type 6H-SiC with a Si terminated (0001) surface (work function = 4.8 eV), and obtained good ohmic contact. This may be because TiN has a smaller work function than that of SiC, or because of the formation of an amorphous Si-N interface layer between the TiN and the SiC, which was involved in the ion-assisted reactive evaporation [Glass et al. 1992]. As-deposited titanium/platinum ohmic contacts were formed on n-type 3C-SiC by Shor and Weber [1992]. Contact resistivity ranged from 2.5×10^{-4} to 9×10^{-5} ohm-cm^2. One-hour sintering at 650°C caused the contact resistivity to decrease by roughly a factor of two. However, after two hours at 650°C, most of the titanium/platinum contacts failed. Jacob et al. [1994] obtained ohmic characteristics in tungsten and molybdenum contacts on 3C- n-type SiC. The contact resistance of tungsten/SiC was about 0.8 ohm-cm^2 before sintering and 0.66 ohm-cm^2 for the sample annealed at 900°C for 30 min. The molybdenum/SiC contact had a high contact resistance of 1.8 ohm-cm^2 in the as-deposited condition. After sintering at 900°C for 30 min, the contact resistance dropped to 0.25 ohm-cm^2. Molybdenum, tantalum, titanium, and zirconium contacts on n- and p-type 6H-SiC were also reported to display ohmic characteristics in the as-deposited state on degenerate epilayers [Petit et al. 1994].

Though ohmic contacts on SiC can be obtained in as-deposited condition, ohmic contacts with low contact resistivity and good thermal stability usually require high-temperature sintering in which silicides and/or carbides are formed. Metals with high melting points and high chemical stability, such as tungsten, molybdenum, titanium, nickel, and chromium, have been widely employed for sintered ohmic contacts. Tungsten contacts were found to be both physically and chemically stable on n-type 3C-SiC at temperatures up to approximately 900°C [Jacob et al. 1994, Geib et al. 1989]. AES data indicated a thin layer of WC and WSi_2 formed during the deposition process. However, no additional reaction was observed after annealing at 850°C for 30 min in an ultrahigh vacuum. The electrical measurements indicated that the tungsten/SiC contact was ohmic and unaffected by vacuum annealing at temperatures up to 900°C. The contact resistance was found to be about 0.24 ohm-cm^2 at 23°C, dropping to 0.08 ohm-cm^2 at 900°C.

The specific contact resistance of tungsten contacts on n-type 3C-SiC obtained by M. I. Chaudhry et al. [1991] prior to heat treatment was of the order of 1.5×10^{-2} ohm-cm^2. The contact resistance decreased to 6.1×10^{-3} ohm-cm^2 after sintering at 300°C for 30 min. The decrease was attributed to the dissolution of the natural oxide at the tungsten/SiC interface during subsequent annealing at 300°C. McMullin et at. [1992] reported a low contact resistivity of 8×10^{-4} ohm-cm^2 for tungsten/gold contacts on n-type 3C-SiC after sintering at 800°C for 1 h. The tungsten/gold contact also demonstrated good thermal stability when subjected to thermal cycles at 600°C for approximately 80 h. Similar tungsten/platinum contacts on n-type 3C-SiC illustrated a contact resistivity of 1.4×10^{-4} ohm-cm^2 after sintering at 650°C for 8 h [Shor and Weber 1992].

The contact resistivity of ohmic contacts on 6H-SiC is sensitive to the crystal-face.

Rastegaeva [1992] reported that specific contact resistance values of tungsten contacts on C-faced n-type 6H-SiC were 2 to 2.5 times greater than for tungsten on Si-faced SiC at the same doping level of $3 \times 10^{18}/cm^3$. The respective resistance values were 2×10^{-3} and 7×10^{-4} ohm-cm^2. Tungsten is also used in many other contact metallization systems as a constituent part [Crofton et al. 1992, 1994], a top layer [Adams et al. 1994, Liu et al. 1995a,b], or a diffusion barrier layer of the metallization [Papanicolaou et al. 1989, Anikin et al. 1992].

Molybdenum forms ohmic contacts on n-type 3C- and 6H-SiC [Jacob et al. 1994, Petit et al. 1994]. Molybdenum reacted with 3C-SiC to form $MoSi_2$ following sintering at 1150°C for 15 minutes. After sintering at 1200°C, Mo_5Si_3 appeared, and the amount of Mo_5Si_3 increased with increasing sintering time [Cho et al. 1994]. Following sintering at 970°C for 15 min, the contact resistivity of a molybdenum/3C-SiC contact was 4×10^{-2} ohm-cm^2. The molybdenum/3C-SiC contact showed good thermal stability. The contact resistivity did not change after sintering at 1200°C for 60 min.

Nickel plays an important role in forming thermally stable, low-contact-resistance ohmic contact on n-type SiC. The Schottky barrier height of nickel contacts on SiC is high and nickel forms good rectifying contacts on both 3C- and 6H-SiC. However, after high-temperature sintering, the contacts change from rectifying to ohmic. Nickel does not react with 3C-SiC below 580°C. When the sintering temperature is above 610°C, nickel begins to react with 3C-SiC and forms polycrystalline Ni_2Si [Cho et al. 1994]. Other nickel silicides such as $NiSi_2$ and Ni_5Si_2 are also reported [Steckl et al. 1994]. A contact resistivity of 2.8×10^{-2} ohm-cm^2 for the nickel contact on 3C-SiC was obtained after sintering at 700°C for 15 min [Steckl et al. 1994]. A similar result was obtained for nickel contacts on n-type 6H-SiC after sintering at 950°C for 5 min [Crofton et al. 1992]. It was also reported that after sintering nickel on 6H-SiC at 1050°C for 5 min, almost all the deposited nickel reacted with SiC and formed Ni_2Si, as identified by X-ray diffraction. At the same time, a low contact resistivity of 10^{-3} to 10^{-4} ohm-cm^2 was obtained [Adams et al. 1994]. Therefore, the low contact resistivity of the nickel contact to SiC is attributed to the formation of Ni_2Si. It is obvious, however, that Ni_2Si is formed by the consumption of silicon in the SiC substrate; thus, the composition of the SiC at the metal/semiconductor interface will be shifted towards a silicon-depleted direction, and isolated graphite carbon atoms would be left behind. These isolated graphite carbon atoms are believed to deteriorate the electric properties of the ohmic contact and the SiC substrate. In order to avoid this problem, a thin silicon layer can be deposited between the metal layer and the SiC substrate [Adams et al. 1994]. Another approach is to employ a carbide-forming element to produce a stable carbide. Both Ni/Ti/W and Ni/Cr/W ohmic contacts on n-type 6H- and 4H-SiC have demonstrated low contact resistivity. Long-term aging tests revealed that Ni/Cr/W contacts yield excellent thermal stability; the contacts were stable in aging at 650°C for 3000 hours [Liu et al. 1995]. AES chemical depth profiles and X-ray diffraction indicated that both silicide and carbide were formed after 1050°C for 5 min [Liu et al. 1995].

In addition to the method of forming silicides or carbides following high-temperature sintering, various compounds can be deposited or grown on SiC. $TiSi_2$ and WSi_2 were reported to be deposited on n-type 3C-SiC by co-sputtering intrinsic silicon and titanium or tungsten. After deposition, the contacts were rapid-thermal annealed (RTA) at 1000°C for 10 s, followed by sintering at 450°C for 10 min to form silicides. After RTA, the contact resistivity of the $TiSi_2$ and WSi_2 contacts were 1.4×10^{-1} ohm-cm^2 and 3.7×10^{-2} ohm-cm^2, respectively. The contact resistivity decreased to 1.1×10^{-4} and 3.0×10^{-4} ohm-cm^2 after sintering at 450°C for 10 min [Chaudhry et al. 1991]. Titanium nitride films were deposited onto the Si-faced n-type 6H-SiC by ion-assisted reactive evaporation in a dual electron-beam evaporation system. The TiN contacts were ohmic in the as-deposited condition and little change was observed after sintering at 450° and 550°C for 15 min [Glass et al. 1991,1992]. Chaddha et al. [1995] used a CVD technique to epitaxially grow a TiC contact layer on an n-type 6H-SiC epilayer. The contact resistivity of the TiC/SiC contacts was 1.3×10^{-5} ohm-cm^2. The contacts were found to be thermally and chemically stable after sintering at 1400°C for 2 h in hydrogen.

Because 3C-SiC has a lower energy gap (Eg ≈ 2.3 eV) than 6H-SiC (Eg ≈ 3.0 eV), it may be easier to make low-resistivity ohmic contacts to 3C-SiC than to 6H-SiC. Dmitriev et al. [1994] utilized a unique technique to obtain low-resistivity ohmic contacts to 6H-SiC. They grew thin 3C-SiC layers (< 2000Å) on 6H-SiC substrates by low-pressure CVD, followed by depositing nickel for n-type contacts, or aluminum/titanium for p-type contacts. The contacts were then sintered using RTA in forming gas at 1000°C for 30 s for n-type contacts and at 950°C for 2 m for p-type contacts. The contact resistivity of nickel contacts to n-type 3C-SiC/6H-SiC grown on the Si-face was < 1.7×10^{-5} ohm-cm^2 and < 6×10^{-5} ohm-cm^2 when 3C-SiC/6H-SiC was grown on the C-face. As a comparison, the contact resistivity of nickel contacts to 6H-SiC (without 3C-SiC layer) was 2×10^{-4} ohm-cm^2 For aluminum/titanium contacts to the p-type 3C-SiC/6H-SiC, the contact resistivity was found to be 2 to 3×10^{-5} ohm-cm^2.

Chapter 4

PASSIVE DEVICE SELECTION AND USE AT HIGH TEMPERATURE

High-temperature electronics research has traditionally focused on the development of active devices and integrated circuits. Some of the technologies and devices developed as a result of this research have been presented in Chapters 2 and 3. However, designing electronic systems for use at temperatures greater than 125°C also requires passive devices and first- and second-level packaging structures that can withstand these temperatures.

Chapter 4 presents the technological issues related to the use of passive devices at elevated temperatures, and chronicles the most recent research aimed at creating passive devices with superior high-temperature performance and quality. The first section focuses on resistors, detailing the effects of elevated temperature of the electrical characteristics and lifetime of various resistors. It reviews several types of commercially available resistors that perform adequately at temperatures to 300°C, including discrete wirewound resistors, surface-mountable thick-film ceramic resistors, and hybrid thick-film paste resistors.

The second section deals with capacitors. While some capacitors, which are stable at elevated temperatures, are commercially available, they are made of insulating materials with low dielectric constants and as a result have low energy densities. Capacitors made with higher dielectric constant insulators exhibit unstable capacitance an high leakage currents at elevated temperatures. This section discusses the trade-offs between high energy density, capacitance stability, and low leakage current and presents the most promising technologies for developing high-energy-density, thermally stable dielectric systems. Both polymer and ceramic-based technologies are highlighted. Since the lack of elevated temperature capacitors is a major constraint in the development of high-temperature electronic systems, this section constitutes the bulk of the chapter.

4.1 Resistors

Resistance to the passage of a current results from the scattering of electrons by thermally generated lattice vibrations, point and line defects, and impurity atoms. Thus, in its simplest form, electrical resistivity can be expressed as the sum of two independent terms, as shown in Equation 4.1.1: one, ρ_L, is the resistivity due to lattice vibrations (or phonons) and another, ρ_I, is the resistivity due to lattice imperfections and impurities. This expression is known as Mattheisen's rule:

$$\rho = \rho_L(T) + \rho_I \qquad (4.1)$$

Only ρ_L is a function of temperature, is given as

$$\rho_L = \rho_{Lo}(1 + \alpha_o(T - T_o)) \tag{4.2}$$

where α_o is the temperature coefficient of resistance, T_O is the reference temperature, usually set to be 23°C, and ρ_{LO} is the resistivity due to lattice vibrations at the reference temperature. This formula provides only an estimate of the actual resistivity, since all materials exhibit some deviation from the rule due to the temperature dependency of the impurity- and imperfection-related term [Ho 1983].

The dependence of resistivity on temperature is the most significant effect of temperature on resistors up to 200°C. The values for the temperature coefficient of resistance for several common resistor materials in a variety of resistor categories are given in Table 4.1. While metals have positive coefficients, it is possible for resistors to have both positive and negative coefficients, based on their materials of construction. For continuous use at an elevated temperature, this fixed change in resistance is not a concern. If properly quantified, it should not cause failure of the system or parameter drifts, since it can be addressed by careful circuit design that balances the temperature characteristics of several resistors to create a network unaffected by temperature [Pecht 1994]. However, instabilities and drifts in resistance that occur at a fixed temperature over time may alter circuit function. Other elevated temperature concerns and temperature-related failure mechanisms include noise, thermal stress, interdiffusion, and oxidation. These depend on the type of resistor used.

4.1.1 Metallized film, carbon film, and carbon composition resistors

The most common resistors in use today are made of metallized film and carbon film, which have replaced carbon composition resistors. Carbon composition resistors have tolerance values and noise levels that are too high for modern use. The resistance of carbon composition resistors changes during temperature or power cycling because of the difference in the coefficients of thermal expansion of the conducting and non-conducting components [Sinclair 1991]. The ultimate temperature limit is imposed by the melting or decomposition temperature of their binder, which is usually a phenol-formaldehyde resin [Smith 1989].

Metallized and carbon film resistors have more stable resistance values than their predecessors. There is some concern, however, about the use of these resistors at elevated temperatures. The films begin to degrade by sublimation of the metal-film resistive element at temperatures in excess of 165°C [Sinclair 1991, Smith 1989]. In addition, softening of the epoxy coating on these devices can cause separation between the encapsulant and the lead end-cap, allowing the ingress of moisture and halogenic corrodants [Smith 1989].

4.1.2 Wirewound resistors

Wirewound resistors have traditionally been used for high-temperature and high-power applications because they do not undergo the degradation typical of film resistors and because they have a more controllable thermal coefficient of resistance [Sinclair 1991]. These devices are limited only by the degradation in the insulation resistance of their coatings and by the thermal cycling fatigue of their coatings, which may fracture and produce an ingress site for moisture and corrodants. Vitreous enamel coatings perform adequately up to 300°C or higher. Use of these resistors allows standard discrete component and board technology to be used up to 200°C. The drawback to these resistors is their relatively large size.

Table 4.1 Temperature Coefficients of Resistance for Common Resistor Materials [Pecht and Lall 1994] [Elshabini-riad 1991] [Heidler 1969]

Resistor Types	*Max Operating Temp (°C)*	*Temp. Coefficient of Resistivity (ppm/ K)*
Wire-wound resistor		
Precision	145	10
Power	275	260
Metal-film resistor		
Precision	125	50-10
Power	165	20-100
Composition resistor		
General Purpose	130	1500
Deposited resistors		
Thin-film		
Tantalum	>200	± 100
Tantalum Nitride	>200	- 85
Titanium	>200	± 1000
SnO_2	>200	-1500 to 0
Ni-Cr	>200	± 100
Cermet (Cr-SiO)	>200	± 150
Thick-film		
Ruthenium-silver	>200	± 200
Palladium-silver	>200	-500 to 150

4.1.3 Thick- and thin-film resistors

Deposition and patterning of resistive films directly on ceramic substrates combines miniaturization with high reliability at high temperatures. Resistor films fall into two categories - thin-film resistors and thick-film resistors.

Thin-film resistors are produced by depositing metals, metal alloys, and metal compounds using thin-film technologies such as vacuum evaporation, sputtering, reactive sputtering, and chemical vapor deposition. Oxidizing the surface of these films produces a coating layer that

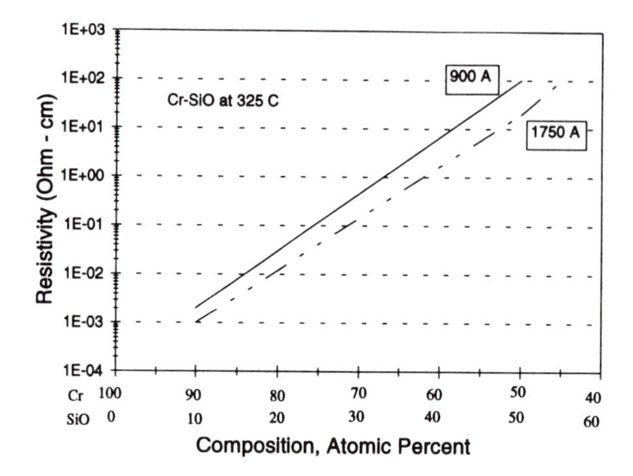

Figure 4.1 The resistivity of Cr-SiO cermet as a function of the Cr composition [Licari 1970].

seals the metal underneath from corrosion and drift. Thin-film resistors are chosen over their thick-film counterparts for their superior resolution, stability, high-frequency performance, small size, and lower TCR values. Commonly used materials are tantalum, tantalum nitride, nickel-chromium, titanium, and cermets. Tantalum, tantalum nitride, and titanium can be used at highly elevated temperatures because of their refractory nature. However, titanium is often not used because of its high TCR. Nickel-chromium and cermets are limited to use at lower temperatures, but are still effective at temperatures well in excess of 200°C. Additionally, they provide the opportunity to tailor the resistivity and TCR by controlling the composition, as shown in Figure 4.1 for the Cr-SiO cermet. SiO also moderates the susceptibility to damage at high humidity for the chromium resistors. SiO or polyimide perform a similar function for the nickel-chromium resistors. Anodization performs a similar function for tantalum nitride resistors [Pecht 1994].

Thick film resistors are preferred in low cost, high volume applications and those requiring high resistivities. Formulations are based on palladium, ruthenium, iridium, and rhenium. Three common materials are ruthenium-silver, palladium-silver, and ruthenium oxide. The sensitivity of the palladium based materials to the firing profile has caused the ruthenium based materials to become more used. Ruthenium oxide is the most popular for its oxidation resistance. It can be heated in air up to 1000°C without any degradation in performance. In addition, resistors made of this material exhibit good stability, low TCR, and low noise.

The film thickness in a film resistor could conceivably be on the same order as the mean free path of the electrons, which would require that a term be added to the resistivity equation to account for surface scattering. This term is given below.

$$\alpha = \frac{\alpha_o}{\ln(1/k) + 0.423} \tag{4.3}$$

where k is the ratio of the film thickness to the mean free path of the electrons, and p is a parameter that represents the fraction of electrons elastically scattered at the surface as opposed to those diffusely scattered at the surface.

In actuality, few applications use film resistors thin enough (< 30 nm) to require this correction and it is sufficient to assume that $\alpha = \alpha_0$. In most cases, the increase in resistivity is linear with temperature up to 200°C. However, in those where it is not, the temperature dependent TCR should be used.

Resistors are subject to Johnson noise, current flow noise, and noise due to fractured bodies, end caps, and leads. Johnson noise, or thermal noise, is a function of temperature but is independent of frequency, while current noise is inversely proportional to frequency. As a result, thermal noise is a particularly large component of the noise for high temperature, high frequency applications. The magnitude of the thermal noise is given as

$$E_{RMS} = \sqrt{4kRT\Delta f} \tag{4.4}$$

where E_{RMS} is the root mean square magnitude of the noise, Δf is the bandwidth over which noise is measured, T is the temperature, R is the resistance, and k is a constant.

Other high-temperature concerns include the generation of stress at the resistor-to-substrate interface during thermal or power cycling as a result of the mismatched coefficients of thermal expansion. Assuming that the film thickness is much less than the substrate thickness, the stress on the resistive film can be approximated by the following:

$$\sigma_f = (\Delta\alpha\Delta T)\frac{E_f}{1-\nu_f} \tag{4.5}$$

where E_f is the elastic modulus of the film; ν_f is Poisson's ratio for the film; $\Delta\alpha$ is the difference in the coefficients of thermal expansion of the film and the substrate, α_f - α_s, and ΔT is the thermal cycle magnitude.

Nickel-chromium resistors are also subject to interdiffusion and oxidation at highly elevated temperatures (400° to 500°C), as shown in Figures 4.2 and 4.3. However, these effects are not noticeable at temperatures up to 200°C [Paulson 1973]. At these temperatures though, the small decreases in resistivity (<2.5%) occurring during 1000 hours of aging at 210°C are significant [Faith 1975]. These must be accounted for in the design tolerances.

Nevertheless, thin-film and thick-film resistors deposited on silicon and ceramic substrates are the best way to achieve high-temperature performance and miniaturization. They are superior to discrete thin-film resistors and carbon composition resistors in their high-temperature performance, and superior to wirewound resistors in their size and ability to be integrated with other components. For this reason, using these resistors for elevated temperature applications is standard design practice for many manufacturers [Torri 1995].

In conclusion, it must be noted that resistors are high power-dissipating components. Therefore, even if the resistor does not fail under the combination of the ambient temperature and the heat it dissipates during operation, it can cause failure in other components. It is good design practice to locate resistors sufficiently far away from capacitors and transistors to minimize the heating of these other devices. Thermal analysis simulations are useful in making the tradeoff between space savings and the reliability degradation resulting from tight placement.

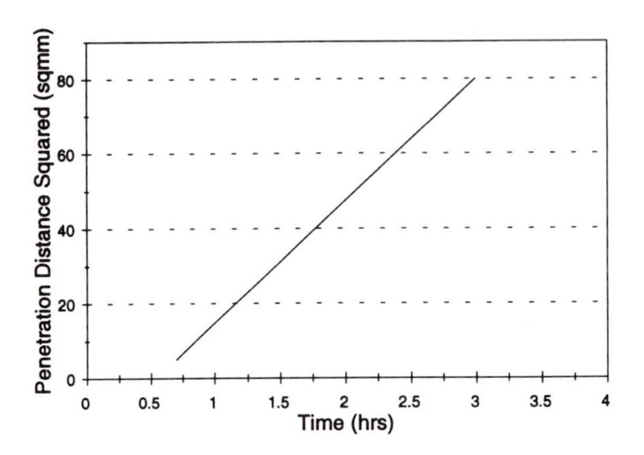

Figure 4.2 Penetration of Al into a NiCr resistor at 525^{0}C [Paulson 1973].

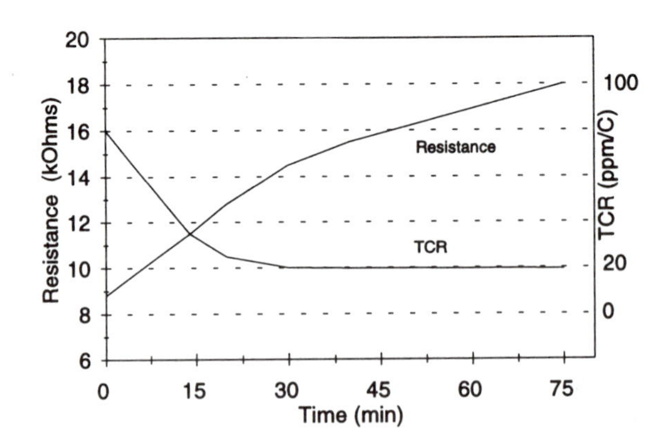

Figure 4.3 Effects of oxidation on a NiCr thin-film resistor [Paulson 1973].

4.2 Capacitors

Many military and commercial systems today require high-temperature electronics to run actuators, high-speed motors, and generators. This is partly due to the greater emphasis on reducing system size and weight with ever higher performance requirements. Future applications will demand military system temperatures in the 300°C+ range with still greater emphasis on increased performance. These goals will be achieved only with the evolution of improved high temperature power electronics. Of all of the passive devices that will be required to satisfy these needs for complete high-temperature system applications, perhaps none is as problematic as the capacitor. Because both the capacitance (due to temperature-dependent values of the dielectric constant) and the dissipation factor of any given capacitive component will often change significantly with increasing temperature, capacitors were identified years ago as a challenging passive component requiring development for reliable high-temperature operation. This is especially so of large devices with values of several microfarads or more, like those required for power conditioning needs associated with electric motors and power switching applications. Innovations on all fronts, including materials, device designs, and packaging, are being pursued to meet these needs.

A number of references [Grzybowski, 1992] also note, however, that unless a capacitor has been carefully designed for long-term operation at elevated temperatures, high-temperature failures observed in commercially available components will most often be related to the device's packaging and contacts technology, rather than the dielectric material employed. Failures evidenced included separation of the encapsulant due to lack of hermeticity or mismatched coefficients of thermal expansion (CTE).

This section considers the two primary capacitor constructions (film wound and multilayer ceramic or glass dielectric), reviews a number of dielectric materials used in both capacitor construction types, and identifies some boundaries on the use of various materials for elevated temperature environments. Finally, it presents a method for characterizing and modelling capacitors at very high temperatures. Table 4.2 summarizes the dielectric materials that have been investigated for use in the fabrication of wound capacitors, lists some commercial sources of the materials, and catalogs some of the more important properties and advantages of each material for harsh environment devices.

4.2.1 Polymer film capacitors

For many years, Mylar polyester has been used for energy-storage capacitors. The use of polyester films, however, has been limited by the relatively low glass transition temperature and, hence, polymer dielectric wound capacitors have been viewed as weak candidates for high-temperature applications exceeding 200°C. For less stressful applications to 150°C, however, some of the previously accepted limitations of polyester film capacitors might be related more to the way in which they are packaged and manufactured than to the basic properties of the polyester film. This deficiency in packaging capability arises often when considering devices for harsh environment applications. Some research has been done to develop a process for producing polyester capacitors that would operate at 150°C without derating [Bauer and Clelland, 1990] by increasing the film stabilization temperature to well above 150°C. Tests were performed on 1.0 μF surface-mounted devices at 150°C for up to 2000 h at 100% of rated voltage. These tests revealed a 1.2% drop in the mean capacitance distribution, with no significant spread in the capacitance values. The mean value of the dissipation factor distribution dropped by 13%, and a significant broadening in the distribution was exhibited. This ability to increase the operating temperature range of polyester film capacitors extends the available options to applications where polyester film is the most cost-effective dielectric.

Scientists at NASA Lewis Research center have characterized many dielectrics for suitability in elevated temperature applications. Evaluations of Nomex 410 and 418 aramid papers, supplied by Dupont, indicated that the dielectric constant and the dissipation factor of these materials was relatively constant to at least 200°C [Ahmad et. al. 1991]. Voltex 450, another aramid paper composed of aramid fiber and neoprene binder, supplied by Lydall, Inc., was evaluated and found to be considerably less stable with temperature. The dielectric constant of the Voltex 450 began to change quite dramatically at 150°C. At this temperature, the dielectric constant had already doubled in value and the dissipation factor had increased by a factor of 10. At 200°C, this material began to char [Ahmad and Myers, 1989]. Also characterized was polybenzimidazole-PBI, a linear thermoplastic polymer supplied by Hoechst Celanese. This material also exhibited a dramatic increase in dielectric constant, though not until a temperature of 250°C was reached. At this temperature the dielectric constant had doubled and the dissipation factor had increased by a factor of 10. This polymer film, however, did maintain its physical integrity to temperatures as high as 300°C.

This work concluded that the PBI film maintained its overall integrity when exposed to temperatures as high as 200°C. At higher temperatures, the film exhibited an increase in its dielectric properties. It was also suggested that improvements in manufacturing and processing might yield a polymer film with electrical properties that would be more stable with temperature.

Other recent research activities have focused on high-temperature polymer film dielectrics such as Kepton [Buritz, 1989] and FPE polyimide [Harris, 1989]. Although FPE polyimide film showed promising results, it was not readily available, both because it was proprietary to the 3M Corporation and thicknesses below 11μm had not been achieved with large-scale film casting equipment. Kepton, however, was plagued by a fabrication process that was complex, time-consuming, and labor-intensive. Coupled with minimal reliability, low-energy density, and high dissipation factor, these devices met with limited success as 200°C components.

Experiments have also been conducted on thick (25μm) samples of other similar materials, including a Dupont polyimide (PI) film, Teflon Perflouroalkoxy (PFA) (also supplied by Dupont), and Poly-P-Xylene (PPX), prepared using Gorham's method and supplied by Nova Tran [Suthar et al. 1991]. Data for relative permittivity vs. temperature was obtained at 60 Hz with 200 V electrical stress. The relative permittivity of the PFA film was quite stable to temperatures as high as 200°C. The relative permittivity of the PI film displayed a negative coefficient that had fallen by approximately 10% at 200°C. The relative permittivity of the PPX film displayed a positive coefficient that had risen by approximately 10% at 200°C. PPX and PFA films displayed increasing trends in their dissipation factor. This increase in dielectric loss in polymers is generally attributed to the increase in free carrier concentration, which often accelerates the breakdown phenomenon, resulting in lower dielectric strength. The PI film, however, evidenced a slight decrease in dissipation factor as the temperature increased. All three materials underwent a reduction in voltage breakdown as temperature was increased. This effect was very slight, however, in the PPX film, which remained relatively stable with an increase in temperature of up to 250°C. The results of these investigations indicated that PFA, with its stable mechanical and thermal properties, combined with lower dielectric loss, was the most viable high-temperature dielectric. It is also recommended as a possible replacement for PI.

Overton et. al. [1993] have reported on characterization efforts that evaluated the effects of thermal aging on metallized Teflon polymer films (supplied by Component Research Company) and arrived at similar conclusions favoring this material for operation at 200°C. The capacitance value of the components tested was 1.0 μF. These devices were

Table 4.2 Dielectric Film Candidates for Elevated-Temperature Wound Capacitors

Dielectric Material	Commercial Source	Properties and Advantages
Polysilseqioxane	David Sarnoff Labs	Good electrical properties up to 250°C, superior to Kapton and Tefzel, can dip or spray coat
Teflon Perflouroalkoxy (PFA)	Dupont	Good mechanical and electrical properties to temperatures as high as 200°C
Polyimide (PI)	Dupont	Small variations in dielectric loss to temperatures as high as 200°C
Nomex 410, 418	Dupont	Aramid papers of synthetic aromatic polyamide polymer; chemically and thermally stable to >220°C; radiation - resistant; 418 grade contains 50% inorganic mica platelets and is designed for high voltage
Diflouro-PBZT Tetraflouro-PBZT	Foster-Miller	High-temperature stability; low dielectric constant
PBO	Foster-Miller, Dow Chemical	Very high temperature stability, 300 to 350°C, significantly exceeding the performance of Kapton and Tefzel
PBO-flourinated IPN	Foster-Miller	High-temperature stability combined with resistance to flash over
Organo-ceramic hybrid nano composites	Garth Wilkes, VPI	Resistant to ionizing radiation; high thermal stability to greater than 200°C
Polybenzimidazole-PBI	Hoechst Celanese	Linear thermoplastic polymer; excellent thermal stability and strength retention >300°C
Flourinated PBO-PI	Hoechst Celanese	Combines possibility of polyimides with high-temperature properties of LCP's
Flourinated polyimides	Hoechst Celanese Ube/ICI, Dupont	Readily available from Ube/ICI and Dupont; thermal stability exceeds Kapton and Tefzel
Voltex 450	Lydall, Inc.	Paper composed of aramid fiber and neoprene binder; low water absorption and high dielectric strength; thermally stable >200°C
Thermoplastic PBO with hexaflourinated moieties	Material lab, WRDC	Thermally processible; high-temperature stability, T_g > 380°C

Table 4.2 (contd)

PQ-100 polyquinolines	Maxdem	Thermally processible; available in a number of configurations; high purity
Polysiloxaneimides	McGrath, VPI	Resistant to ionizing radiation; high thermal stability
Poly-P-Xylene (PPX)	Nova Tran	Stable dielectric strength to temperatures as high as 250°C; may lose some of its good mechanical properties when exposed for long periods to elevated temperature
Flourocarbon-hydrocarbon polymers	Tefzel, Dupont	Readily available; high-quality films; moderate thermal stability
FPE Proprietary aromatic polyester	3M	Readily available high-quality aromatic films useful up to 250°C

aged in an air ambient for 2000 h (12 weeks) at 200°C. Components were gradually brought to room temperature on a weekly basis for data taking. After 12 weeks, at a test frequency of 50 Hz, the Teflon capacitors evidenced very little change in capacitance as a function of thermal aging. At a test frequency of 20 kHz, these components underwent a gradual decrease in capacitance value over time. This decrease reached a maximum of approximately 2% of the original value at the end of the twelfth week. The dissipation factor of the capacitors tested increased slightly and gradually at 50 Hz throughout the testing. At 20 kHz, the dissipation factor increased by more than an order of magnitude. This increase, however, was not attributed to the polymer film but to the resistance of the termination and lead structures. Once again, careful design of the component's packaging could greatly reduce the effect of frequency on the capacitor's dissipation factor and produce a superior high-temperature passive device.

In addition to polymer films, other innovative alternatives include a solid-state double-layer capacitor employing a solid electrolyte [Bullard 1989]. Although the feasibility of such devices for 200°C operation was demonstrated, they suffered high leakage currents, very low voltage (0.65 V/cell), and high ESR (0.2 cm^2/cell).

Although a number of dielectric films possess glass transition temperatures that exceed 350°C, and may thus retain their mechanical properties at operating temperatures of 300°C, unsatisfactory electrical stability of the films precludes their use for fabricating capacitors for operation above 150°C to 200°C. Some extraordinary materials show the promise of demonstrable capabilities in the 250°C to 300°C temperature range, but they have yet to prove their worth in real-life applications. In addition, no manufacturer's characterization data sheets have been published on devices that exceed temperatures of 140°C to 150°C. Also, no manufacturer has expressed any confidence in the development of a series of high-temperature wound-film components for harsh environment applications. For these reasons, the most promising dielectric candidates for high-temperature capacitors may be glass or ceramic multilayer devices.

4.2.2 Ceramic and glass multilayer capacitors

Long-term operation of capacitors fabricated with PLZT (lanthanum modified lead zirconate titanate) with X7R dielectrics has been demonstrated [Maher et. al., 1986]. Following 10,000 h of life testing at 150°C at 100 VDC, no evidence of degradation in the PLZT dielectric was observed. Cygan and McLarney [1991] of Olean Advanced Products have evaluated two other types of ceramic materials for high-temperature applications; these two dielectric formulations were standard NPO (K66) and high-fired X7R (K2000). These ceramics showed minimal reduction in insulation resistance at evaluation temperatures up to 200°C.

Grzybowski et. al. reported on the characterization and modeling of both ceramic and glass dielectric multilayer capacitors at elevated temperatures [Grzybowsky 1992, Grzybowski and Beckman 1993]. This work showed that components fabricated with glass or glass-K dielectrics exhibit excellent properties and useful operating characteristics to temperatures exceeding 300°C. Glass dielectric capacitors are most often used where low noise, high stability, high reliability, and high Q are important to the circuit being designed. Glass dielectrics possess a noncrystalline structure that boasts minimal aging effects, zero voltage coefficient, low loss, zero piezoelectric noise, and a temperature coefficient (TC) that does not exhibit hysteresis and retraces to within ±5 ppm of its TC curve [Demcko et al. 1989]. Capacitors fabricated with these glass dielectrics have proven to be reliable and very stable in harsh environments. A fair criticism, however, may be that these devices possess a poor volumetric efficiency that could make their incorporation in high energy-density applications difficult.

For low level signal conditioning applications, Grzybowski and Beckman [1993] demonstrated empirically that some commercial capacitors employing common ceramic dielectrics, such as NPO, looked like promising candidates for reliable long-term operation at high temperatures. The experiments included data collection and device characterization to temperatures as high as 600°C. Figure 4.4 permits an easy comparison of the capacitance vs. temperature characteristics of four devices fabricated from different glass and ceramic dielectrics to temperatures as high as 500°C. How these experiments were configured and the test hardware utilized to obtain the characterization data at these high temperatures is covered in Chapter 9.

Of the components shown in Figure 4.4, two dielectric types possess positive thermal coefficients - those fabricated with glass or NPO (COG) dielectrics. The other two dielectric types possess negative thermal coefficients - those fabricated with glass-K or X7R dielectrics. Not unexpectedly, large variations in the component values with temperature can have a significant impact on a circuit's performance. Therefore, this study will emphasize the SPICE electrical circuit simulation results of a real application, the approach used to model the behavior of these ceramic capacitors within high-temperature circuit simulations, and a sample of the results obtained to date.

We look specifically at a number of devices fabricated with both NPO and X7R dielectrics, all of which are commercially available from Maxwell's Sierra Capacitor/Filter Division. The ceramic multilayer capacitors evaluated were designed for continuous operation to 200°C, and the materials used to package these ceramic devices were chosen to operate in environments similar to those for the AVX Glass and Glass-K dielectric devices in reference [Grzybowski 1992]. The ceramic capacitors were empirically evaluated in the same way as the AVX Glass dielectric components. The resulting data was curve-fitted and incorporated within SPICE circuit models. Simulation results obtained were compared to those obtained utilizing components made from glass and glass-K dielectrics. The ultimate goal is to develop a design rule set and modeling capability so that these components, and others, may be used routinely for high-temperature circuit applications.

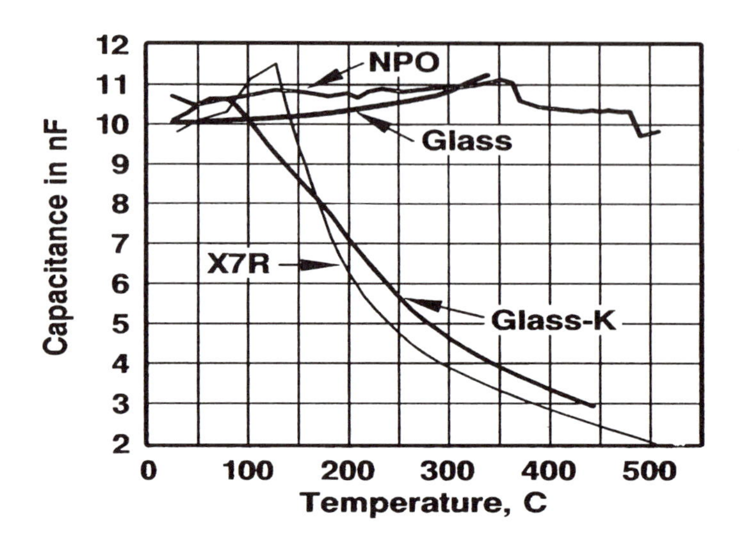

Figure 4.4 Comparison of capacitance vs. temperature for four 10 nF fabricated from different glass and ceramic dielectrics.

One caveat must be borne in mind. All of the data in this example represent short-term excursions to very high temperatures - on the order of several hours. The significant effects of long-term electrical, environmental, and mechanical stress and aging comprise a key aspect of harsh environment electronics and packaging development.

The C^3 capacitors manufactured by Sierra Aerospace have the ceramic capacitor pellet contained within a ceramic case of the same material. The case and capacitor are fired at the same time to yield a monolithic unit [Stevenson 1991]. This attention to the device's packaging is especially important for components being considered for high-temperature applications. Many more ceramic components would perform very well if the packaging materials were chosen so that their TCEs could be matched to the dielectric pellet and the device would remain hermetically sealed in the harsh environment application. These characteristics, bolstered by published test data [Stevenson 1991] regarding C^3 device performance at temperatures as high as 260°C, make these devices an excellent selection from among the many commercially available components for high-temperature characterization.

To obtain the desired high-temperature characterization data, all of the capacitors were fixtured within a high-temperature test chamber and interfaced to the appropriate instrumentation. A set of socket fixtures designed to operate to temperatures as high as 600°C were used (as described at length in [Grzybowski 1991] and in Chapter 9 of this book.) Capacitance and dissipation factor data were acquired using a Hewlett-Packard 4275A Multi-Frequency LCR Meter at a frequency of 1 MHZ. Data points were recorded every 25°C from room temperature to the maximum temperature of 500°C.

Once the capacitance and dissipation factor data were obtained from representative samples of a given component value and voltage rating, the spread of values at each temperature point was averaged. A curve fit was performed on the resulting distribution to render the best polynomial or exponential equation describing the average data obtained. In all samples with an NPO dielectric, the curve representing the change of capacitance with temperature was fit to an equation of the form

$$C = a + bT^2 + cT^2\sqrt{T} + d\,(e^{-T}) \tag{4.6}$$

and the curve representing the change in dissipation factor with temperature was fit to an exponential equation of the form

$$C = a + bT^2 + cT^2\sqrt{T} + d(e^{-T}) \tag{4.7}$$

The curve representing the change in dissipation factor with temperature was fit to an equation of the form

$$DF = a + be^{-T/c} \tag{4.8}$$

Samples with an X7R dielectric were fit to an equation of the form

$$C = \frac{a + cT + eT^2}{1 + bT + dT^2} \tag{4.9}$$

$$DF = a + bT\sqrt{T} + cT^2 + dT^3 \tag{4.10}$$

Figures 4.5 and 4.6 provide sample curve fits of capacitance and dissipation factor data, respectively, for a number of X7R dielectric capacitors. The curves for the samples shown represent typical curve fits for the data spreads obtained during testing. Similarly, Figures 4.7

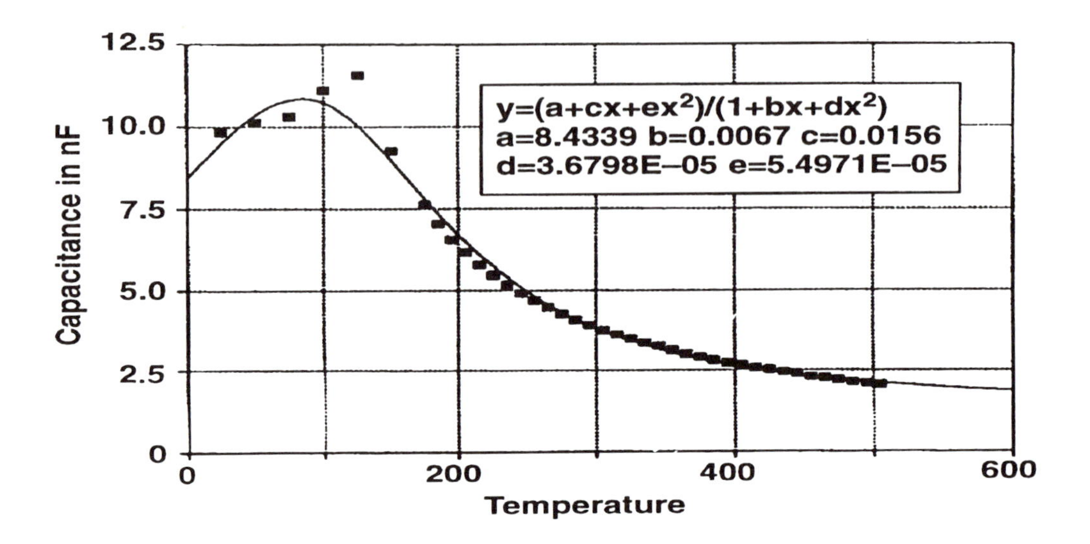

Figure 4.5 Average curve fit for capacitance vs. temperature for a Sierra 10 nF, 100 VDC, X7R component (TRR06D103KG).

and 4.8 provide sample curve fits of capacitance and dissipation factor data, respectively, for a number of NPO dielectric capacitors. The curves for these samples also represent typical curve fits for the data spreads obtained during testing.

Figure 4.4 showed a plot of the capacitance vs. temperature data obtained from typical capacitors fabricated with NPO and X7R dielectrics compared to those obtained from

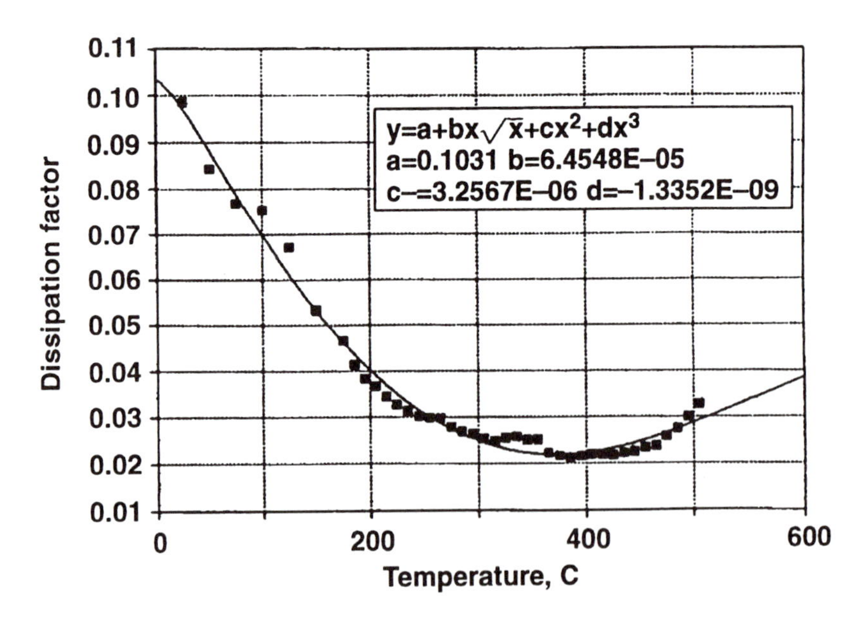

Figure 4.6 Average curve fit for dissipation factor vs. temperature for a Sierra 10 nF, 100VDC, X7R component (TRR06D103KG).

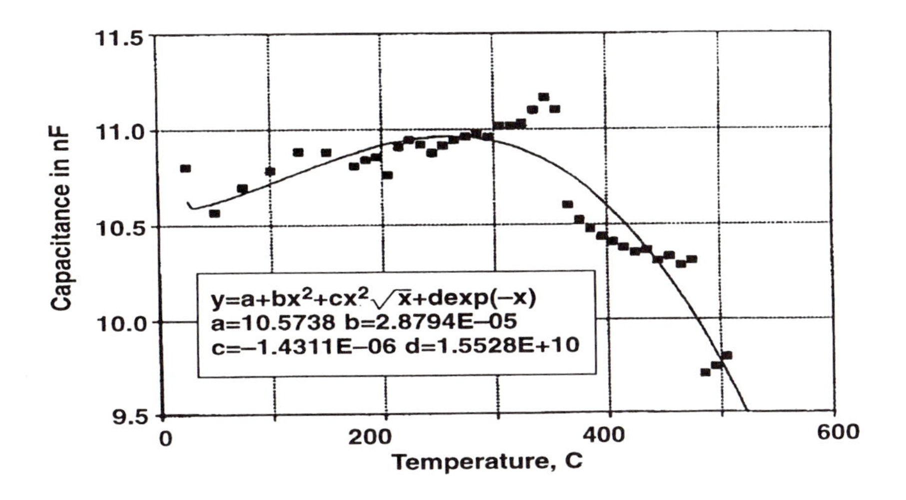

Figure 4.7 Average curve fit for capacitance vs. temperature for a Sierra 10 nF, 50 VDC, NPO component (ACR05D103KG).

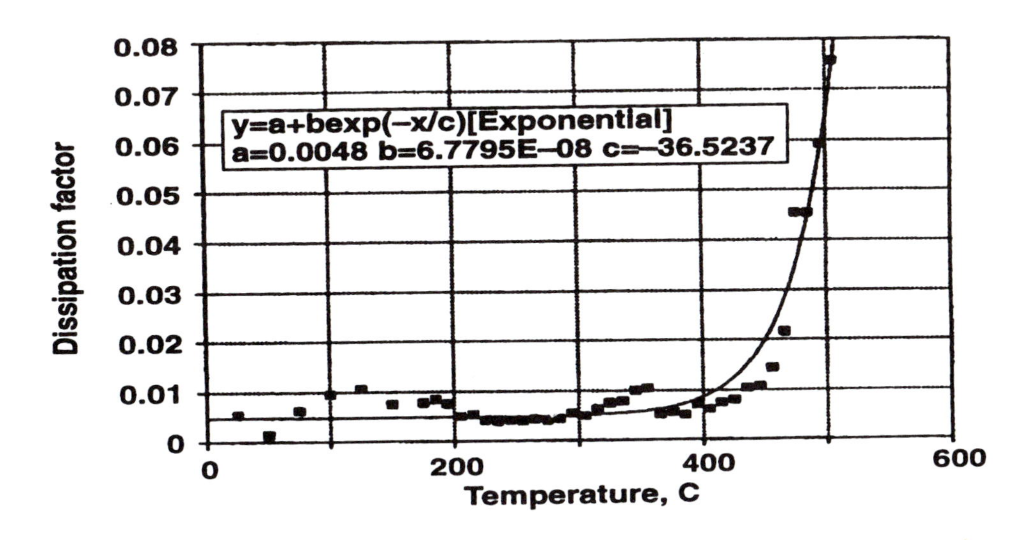

Figure 4.8 Average curve fit for dissipation factor vs. temperature for a Sierra 10 nF, 50 VDC, NPO component (ACR05D103KG).

capacitors fabricated with the glass and glass-K dielectrics discussed in reference [Grzybowski, 1992]. All capacitors represented in Figure 4.4 were 10,000 pF, 20% devices. From this figure it was clear that there were certain similarities existing between the NPO and glass dielectrics and between the glass-K and X7R dielectrics. Most noteworthy was the fact that the devices fabricated with the glass and NPO dielectrics possessed the most stable values of capacitance as the temperature increased. The best-fit modeling equations and temperature coefficients are, however, significantly different. It should be observed that the glass components have a positive temperature coefficient, modeled by Equation (4.11), while the NPO devices have the more complex parabolic behavior modeled by Equation (4.7).

$$C = \frac{a + cT}{1 + bT + dT^2} \tag{4.11}$$

The curve fit equations used to model the capacitance vs. temperature behavior of the X7R and glass-K dielectric devices were identical in form, as expressed by Equation (4.9). The glass-K capacitors, however, displayed slightly less variance in their capacitance vs. temperature characteristics.

Figure 4.9 is a plot of the dissipation factor vs. temperature data obtained from the same four capacitor types identified in Figure 4.4. While there are some similarities among the four capacitor types, the most outstanding feature is the comparatively large variation in dissipation factor vs. temperature exhibited by the X7R devices. Furthermore, the glass-K capacitors generally had a smaller dissipation factor and were, therefore, possessed less loss to temperatures of at least 400°C. Equations (4.11) and (4.12), below, provide the curve fit equations used to model the dissipation factor vs. temperature for the glass-K and glass dielectric devices, respectively.

$$DF = a + bT + cT^2 + dT^3 + eT^4 \tag{4.12}$$

$$DF = \frac{a + cT + eT^2}{1 + bT + dT^2} \tag{4.13}$$

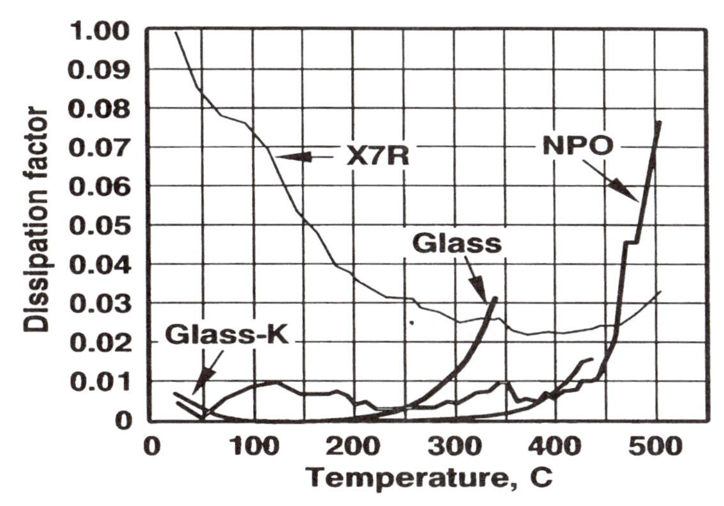

Figure 4.9 Comparison of dissipation factor vs. temperature for four devices fabricated from different glass and ceramic

4.2.3 High temperature device modeling approach and design example

The goal of this effort was to develop a circuit modeling capability for capacitive components at elevated temperatures. This will serve as a first step toward the development of a routine high-temperature circuit design procedure. Modeling the C^3 devices using SPICE may be approached in two ways. The first creates a model in which the ideal (room temperature) value of a capacitor with a specific dielectric material is scaled as a function of desired SPICE simulation temperature. The simulation then proceeds, using the scaled value of capacitance calculated. In this way, only the ideal value of capacitance and an identifier indicating which dielectric material is involved would have to be passed to the simulator. This approach yields a useful set of models serving more general, first-pass simulations. The second approach requires the development of models for specific devices or narrow ranges of specific devices. This approach is directed at applications for which simulation results over temperature are important to the reliability or basic operation of the circuit. This is the strategy that will be pursued in the example to follow. It is a more demanding approach, in that it requires a knowledge of the constants in the curve fit equation for the particular capacitor specified.

The application circuit selected to aid in the comparison of the effects of elevated temperature on capacitive components was a fifth-order elliptic filter. This circuit was used in reference [Grzybowski, 1992] to model glass and glass-K capacitor characteristics within a SPICE simulation and will, again, serve to compare the earlier results to those obtained for the NPO and X7R capacitors. The schematic for this example, shown in Figure 4.10, illustrates an easily implemented way of incorporating a curve-fitted equation, such as one of those described above, within a high-temperature SPICE circuit simulation. This example will also confirm the importance of accurately modeling the high-temperature behavior of passive discrete components used in any design expected to operate at elevated temperatures.

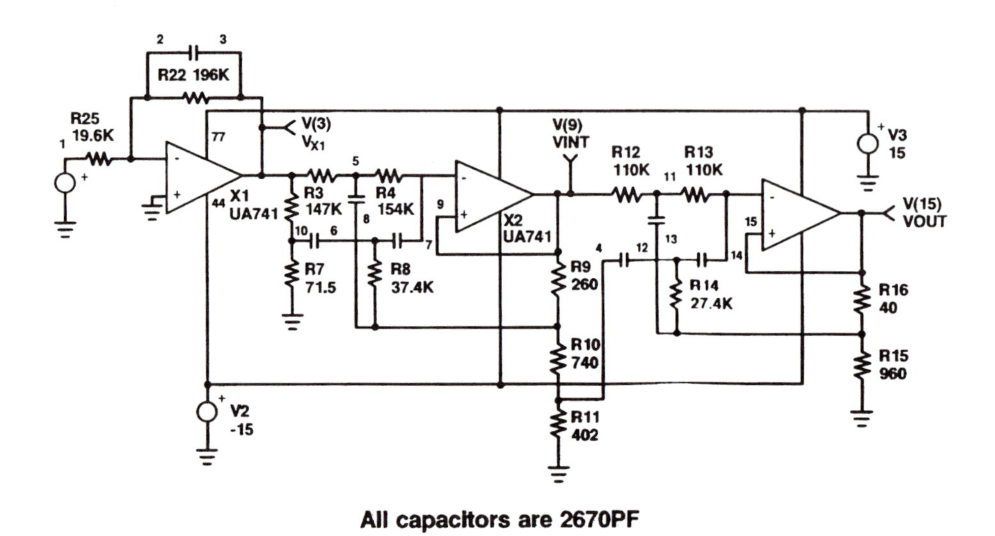

Figure 4.10 Schematic of a fifth-order elliptic filter used to compare SPICE simulation results at elevated temperatures

There will be a large variation in a circuit's response as the ambient temperature changes [Grzybowski, 1992]. In order not to clutter this design example with unrelated high-temperature circuit effects, only the capacitors will be given temperature coefficients in the simulations that follow, all other components will be treated as ideal. However, it must be understood that every component, both passive and active, as well as the methods of component attachment and system packaging, will have a definite impact on the application's overall long-term operational lifetime and reliability. All of the capacitors in this circuit have the same starting room temperature value of 10,000 pF.

The parameter card (.PARAM) in SPICE will be used to pass the value for the temperature at which the simulation should take place to the simulator in the code containing the capacitor definitions. For example, to use Equation (4.9) for an X7R dielectric device within a 300°C circuit simulation, the following lines of code will be required:

```
.PARAM = 300

C12 6 1 {(aT + cT^2 + eT^3)/(1 + bT^2 + dT^3)}
```

A SPICE circuit simulation was run for capacitors representing each of the four dielectric types, with the model for the appropriate dielectric inserted. The output voltage obtained from each of the four simulations for the elliptic filter example in response to a step input at 300°C, are shown in Figure 4.11. From this figure the similarities between the NPO

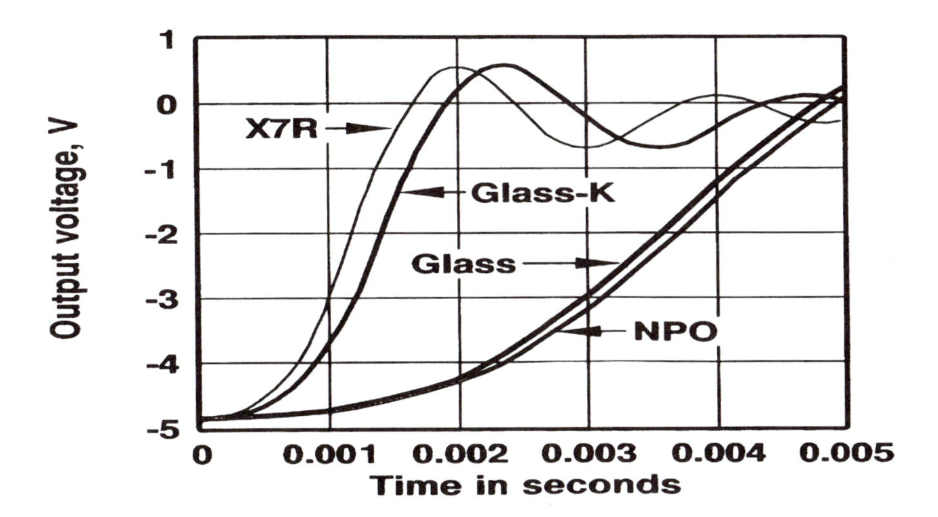

Figure 4.11 Comparison of simulated elliptic filter response, using capacitors of four different dielectric types at 300 °C

and glass dielectrics and between the X7R and glass-K dielectrics are again evident. This example makes it clear that, although all simulations were run with 10,000 pF capacitors, at high temperatures the circuit response can vary drastically. These results illustrate the significance temperature-dependent capacitance variations - both within a given class of dielectrics and among different dielectric types - can have in circuit designs.

At the beginning of this section, concerns were raised regarding the lack of large value-capacitors for elevated-temperature energy storage applications. The author has recently evaluated a number of higher capacitance multilayer ceramic components fabricated by Olean Advanced Products. Included in these device splits were capacitors employing an X7R dielectric with room temperature ratings as high as 10.0 μF at 100 VDC and those employing NPO dielectric with room temperature ratings as high as 0.33 μF at 100 VDC. In addition to being characterized at temperatures as high as 500°C, some of these devices are undergoing long-term life testing and thermal cycling from room temperature to 500°C.

Figure 4.12 shows capacitance vs. temperature characterization results obtained from a selection of 10 μF X7R devices. These capacitors are actually fabricated by stacking two 5 μF devices together and soldering metal leads to both sides of the stack. Because of the limitations on the melting temperature of the solder used to attach these leads in place, the collection of characterization data proceeded only to 300°C, as represented in this figure. Based on earlier data earlier for capacitors fabricated with X7R dielectrics, it should come as no surprise that there is a significant negative thermal coefficient of capacitance associated with these devices. From Figure 4.12 it is evident that at 200°C, these components have derated to only 60% of their room temperature value. At 300°C, the capacitance value has

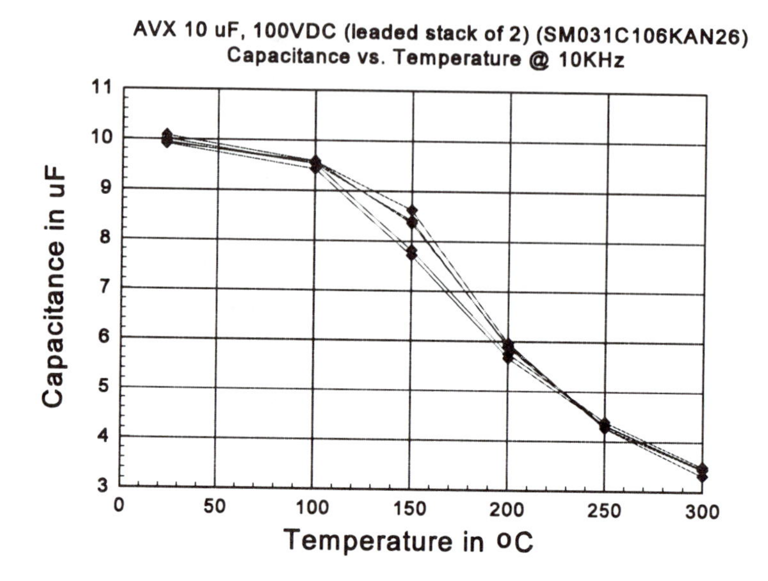

Figure 4.12 Capacitance vs. temperature data obtained from a selection of 10 μF X7R multilayer ceramic capacitors

dropped to only 35% of the starting room temperature value. These large variations in capacitance vs. temperature alone make devices fabricated from X7R dielectrics poor choices for all but the most rudimentary filtering functions at elevated temperatures. Data for the dissipation factor of these leaded stacks of two 5 μF components vs. temperature is provided in Figure 4.13.

In striking contrast to the X7R characteristics, however, is the high-temperature behavior of large-value components fabricated with NPO dielectrics. The greatly reduced volumetric efficiency of these devices is their main drawback for elevated temperature applications. Figure 4.14 shows a set of capacitance vs. temperature characterization results obtained from a selection of 0.33 μF NPO devices. The fact that the capacitance value of these components with temperature remains essentially constant to 500°C makes this a remarkable presentation, indeed. Data for the dissipation factor of these 0.33 μF NPO components vs. temperature is provided in Figure 4.15. Clearly, most of the devices tested exhibit modest increases in dissipation factor until approximately 450°C. Moreover, two devices in this family of curves exhibited abnormally high losses at lower temperatures when compared with other devices undergoing the same testing. It is possible that these two devices had been fixtured such that additional loss mechanisms were artificially introduced.

As noted, the effects of long-term electrical, environmental, and mechanical stress and aging that seriously affect high-temperature/harsh environment electronics and packaging development will assuredly be significant in any such application. It must also be acknowledged that, because high-temperature (300°C+) electronics and packaging is in its infancy as a technology and as an industry, there is a conspicuous lack of long-term or reliability data. The hurtles include several issues. How should accelerated testing for high-temperature components be performed? How does the physics of failure for any given

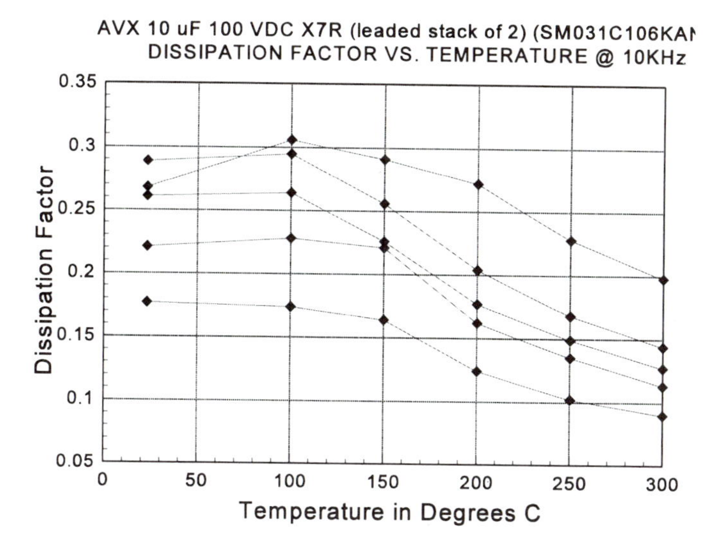

Figure 4.13 Dissipation factor vs. temperature data obtained from a selection of 10 μF X7R multilayer ceramic capacitors

component or packaging scheme relate to the product's predicted lifetime or reliability? How should the basic materials for any given component or packaging scheme be selected? What market has a large enough volume production need to produce the statistically significant data that may be required to answer these questions? Researchers are working to solve these problems, but it is going to take some time.

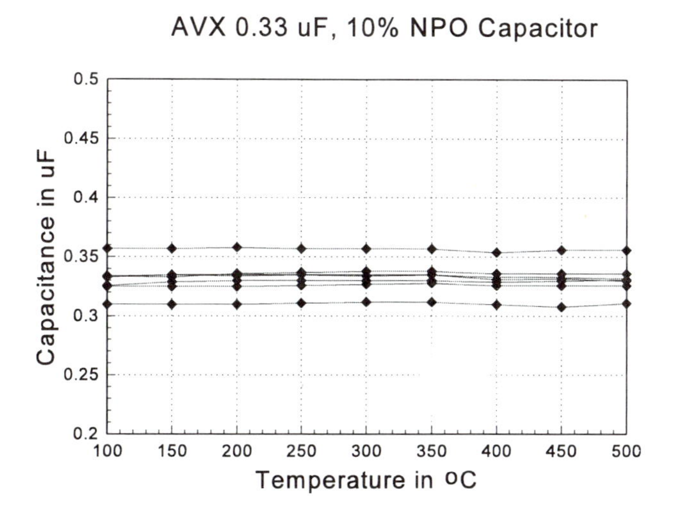

Figure 4.14 Capacitance vs. temperature data obtained from a selection of 0.33 μF NPO multilayer ceramic capacitors

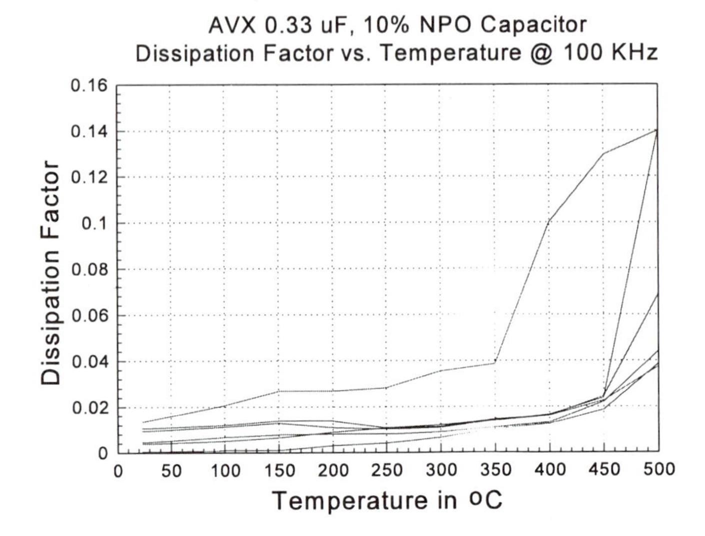

Figure 4.15 Dissipation factor vs. temperature data obtained from a selection of 0.33 μF NPO multilayer ceramic capacitors

As an example, Figure 4.16 shows the time variation of the dissipation factor obtained from two of the 0.33 μF capacitors considered earlier. These two devices were ramped from room temperature to 500°C in a short two hours, then soaked at 500°C for 200 hours. The temperature was then ramped down to room temperature again. Dissipation factor data was obtained periodically throughout the testing. The capacitors then underwent six cycles of being ramped to 500°C and soaked for 450 hours and ramped back to room temperature. At the end of each cycle, the test fixture was outfitted with new interconnects and the contacts to the capacitors were abraded to remove any oxide build-up. Although this life test was still ongoing at the time of this writing, Figure 4.16 provides substantial insight into the effects of cyclic fatigue and aging on the two components. The most obvious point is that there is a dramatic effect of time at temperature on the loss behavior of these components at the test frequency of 100 KHz. The dissipation factor is steadily climbing with time throughout each of the cycles. The difference in the absolute value of the dissipation factor in each cycle may be attributed to the replacement of the interconnects and the cleaning and reseating of the capacitor contacts at the end of each cycle. Throughout the entire 2500+ hours of testing represented by this data, however, the capacitance values of these components remained virtually unaltered from their starting values of approximately 0.33 μF. These devices have performed remarkably well under these adverse conditions, with no signs of permanent long-term change or damage. This is not always the case, and the point is that long-term reliable operation of passive components, such as capacitors, will have to be modeled much more carefully than are the traditional 125°C rated components of today.

4.2.4 Polarized/Electrolytic capacitors

There are fundamental incompatibilities between the materials science associated with constructing components for high-temperature operating environments and the physics associated with polarized capacitors. Characterization data for commercially available wet slug and hermetically sealed tantalum capacitors, along with aluminum electrolytic devices is seldom provided above 100°C. The maximum temperature at which these devices are rated rarely exceeds 85 °C. The Ta_2O_5 dielectric used to fabricate tantalum capacitors is prone to breakdown when exposed to elevated temperatures. This problem is compounded by isolated hot spots induced by surge currents. This effect can induce combustion of the dielectric in solid tantalum capacitors. In addition, all of these components can fail catastrophically, and often quite spectacularly, should the packaging fail and the case rupture or explode. That these types of capacitors are so unsuitable for use at high temperatures is an unfortunate problem that still needs to be solved for harsh-environment applications that require high energy-density components and large-value capacitors.

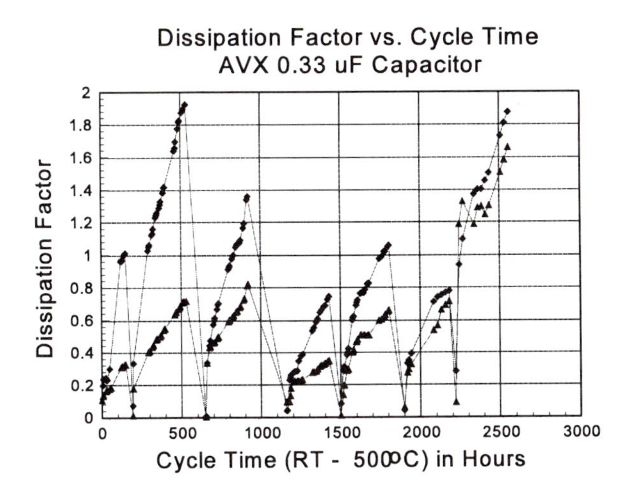

Figure 4.16 High-temperature thermal cycling of two 0.33 μF NPO multilayer ceramic capacitors

Chapter 5

FIRST-LEVEL PACKAGING CONSIDERATIONS FOR THE USE OF ELECTRONIC HARDWARE AT HIGH TEMPERATURES

Chapter 2 noted that, with proper design, integrated circuits manufactured from silicon and other common semiconductor materials could withstand ambient temperatures of 200°C. This chapter presents the fundamental material-related limitations of high-temperature operation of electronic systems caused by first-level packaging materials. Conventionally, ceramic and metal hermetic packaging technology have been utilized for high-temperature and high-power dissipation applications. Alternatively, this chapter discusses the use of both plastic and hermetic packaging technology at elevated temperatures. It concludes by summarizing high-temperature packaging concerns and listing the materials acceptable for different temperature ranges.

The chapter focuses on the temperature limitations and dependencies of key thermal, mechanical, and electrical properties up to 200°C for the materials used in wires and wirebonds; die attach materials; leadframe materials; plastic encapsulants; hermetic case materials; lid seals; and lead seals.

5.1 Wires and Wirebonds

Wirebonding is a method of interconnecting the die to the leads of the package. It is accomplished by ultrasonic, thermosonic, or thermocompression bonding of a metal wire from a bond pad on the die to either a bond pad on a substrate or directly to a lead. Two temperature-related factors are important to the reliability of a wirebonding system. The first is the compatibility of the wire and the bond pad; the second is the integrity of the wire against fatigue and fracture.

Table 5.1 lists common wire/wirebond material combinations, together with the maximum use temperature for each and the reason for the limitation. These reasons relate to the compatibility of the bonding materials and fall into three categories: (1) interdiffusion, which causes void formation and, in turn, a decrease in the bond strength and an increase in the contact resistance; (2) intermetallic formation, which creates brittle phases that weaken the bond; and (3) melting.

If two dissimilar metals must be bonded, they should have interdiffusion constants large enough to permit a strong bond, but small enough to prevent the formation of excess intermetallics in periods less than the operating life of the device. Intermetallic layers have complex crystal structures, and thus are usually brittle and more susceptible to flexure damage than pure metals. Because interdiffusion governs intermetallic growth, intermetallics are also associated with Kirkendall voiding and volumetric expansion in the bond region. When exposed to shear stresses generated between the wire, the substrate, and the bond as a result of

Table 5.1 Temperature Guideline Table for Wirebonds

Wirebond	Maximum temperature	Reason	References
Al - Au	175°C	Forms brittle intermetallic phases which reduce bond strength and conductivity.	Newsome et al. 1976
Al - Ag	175°C	Interdiffusion creates excessive voids that decrease the bond area and strength. Also limited by chloride corrosion susceptibility.	James 1977 Jellison 1975
Cu - Al	200°C	Forms brittle $CuAl_2$ intermetallic phases that lower shear strength.	Atsumi et al. 1986 Onuki et al. 1987
Cu - Au	300°C	Interdiffusion creates excessive voids that decrease the bond area and strength.	Hall et al. 1975 Pitt and Needes 1982
Al - Ni	300°C	Interdiffusion creates excessive voids that decrease the bond area and strength.	Harman 1989
Al - Al	660°C	Melting temperature.	Weast 1979
Au - Au	> 660°C (up to 1064°C)	Melting temperature	Jellison 1975 Weast 1979

differential thermal expansion, brittle intermetallics often fracture, causing failure at the bond site [Pecht 1994].

The most common example of a system limited in its high-temperature use by intermetallic formation is gold-aluminum, which is present when gold (Au) wire is bonded to aluminum (Al) bond pads on the chip, or when aluminum wire is bonded to gold bonding fingers on a leadframe or substrate. The intermetallic formation process is governed by the diffusion of gold into aluminum, with Au_5Al_2 forming before $AuAl_2$, the final phase in aluminum-rich systems. In gold-rich systems, it is also possible to form AuAl, Au_2Al, and Au_4Al. The higher diffusivity of the gold leads to voiding on the gold side of the interface. Below 175°C, intermetallic growth in the gold-aluminum system is not dominantly dependent on steady-state temperature [Pecht 1994]. At these temperatures, interdiffusion and intermetallic formation are governed by the defect density. Thick-film metallizations and poorly welded joints with many grain boundaries, vacancies, dislocations, and free surfaces have higher rates of interdiffusion than bulk metals, and therefore often fail due to interdiffusion and intermetallic formation mechanisms at times and temperatures less than those predicted from lattice diffusion alone. As a result, various studies have placed the point at which interdiffusion causes a problem as low as 125°C and as high as 175°C. While intermetallics have been observed at temperatures as low as 75°C [Newsome et al. 1976], the related decrease in bond strength that results from Kirkendall voiding and residual stress was not observed until 125°C in one study of aluminum wires on gold bond pads [Takei and Francombe 1968] and until 160°C in another [Newsome et al. 1976]. Above 175°C, the rate of intermetallic formation accelerates greatly. This is shown in Figure 5.1 [DM Data 1990] as an increase in the resistance in the interconnect, which is another damaging effect of intermetallic growth. At these temperatures, intermetallic growth becomes a function of the lattice interdiffusion constant and thus the steady-state temperature, as described by the parabolic relationship

$$x_l = k_c \ t_1/2 \tag{5.1}$$

where x_l is the intermetallic layer thickness, t is the time, and k_c is the interdiffusion constant, given by

$$k_c = c_x e^{[-E_a/K_B T]}, \quad T > 150°C \tag{5.2}$$

where c_x is a constant, E_a is the activation energy for layer growth, and T is the steady-state temperature. The value of k_c changes for each intermetallic phase. For gold-aluminum, the value of E_a is 0.69 eV and the value of c_x is 5.2 x 10^{-4} cm^2/s [Philosky 1970, 1971].

The most recent studies found that aluminum wirebonded to gold over nickel bond pad

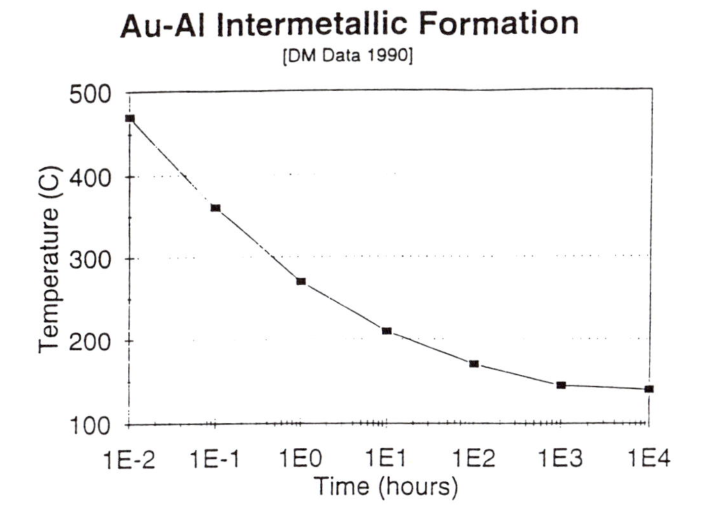

Figure 5.1 The effect of temperature on the growth of Au-Al intermetallics [DM Data 1990]

fingers can last 460 hours at 400°C and over 2500 hours at 250°C. However, this extended life is attributed to diffusion of the gold into the nickel rather than the aluminum, creating a more stable aluminum-nickel bond. This study concluded that a gold-aluminum system is acceptable to 125°C, questionable to 200°C, and unacceptable above 200°C, while a gold-gold system is suitable up to 400°C, where bond life exceeded 3000 hours [Grzybowski 1994].

Other wirebond systems have smaller interdiffusion coefficients. This slows the rate of intermetallic formation and elevates the allowable use temperature. Copper-aluminum systems exhibit a decrease in shear strength above 200°C [Atsumi et al. 1986, Onuki et al. 1987]. Copper-gold systems do not show void-related failures resulting from a decrease in bond strength at temperatures less than 250°C [Pitt and Needes 1982]. Aluminum wires bonded to nickel coatings retain their bond strength at temperatures up to 300°C, and have therefore been widely used in discrete power devices as an alternative to gold wires bonded to aluminum [Harman 1989]. Aluminum-aluminum bonds can be used to temperatures near the melting point of 660°C, because there is no potential for intermetallic formation and galvanic corrosion; gold-gold bonds are stable to even higher temperatures (>1000°C) for the same reasons.

Failure mechanisms identified for wire integrity include wire fatigue and wirebond fatigue. These mechanisms are not dependent on steady-state temperature, but do depend on temperature changes and on the thermal dependencies of the elastic modulus and the coefficient of thermal expansion (CTE) of the wire. The temperature dependency of these properties is shown in Figures 5.2 and 5.3 for gold, aluminum, copper, and silver wire materials [Smithells 1955, Leksina and Novikova 1963]. Both the elastic modulus and the coefficient of thermal expansion for metals are dependent on temperature. This is because, as the temperature is increased, the lattice vibrations increase and the lattice sites move farther apart. This results in a microscopic thermal expansion of the material, defined as

$$\alpha_{CTE} = \left(\frac{a_f - a_i}{a_i}\right)\left(\frac{1}{\Delta T}\right) \tag{5.3}$$

where α_{CTE} is the coefficient of thermal expansion, a_f is the final lattice parameter, a_i is the initial lattice parameter, and ΔT is the change in temperature. As the atoms move farther apart, the force required to separate them further decreases. This results in both a decrease in the elastic modulus with temperature, with the modulus being proportional to a^{-4}, and to an increase in the coefficient of thermal expansion with temperature.

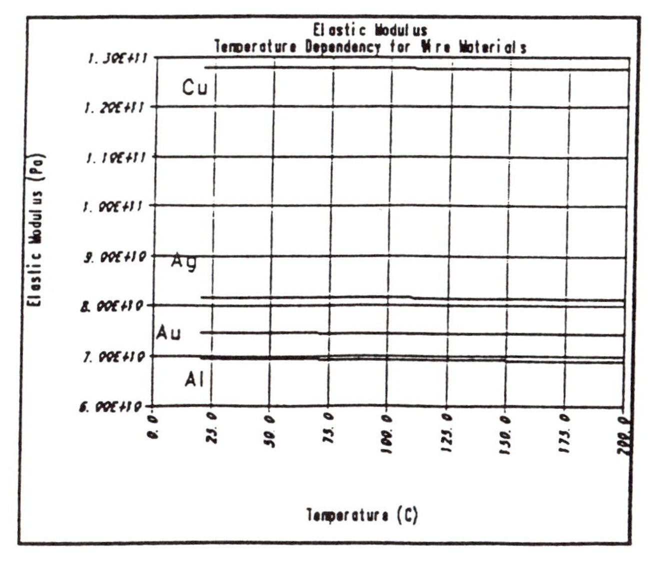

Figure 5.2 The elastic modulus of wire materials as a function of temperature [Pecht 1994].

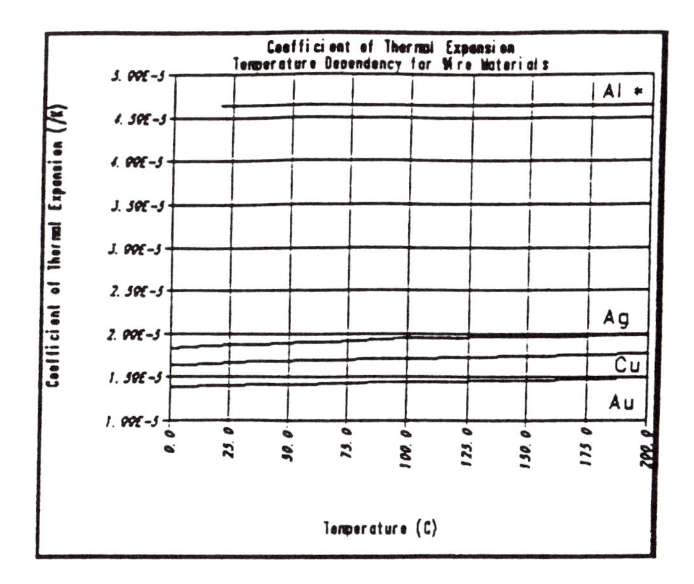

Figure 5.3 The coefficient of thermal expansion of wire materials as a function of temperature [Pecht 1994, Leksina and Novikova 1963].

For the materials studied here, the temperature dependence of the elastic modulus and the coefficient of thermal expansion are much smaller than the differences in the values for different materials, so what is more important is selecting the proper material to alleviate the effects of larger temperature changes. In this case, the best options are materials with a small coefficient of thermal expansion and a low modulus to minimize fatigue. This makes gold appear to be the best choice, as the CTE of aluminum and the modulus of copper are too high. Extensive design guidelines detailing the quantitative relationships between the fatigue life and the materials, wire diameters, and ΔT's are available in Pecht [1994].

Finally, as the main purpose of the wire is to conduct signals [Pecht 1994], an increase in resistivity at elevated temperatures is a concern because it increases signal propagation delay and causes losses in signal fidelity. The temperature dependencies of the electrical resistivity and thermal conductivity are shown in Figures 5.4 and 5.5 for gold, aluminum, copper, and silver wire materials [Matula 1978, Smithells 1955, Holman 1986]. No attempt is made to choose a cut-off temperature for the use of these materials based on their increase in electrical resistivity, since the acceptable value is based on the circuit design.

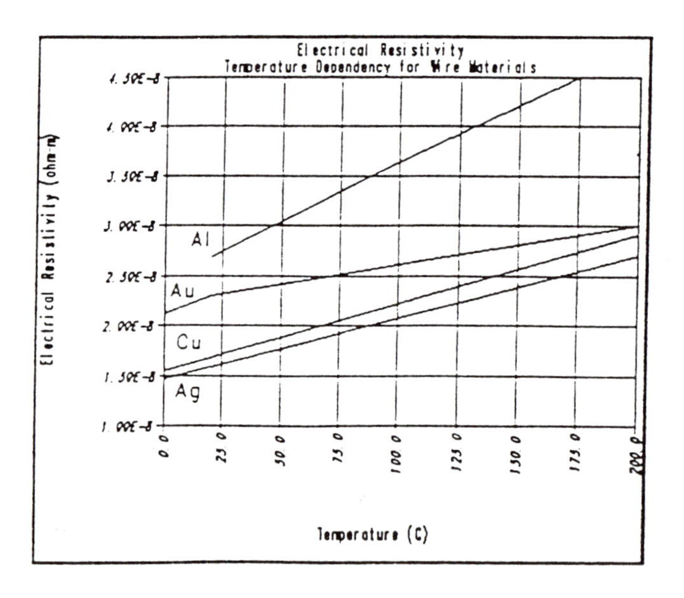

Figure 5.4 The electrical resistivity of wire materials as a function of temperature [Matula 1978, Smithells 1955].

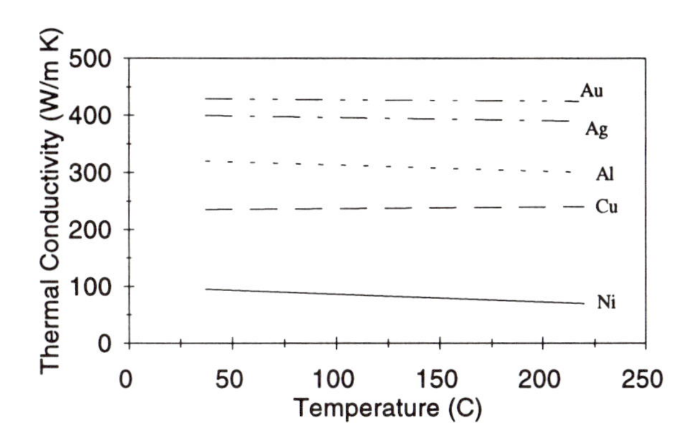

Figure 5.5 The thermal conductivity of wire materials as a function of temperature [Holman 1986].

The electrical resistivity increases with temperature because of increased scattering of the electrons by lattice vibrations (or phonons). In its simplest form, electrical resistivity can be expressed as the sum of two independent terms, as shown in the equation below — one, ρ_L, is the resistivity due to lattice vibrations (or phonons), and the other, ρ_I, is the resistivity due to lattice imperfections and impurities:

$$\rho = \rho_L(T) + \rho_I \tag{5.4}$$

This expression is known as Mattheisen's rule. Only ρ_L is a function of temperature which is given as

$$\rho_L = \rho_{LO}(1+\alpha_0(T-T_0)) \tag{5.5}$$

where α_o is the temperature coefficient of resistance; T_O and ρ_{LO} is the reference temperature, which is usually set to be 23°C, and the resistivity due to lattice vibrations at the reference temperature. These expressions allow an estimate to be made of the resistivity of a dilute alloy solid solution, such as Ag-1% Si or Al-1% Mg wire, at any temperature, using the resistivity of the alloy at room temperature as ρ_o and the temperature coefficient of resistance for the pure element as α_o. However, this formula provides only an estimate of the actual resistivity since all materials exhibit some deviation from the rule due to the temperature dependency of the impurity- and imperfection-related term [Ho et al. 1983]. As a large fraction of the heat is also conducted in metals by the electrons, thermal conductivity decreases with temperature because of the same temperature-related increase in scattering.

5.2 Die Attach

The primary function of a die attach material is to secure a semiconductor chip to a leadframe or substrate, and to ensure it does not detach or fracture over an operational lifetime that may include many power and temperature cycle excursions. The maximum use temperatures for die attaches are those at which the die attach materials melt or soften to a level at which they no longer keep the die in place during operation. These temperatures are given in Table 5.2 for various common die attach systems.

Table 5.2 Temperature Guideline Table for Die Attach Materials [Manko 1992, Pecht 1994, Feinstein 1989, Khatchatourian 1995, Olson 1995]

Die attach material	Maximum use temperature	Reason
Au80Sn20	280°C	eutectic melting point
Au88Ge12	356°C	eutectic melting point
Au97Si3	363°C	eutectic melting point
Sn96Ag4	221°C	solidus
Sn63Pb37	183°C	eutectic melting point
Sn60Pb40	183°C	solidus
Sn62Pb36Ag2	179°C	solidus
Sn50Pb50	183°C	solidus
Sn5Pb95	308°C	solidus
Sn95Sb5	235°C	solidus
Sn92Sb8	236°C	solidus
Sn65Ag25In10 (J-alloy)	236°C	liquidus
Pb50In50	180°C	solidus
Pb92In5Ag3	300°C	solidus
Epoxy Novalac	200°C	softening/ depolymerization
Silicone Epoxy	260°C	softening/ depolymerization
Silver filled Epoxy	200°C	softening/ depolymerization
Polyimide	300°C	glass transition temperature
Silver filled Glass	450°C	softening point

The utility and reliability of die attach materials may be compromised at temperatures below the solidus, melting, or decomposition temperature, as a result of failure mechanisms related to the thermal stresses induced by temperature cycling, power cycling, and initial die bond cooldown. Stiffer die attach materials, such as silver-filled glasses and gold-based eutectics, concentrate the thermal stress in the die, which can potentially cause die fracture and fatigue. Peeling stresses at the edge of the die can cause horizontal crack propagation and die lifting, while tensile stresses in the center of the die can cause vertical crack propagation, leading to die cracking. The defects causing crack initiation develop during such manufacturing steps as ingot growth, wafer slicing, and die separation. MIL-STD-883 lists the maximum allowable initial crack sizes that can be detected during visual inspection by a die manufacturer.

Die fracture and fatigue mechanisms depend primarily on the modulus of the die attach (which determines the fraction of the stress imparted to the die); on the difference in the coefficients of thermal expansion of the die and substrate (which determines the total thermal stress); and on the fracture toughness and fatigue constants of the die. Figures 5.6 and 5.7 illustrate the temperature dependence of the elastic modulus and tensile strength of gold eutectic die attaches. Figures 5.8 to 5.10 illustrate the temperature dependence of the coefficients of thermal expansion of silicon dies, ceramic substrates, metal substrates, and leadframes.

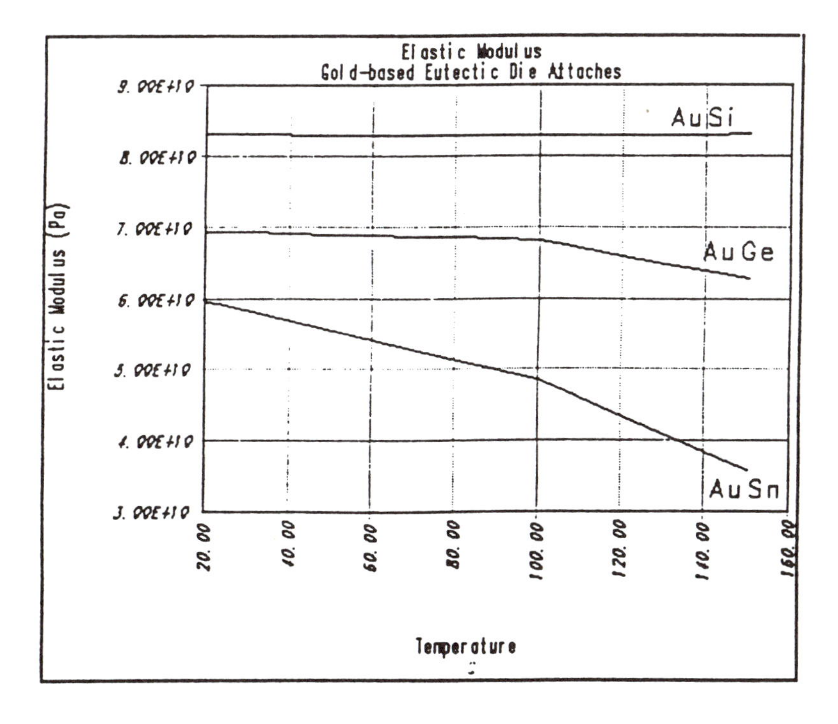

Figure 5.6 The elastic modulus of gold-based eutectic die attach materials as a function of temperature [Olson and Berg 1979, Yost et al. 1990].

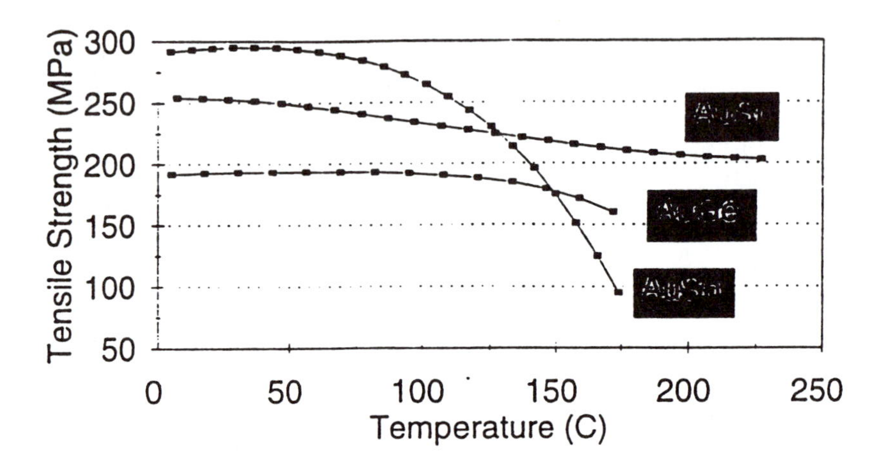

Figure 5.7 The ultimate tensile strength of gold-tin eutectic die attach materials as a function of temperature [Olson and Berg 1979].

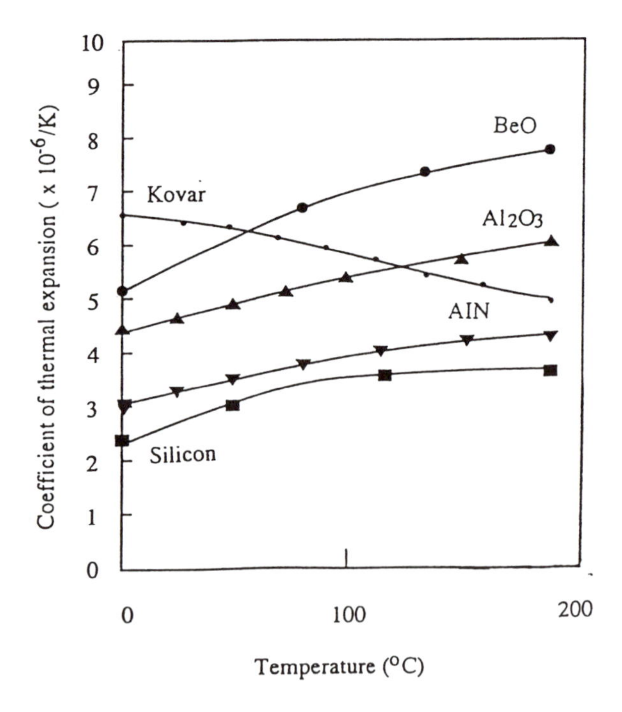

Figure 5.8 The coefficient of thermal expansion of ceramic case materials as a function of temperature [Charles and Clatterbaugh 1994].

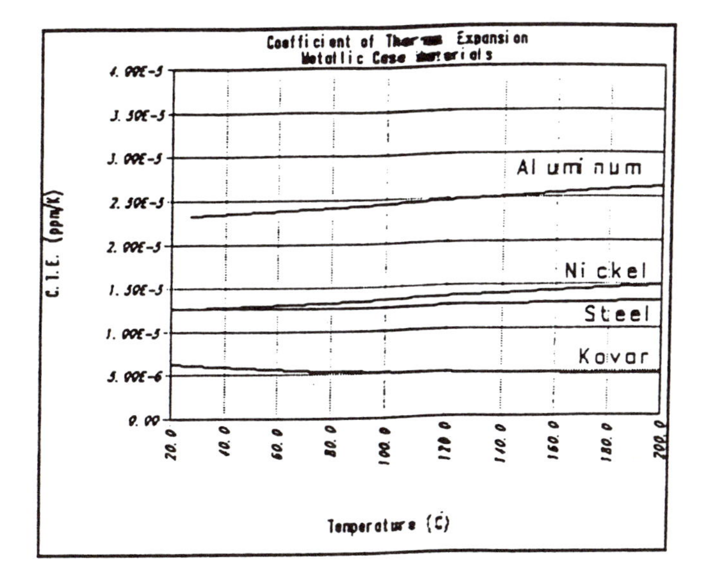

Figure 5.9 The coefficient of thermal expansion of metal case materials as a function of temperature [Holman 1986, Charles and Clatterbaugh 1994].

The more flexible die attach materials, such as epoxies, polyimides, and lead-tin solders, impart less thermal stress to the die, but are therefore more susceptible to failure by die-attach fracture and fatigue. As before, thermal stresses are a function of the modulus of elasticity of the attach material. In addition, the tensile strength and shear strength of the attach are important since they determine the cohesiveness of the assembly. The variation in these parameters with temperature for soft solders is given in Figures 5.11 to 5.13. The decrease in

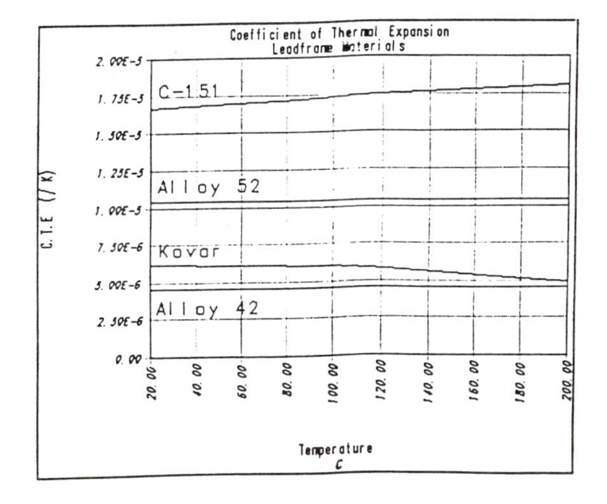

Figure 5.10 The coefficient of thermal expansion of leadframe materials as a function of temperature [Alloys Digest 1982, Pecht and Agarwal 1994].

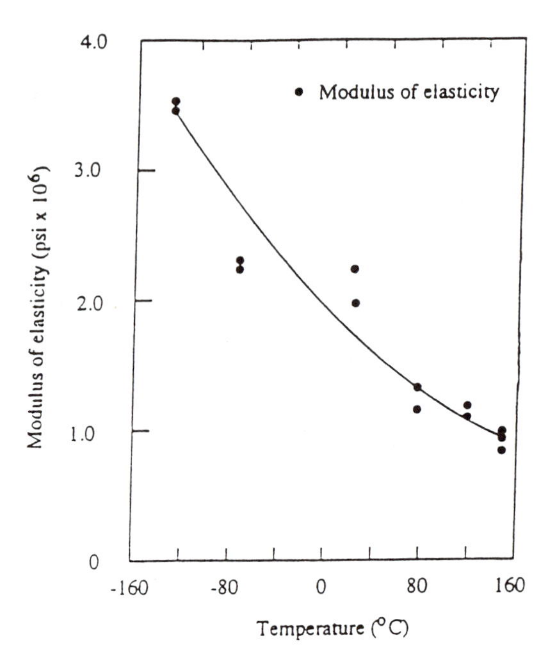

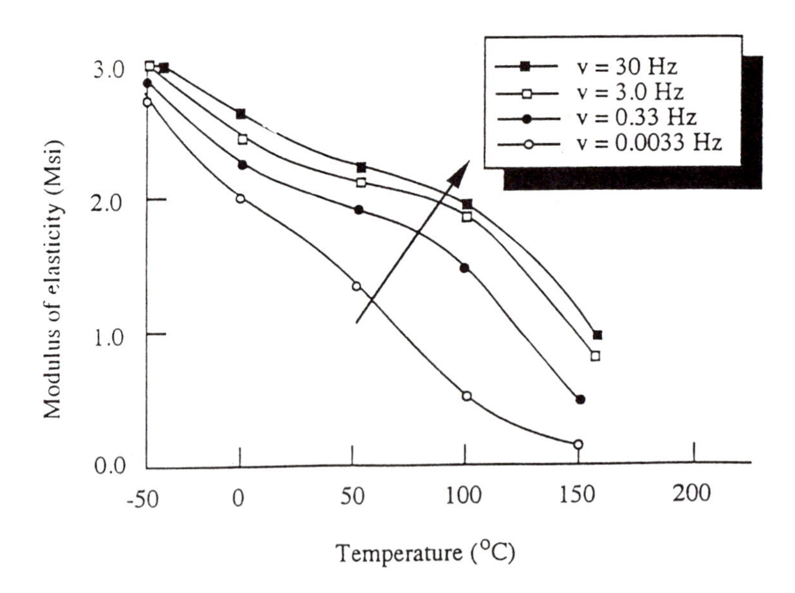

Figure 5.11 The variation of the modulus a) for Sn63Pb 37 with respect to temperature and b) for Sn60Pb40 with respect to temperature and frequency [Vianco 1993, Dasgupta 1991].

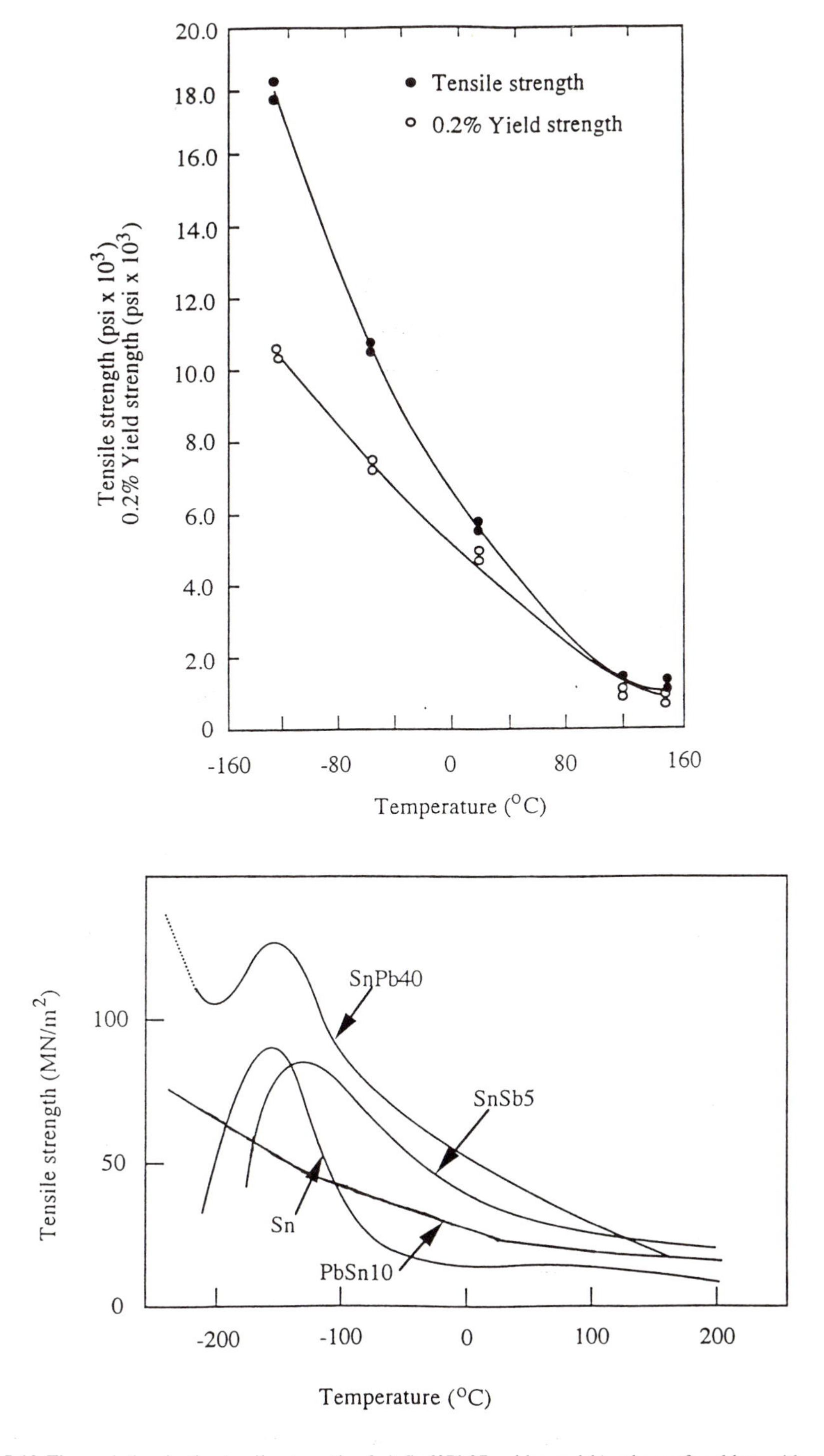

Figure 5.12 The variation in the tensile strength of a) Sn63Pb37 solder and b) other soft solders with respect to temperature [Vianco 1993, Dasgupta 1991].

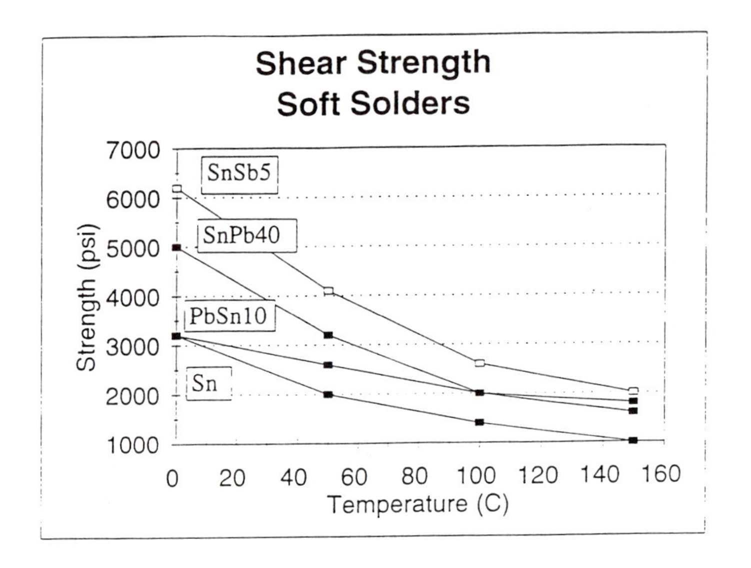

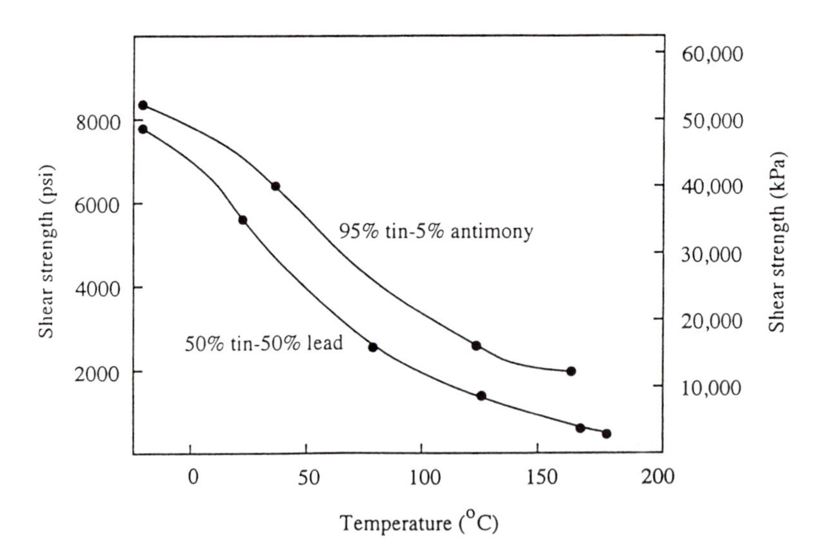

Figure 5.13 The variation in the shear strength of soft solders with respect to temperature [Thwaites 1972, Rahn 1993].

the elastic modulus for both stiff and flexible materials causes less stress to be placed on the die at elevated temperatures; however, the decreases in strength lower the cohesiveness of either assembly. This particularly affects materials that are more flexible to start. Voids in the die attach are important, as they serve as crack initiation sites and reduce the effective heat transfer of the package. A random distribution of small voids has little effect on the heat transfer, whereas one large void can create a temperature gradient high enough to degrade package performance. Voids are formed by poor wetting of solders, outgassing of polymers, trapped air in the attachment, or attachment shrinkage during solidification [Pecht 1994].

Extensive guidelines for the design of die attach materials that can be used to further quantify the reliability effects of temperature cycling and temperature-related parameter changes on fatigue life are given in Pecht [1994] or Hu et al. [1993].

Die attach materials are required to perform other functions as well. They must have good thermal conductivity to conduct heat from the chip to the mounting platform and then out of the package; and in many cases, particularly in power devices, they must function as the ohmic contact that makes the drain connection from the back of the die to the header. Die attaches are therefore often divided into three categories reflecting their thermal and electrical conductivity, and into two categories reflecting their stiffness. Table 5.3 shows several common die attach families in their appropriate categories. Room temperature properties for common die attach materials are given in Table 5.4.

Packages intended for room-temperature applications tend to use the more flexible tin-lead solders or polymer-based systems as die attaches. They are low in cost, and their low moduli and high ductilities place little stress on the die. Soft solder attachments are preferred for high-power circuits, because of their superior electrical and thermal conductivity. The

Table 5.3 Classification of Die Attach Materials

	High modulus (stiff)	Low modulus (flexible)
Electrical/Thermal Conductive	gold eutectics silver-filled glass	silver-filled polyimides silver-filled epoxies tin-lead-based solders indium-based solders
Thermal Conductive	tin oxide-filled glass indium oxide-filled glass beryllia-filled glass	AlN-filled epoxy alumina-filled epoxy
Non-conductive	unfilled glass	unfilled epoxies epoxy silicone polyimides

Table 5.4 Room Temperature Properties for Attach Materials [Pecht 1994, Manko 1979, Goodnan 1986, Wright 1977, ASM 1990, 1993]

Material	Tensile strength (MPa)	Shear strength (MPa)	Elastic modulus (GPa)	Thermal conduct. (W/mK)	CTE (ppm/K)	Electrical resistivity (μΩ-cm)
Sn63	35.4-42.2	28.5	14.9	50	24.7	14.5
Sn60	18.6-28.0	24-33	20	50-65	23.9	17
Sn50	-	24.2	26	46	23.4	15.8
Sn5	23.2	14	23.5	35	28.7	21.4
Sb5	56.2	32	50	-	27	15
AuSn	198	185	69	251	16	36
AuSi	255-304	220	69.5	293	11	775
AuGe	233	220	63	44	13	29
Epoxy novalac	55-83	26.2	3	0.017	35	$1x10^{13}$ - $1x10^{16}$
Epoxy conductive	4-34	30	2.7-3.3	0.17-1.5	41	100

electrical and thermal conductivities and their temperature dependencies for selected solders are shown in Figures 5.14 to 5.15. Common soft solders in use in power devices are 95Pb5Sn, 65Sn25Ag10Sb (J-alloy), 60Sn40Pb, and Pb92Ag5In3 [Pecht et al. 1995]. However, as discussed above, soft solders are subject to low-cycle fatigue under thermal cycling; they experience creep damage at temperatures below 200°C; and as was shown in Table 5.2, some compositions can even melt at temperatures below 200°C. Table 5.5 shows the flow properties for solders in various temperature ranges [Condra 1993].

Table 5.5 Flow Behavior of Solder Die Attaches [Condra 1993]

Flow behavior	Sn63 temperature range	Homologous temperature (fract. of melt temp in K)
Solid linear elastic	-65°C to 10°C	0.45 - 0.62
Elastic-plastic	10°C to 40°C	0.62 - 0.69
Visco-plastic (elastomeric)	40°C to 80°C	0.69 - 0.77
Viscous (paint-like)	75°C to 100°C	0.77 - 0.82
Navier-Stokes (honey-like)	100°C to 125°C +	0.82 - 0.87

Solventless conductive epoxies are chosen for 80% of plastic-encapsulated microcircuits [Kearny 1988] because they can accommodate the strain in packages that typically have high CTE mismatches at the leadframe-to-die interfaces. Long-term bond integrity is maintained in most epoxies for operating temperatures as high as 150°C. Some silver-filled epoxies maintain as much as 40% of their initial shear strength, even after 1000 hours at 200°C. Much shorter lives of 168 hours and 1 hour have been recorded for exposures at 250°C and 350°C, respectively [Feinstein 1989]. Use of epoxy die attaches above their glass transition temperatures, which are usually in the range of 75°C to 125°C, requires that allowance be made for a higher coefficient of thermal expansion, a lower modulus, and lower strength. Table 5.6 lists the glass transition temperatures (Tg), which mark the onset of long-range molecular motions in polymers, for several common die attaches, along with the coefficients of thermal expansion above and below the Tg. High-temperature storage and use may also release ionic chlorine from the epoxy, which can lead to increased metallization corrosion. All these factors limit the use temperature of epoxies to a maximum of 200°C. Polyimides are a high-temperature alternative to epoxies, with higher Tgs (180 to 275°C). However, they cure at higher temperatures, develop higher stresses upon cooling, are more prone to cracking, and use up to 30% more solvents, which can cause voiding during solvent evaporation [Feinstein 1989].

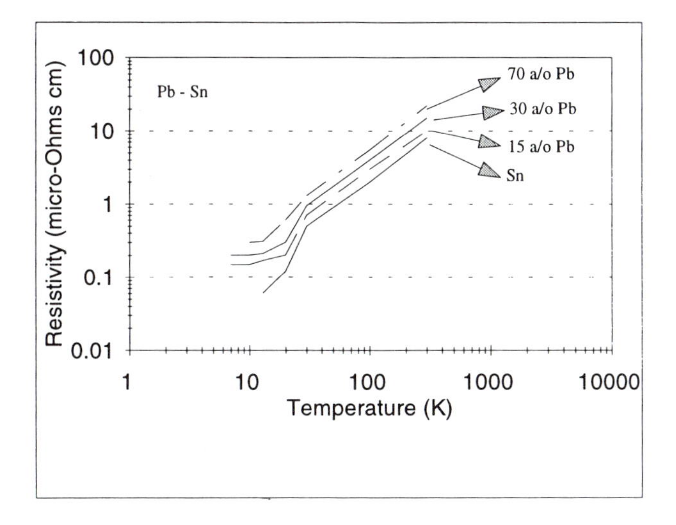

Figure 5.14 The electrical resistivity of lead-tin alloys as a function of temperature [Schroder 1983].

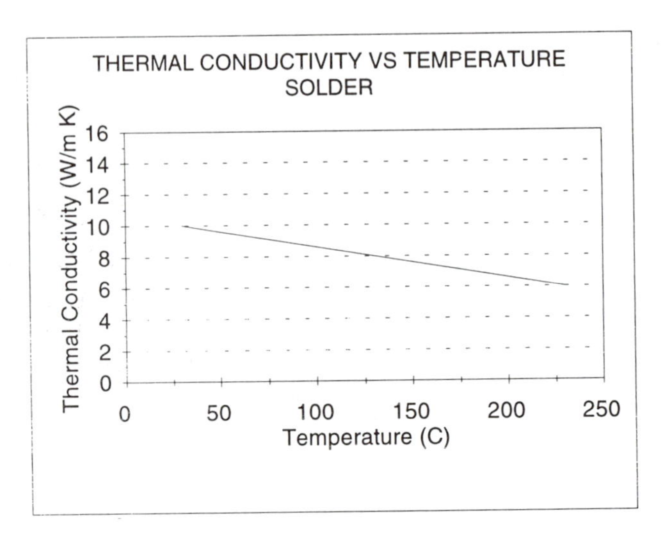

Figure 5.15 The thermal conductivity of solder as a function of temperature [Holman 1986].

Table 5.6 Thermal Expansion of Polymer Die Attaches [Ablestik 1995]

Die attach	Glass transition temperature	CTE below Tg (ppm/K)	CTE above Tg (ppm/K)
Ablebond 979-1AS	230°C	45	98
Ablebond 84-1LMI	103°C	55	150
Ablebond 84-1LMIS	127°C	46	240
Ablebond 8350T	85°C	65	45
Ablebond 8350M	85°C	30	140
Ablebond 8360	88°C	45	200
Ablelock P1-5500 (polyimide)	170°C	35	150

Packages intended for elevated temperature applications tend to use stiffer die attach materials, such as gold eutectics or silver-filled glasses. These systems have the higher moduli and higher flow stresses needed to provide fatigue and creep resistance at elevated temperatures. However, they transfer more stress to the die, often causing failure to occur by die fracture. The gold eutectic metallurgical attachment is preferred for high-power circuits, because of its superior electrical and thermal conductivity. In cases where gold eutectic is not practical because of cost or rework issues, or in cases where high conductivity is not required, silver-filled glasses can be used at the expense of one order of magnitude of electrical and

Table 5.7 Lead and Leadframe Material Properties [Pecht 1994, Tummala 1989, Alloys Digest 1982]

Material	Tensile strength (MPa)	CTE (ppm/K)	Thermal conductivity (W/mK)	Electrical conductivity (μohm-cm)
Kovar	525-550	4.0-4.7	40	49
Alloy 42	565	5.1-5.8	12	70
C-151	350	17.7	360	1.92
C-155	360	17.7	344	2.00
C-194	410	16.3	262	2.65
C-195	470	16.9	197	3.45

thermal conductivity, while aluminum nitride or beryllia-filled glasses provide one order of magnitude less thermal conductivity, combined with electrical insulation [Pecht 1994]. Silver-filled glasses have several advantages over gold eutectics as well. They are lower in cost, they exhibit lower stress than gold eutectics after 10 thermal cycles of -10°C to 135°C [Loo and Su 1986], they do not fail due to cracked dies after 1000 cycles of thermal shock from -65°C to 150°C [Moghadan 1984]; and they have a lower thermal resistance as a result of the smaller void percentage [Moghadam 1984]. High-temperature packages requiring very little stress to be imparted to the die, such as those with large dies or thin leadframes, can use cyanate ester die attaches, which combine a modulus of rigidity (400 MPa) that is 2.5 times less than epoxy with thermal stability at temperatures up to 300°C [Chien and Nguyen 1994].

5.3 Lead and Leadframe Materials

Leads serve as power, ground, and signal I/O interconnections from the device to the electrical conductors on the mounting platform, which requires that they provide mechanical support, an electrical conduction path, and a thermal conduction path. Therefore, important properties of the base materials for leads are electrical conductivity, thermal conductivity, strength, and the coefficient of thermal expansion. The room-temperature values for these parameters are given in Table 5.7 for six common leadframe materials.

Annealing the leadframe, with a corresponding decrease in strength, is not a concern for temperatures up to 200°C. The hardness of copper leadframe alloys C-194 and C-197, which is directly related to their strength, decreases less than 20% after 1 hour at 350°C, as shown in Figure 5.16 [Crane et al. 1989]. Since annealing is an activated process that can be described by an Arrhenius relationship, this roughly equates to greater than 30,000 hours (3 years) of continuous use at 200°C if only steady-state temperature effects are considered. Copper alloys C-151 and C-192 do not exhibit a decrease in hardness at temperatures below 400°C, which equates to greater than 1,000,000 hours (110 years) of continuous use at 200°C, while iron-nickel alloys, such as Kovar and Alloy 42, do not exhibit a decrease in hardness at temperatures less than 600°C. Further evidence for this is provided by the work of Zarlingo and Scott [1981], which indicated that the yield strengths of copper alloy C-195 and iron-nickel Alloy 42 do not show an appreciable decrease after 100 minutes of exposure to 350°C.

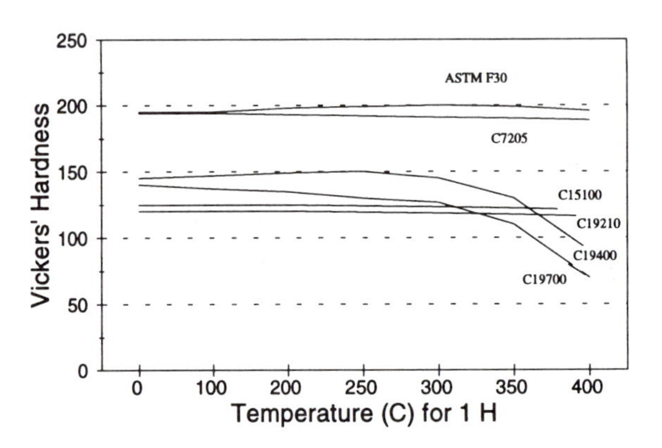

Figure 5.16 The hardness of leadframe alloys as a function of temperature [Crane 1989].

Elevated temperatures do raise some concerns, however. Electrical conductivity decreases with temperature, which can lead to reduced signal fidelity and slower switching time. Thermal conductivity is also decreased, as shown in Figure 5.17, although the effect of temperature is small compared to the differences between leadframe materials. Choosing a material with high thermal conductivity is important for heat dissipation. However, copper alloys that have higher thermal conductivities than the iron-nickel alloys, as shown in Table 5.7, also have larger coefficients of thermal expansion [Pecht 1994]. The higher CTE values create more thermal stress in the die and die attach during temperature cycling, and can lead to die and die attach failure. In hermetic packages, the higher CTE mismatch could cause lead seal fracture. The variation in the coefficients of thermal expansion with temperature are given in Figure 5.10.

There is also some concern with respect to the layers plated on the leads to improve solderability, corrosion resistance, and oxidation resistance. Lead platings usually consist of two layers- a top layer of gold or solder (hot-dipped or electroplated) for corrosion resistance and solderability, and an undercoat of electrolytic nickel as a barrier to the diffusion of the base metal into the top layer. Extended storage at elevated temperatures can contaminate surfaces, reducing solderability and increasing contact resistance. Oxidation and intermetallic growth limit tin and tin-lead plating on copper to 100°C. 110°C can be achieved with tin and tin-lead plating on nickel, because the rate of formation of nickel intermetallics is only 1/3 that of copper-tin intermetallics [Mroczkowski 1995]. Diffusion of base metal, impurities, and hardening agents to the surface limit hard gold over nickel to 125°C, as shown in Figure 5.18 [Antler 1970]. Thicker layers of soft gold over nickel can increase the maximum temperature to 200°C. Plating concerns cease to be an issue after the leads have been soldered onto the board. The formation of tin whiskers and dendritic growth of copper and silver are also processes of concern, but they are related to bias conditions, humidity, ionic concentrations, and driving forces other than temperature [Pecht 1994]. This is also true of corrosion, which is best addressed simply by using gold over sulfamate nickel plating and controlling the porosity.

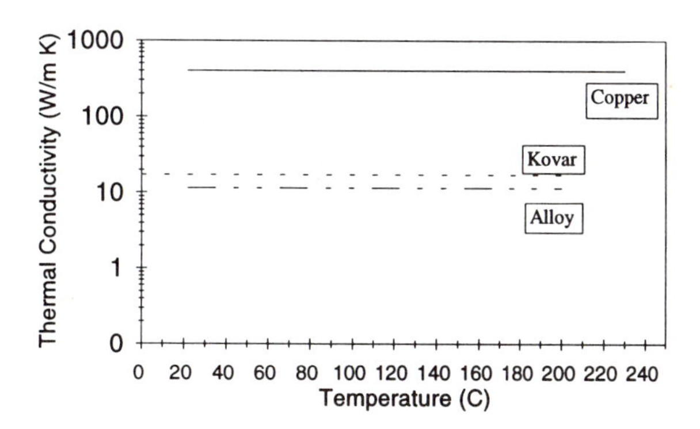

Figure 5.17 The thermal conductivity of leadframe alloys as a function of temperature [Holman 1986, Mattox 1994, Touloukian et al. 1970].

5.4 Plastic Encapsulants

Traditionally, high-temperature systems have been packaged in ceramic and metal hermetic packages, because of the limitations to packaging high-temperature circuits in plastic. The primary limitation is failure mechanisms related to the softening and weakening of the molding compound at raised temperatures. The most widely used molding compounds are thermoset epoxies but several other compounds are in use. Modified polyimides are used for thermal and moisture stability, and have low coefficients of thermal expansion and high purity. Polyurethanes, polyamides, and polyesters are used primarily to encase modules and hybrids intended for use under low-temperature, high-humidity conditions. Thermoplastics are avoided because of the high temperatures and pressures required for molding, and their low purity.

Chosen thermoset epoxies have a low initial viscosity for minimal wirebond and leadframe deformation, but a high modulus and heat distortion temperature in the final cured state to withstand manufacturing and operational temperatures. Novalac epoxies are the primary compounds used today because they exhibit the higher heat-distortion temperatures, glass transition temperatures, and purity of silicones, with the adhesion characteristics of lower Tg epoxies. The novalac epoxies are also preferred for their coefficients of thermal expansion, which at 17-20 ppm/K [Tummala 1989] are well matched to copper alloy leadframes (17 ppm/K). Other systems in common use include the diglycidyl ethers of bisphenols A and F, and the cycloaliphatic epoxides [Pecht 1995].

Circuits intended for high-temperature use are often encapsulated in silicone or placed in high- temperature hermetic packages filled with silicone gel. Not only does silicone effectively stop moisture from reaching the die, it is also resistant to depolymerization at temperatures up to 260 to 280°C, has a lower dielectric constant than epoxies, and is better at absorbing thermal stress. Silicones are limited in their use, however, by their poor adhesion to the leadframe, which provides a path for the ingress of moisture and contaminants, leading to subsequent corrosion [Pecht 1994].

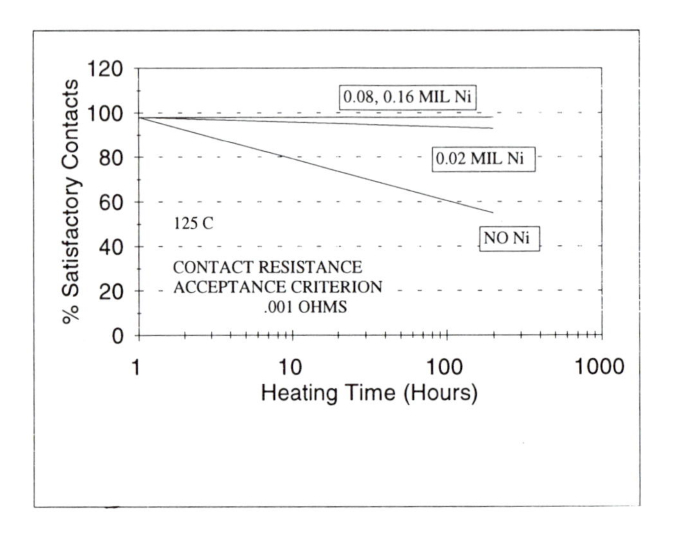

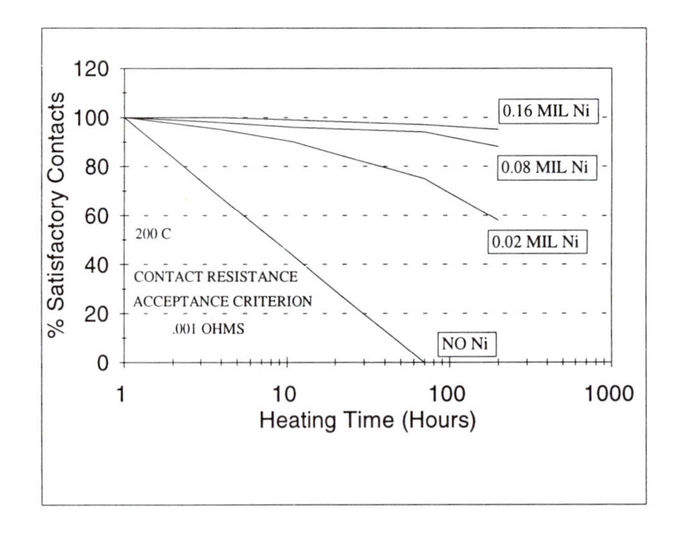

Figure 5.18 The effect of temperature on the resistance of precious metal plated contacts [Antler 1970].

Plastic packages are ultimately limited in their use temperature by depolymerization, which converts the solid polymer into a viscous mixture of monomers, dimers, and low-molecular- weight polymer fragments. This occurs at temperatures between 190°C and 230°C for epoxies and between 260°C and 280°C for silicones [Wong 1989, Khan et al. 1988]. The maximum use temperatures for typical encapsulation resins are given in Table 5.8.

The strength of molding compounds begins to decrease at temperatures well below that needed for depolymerization, as shown in Figure 5.19. The relative high-temperature strength of a molding compound is defined by its heat deflection temperature, which is the temperature at which it can no longer support 264 psi without deforming plastically. The heat deflection temperatures for several common molding resins are given in Table 5.8 and in Figure 5.20. As the molding compound approaches the glass transition temperature, the coefficient of thermal expansion increases, and the modulus decreases as illustrated in Figures 5.21 and 5.22 for a typical semiconductor-grade molding compound. In addition the adhesion strength decreases. As with epoxy die attach, these changes are due to the onset of long-range molecular motions. The glass transition temperature for most epoxy encapsulants ranges from 150 to 180°C [Pecht et al. 1995, Tummala 1989]. Operating above the glass transition temperature increases the thermal strain between the molding compound, die, and leadframe, not only because of the temperature increase, but also because of the increase in the CTE. This can lead to delamination and cracking. Finally, elevated temperature operation reduces the volume resistivity of the molding compound, as shown in Figure 5.23.

Table 5.8 Maximum Use Temperatures for Encapsulants [Pecht et al. 1995]

Material	Maximum temperature	Reason
Epoxies	190°C-230°C	Depolymerization
Aliphatic Amine	125°C	Heat deflection temperature
Anhydride	130°C	Heat deflection temperature
Aromatic Amine	150°C	Heat deflection temperature
High T Anhydride	170°C	Heat deflection temperature
Novalac	195°C	Heat deflection temperature
Dianhydride	280°C	Heat deflection temperature
Silicones	260-400°C	Depolymerization
Polyimides	300°C	Glass transition temperature
Phenolic	205-260°C	Depolymerization

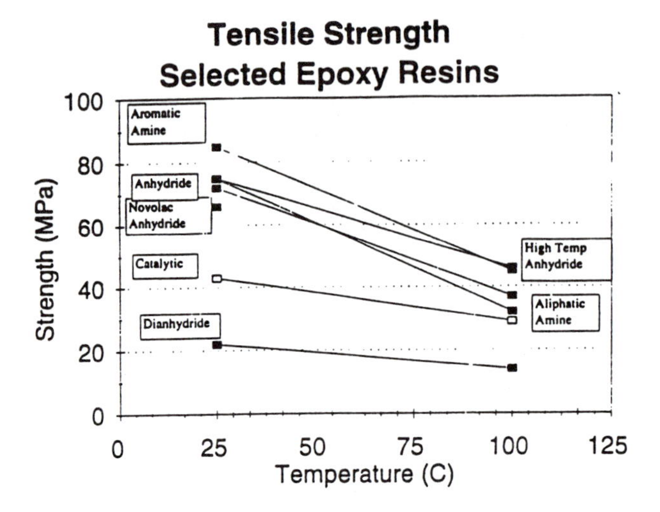

Figure 5.19 The tensile strength of selected epoxy resins as a function of temperature [Pecht et al. 1995].

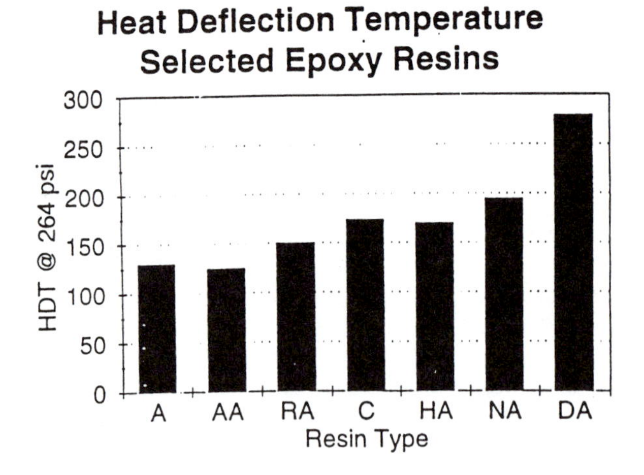

Figure 5.20 The heat deflection temperatures of selected epoxy resins [Pecht et al. 1995].

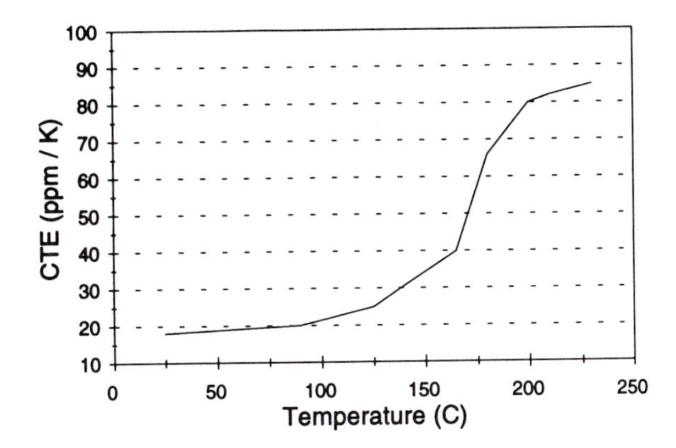

Figure 5.21 The change in the coefficient of thermal expansion of a semiconductor grade molding compound across its glass transition region [Rosler 1989].

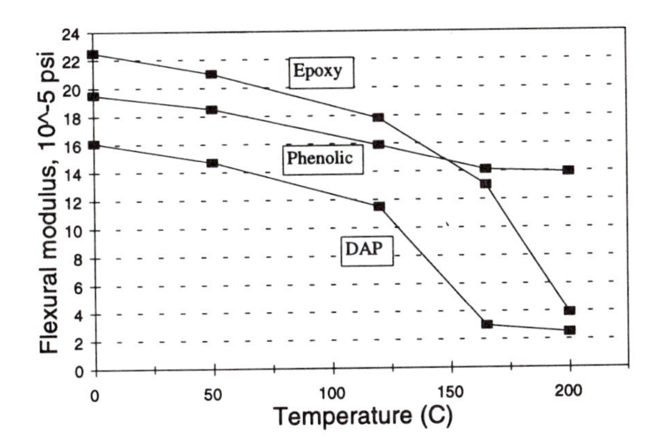

Figure 5.22 The effect of temperature on the elastic modulus of plastic molding compounds [Harper 1970].

These effects can be partially alleviated by using low-stress molding compounds, which are made by maximizing filler loading. Fillers can have profound impact on material properties. For example, adding 68% ground fused silica can increase the thermal conductivity from 50 W/mK to 16,000 W/mK, and decrease the coefficient of thermal expansion from around 40 to 50 ppm/K to 20 to 30 ppm/K. In addition, stress-relief additives can be added to lower the elastic modulus, improve toughness and flexibility, and lower the CTE. The main stress relief agents for epoxy molding compounds are silicones, followed by acrylonitrile-butadiene rubbers, and polybutyl acrylate [Pecht 1995].

As with lead and leadframe materials, the encapsulant material and its material properties can be a major factor in failure mechanisms for which the failure site is not the encapsulant itself.

Cracking of the passivation layer due to thermal stresses imposed by the molding compound accelerates moisture and contaminant ingress. This results in metallization corrosion or surface charge spreading, and can also cause ball-bond liftoff and shearing. Adding phosphorus to the glassivation allows it to stress-relax, but excessive phosphorus (>2% by weight) accelerates corrosion failure due to the formation of corrosive phosphoric acid with ingressed moisture. Elevated temperature cycling can accelerate this passivation cracking. Not only does temperature accelerate cracking of the passivation, it also accelerates the die metallization corrosion that occurs after moisture ingress [Pecht 1995, Intel 1990]. Elevated temperatures in the presence of moisture also promote the release of flame retardants, such as bromides and antimony trioxide, from the molding compound, along with fumes originating from the decomposition of silicone modifiers and the resin blend. One solution to control this is to use softer and more ductile passivation materials for high temperature, such as polyimides [Inayoshi 1979]. Galvanic corrosion of gold and aluminum at the unpassivated pads can also happen at elevated temperatures in the presence of moisture and ionic contaminants. Current efforts at controlling corrosion focus on reducing ionic impurities and moisture ingress in the encapsulant [Pecht 1995].

Finally, exposure of plastic encapsulants to elevated temperatures causes outgassing of monomer and low-molecular-weight polymer molecules. The organic vapors build up and deposit organic films on other components, particularly if the package is sealed in a hermetic multichip module. This phenomenon causes corrosion, facilitates leakage across insulators, and increases the resistivity of contacts [Wilcoxon 1995].

5.5 Hermetic Case Materials and Lids

The purpose of a hermetic case is to provide a platform for mounting devices, interconnects, and, possibly, substrates; to provide mechanical protection for the contents; to act as a path for removing the heat generated by device operation; and to protect the contents from moisture, contamination, and harsh environments. In order to perform these functions, the case should be impermeable to moisture and resistant to corrosion, and have a high thermal conductivity, a high elastic modulus, a coefficient of thermal expansion matched to the substrate, die, and leads, and a high tensile strength. Values of these properties at room temperature for ceramic case materials are given in Table 5.9.

The base materials used to construct hermetic cases are high-temperature metal alloys and ceramics, so there is little degradation in these materials at elevated temperatures. Nonetheless, an understanding of their high-temperature properties is important because they can have a profound effect on the stresses in the package and, consequently, the reliability of other materials used in the package. The primary concerns with using hermetic packages are their high cost and low packaging density, which restricts their use to applications requiring

Table 5.9 Properties of Case and Lid Materials [Pecht 1994]

Material	Tensile strength (MPa)	Flexural strength (MPa)	Elastic modulus (GPa)	Thermal conduct. (W/mK)	CTE (ppm/K)	Electrical resistivity (Ω-m)
BeO	230	250-490	345	150-300	6.3-7.5	10^{11}-10^{12}
SiC	17.24	440-460	412	120-270	3.5-4.6	> 10^{11}
AlN	-	360-490	310-343	82-320	4.3-4.7	> 10^{11}
Al_2O_3 95%	127.4	317	310.3	15-33	4.3-7.4	10^{9}-10^{12}
Kovar	522-552	-	138	15.5-17.0	5.87	50 x 10^{-8}

high reliability in harsh environments, such as space and military equipment. This section will discuss failure modes for case materials, the dependence of case material properties on temperature, and guidelines for materials selection to minimize any high-temperature concerns imposed by these materials on other materials around them.

5.5.1 Ceramics

Ceramic packages are preferred for all applications, except those involving power devices that dissipate greater than 0.75 watt/cm^2, which use metal because of its higher thermal conductivity, and those involving shock environments, which use metal because of its toughness. Ceramics are used because they eliminate the need to seal leads with glass, which is subject to several failure mechanisms. As is the case with leadframes, the choice of ceramic header materials is more a function of how they affect failure at other sites than how they fail themselves, as seen by the maximum use temperatures in Table 5.10 below.

Table 5.10 Maximum Temperatures for Ceramic Case and Lid Materials [Pecht 1991, Coors 1993]

Material	Max use temperature (°C)	Reason
Al_2O_3	>1000	
BeO	>1000	
AlN	>1000	
SiC	>1000	
Cu/W (10/90)	>1000	
Kovar	600	Annealing causes loss of yield strength

Common ceramic case materials include alumina (Al_2O_3), beryllia (BeO), and aluminum nitride (AlN). The advantage of alumina is its availability at relatively low cost. Beryllia is chosen for applications, such as power devices, requiring higher thermal conductivity than that provided by alumina. However, beryllia is more expensive, potentially toxic, and produces greater thermal stresses in the die and die attach at elevated temperatures, because of its higher coefficient of thermal expansion [Pecht 1994]. If beryllia is used for the header (to provide heat dissipation), and alumina is used for the walls, thermal stresses can also occur at the header-to-case wall interface. AlN is a newer material that combines a thermal conductivity near that of beryllia with a coefficient of thermal expansion better matched to silicon than either alumina or beryllia. Its use, however, is still confined to small niche markets. Any designs utilizing these case materials should take into account the variation in the coefficients of thermal expansion and thermal conductivity with temperature, which are shown in Figures 5.24 and 5.25 [Charles and Clatterbaugh 1994, Mattox 1994].

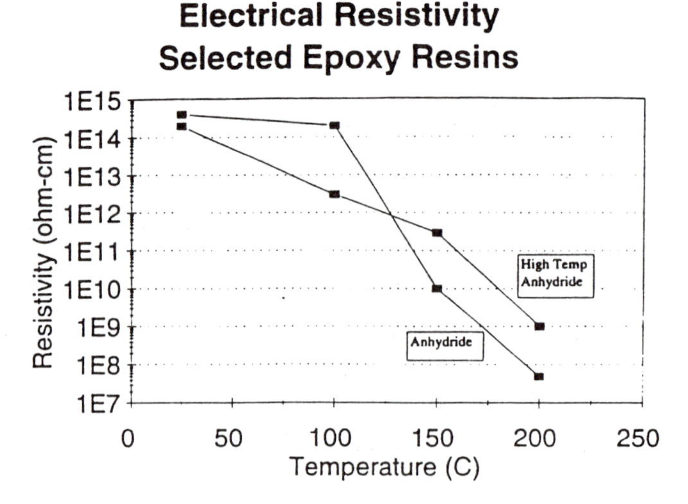

Figure 5.23 The volume resistivity of selected epoxy resins as a function of temperature [Pecht et al. 1995]

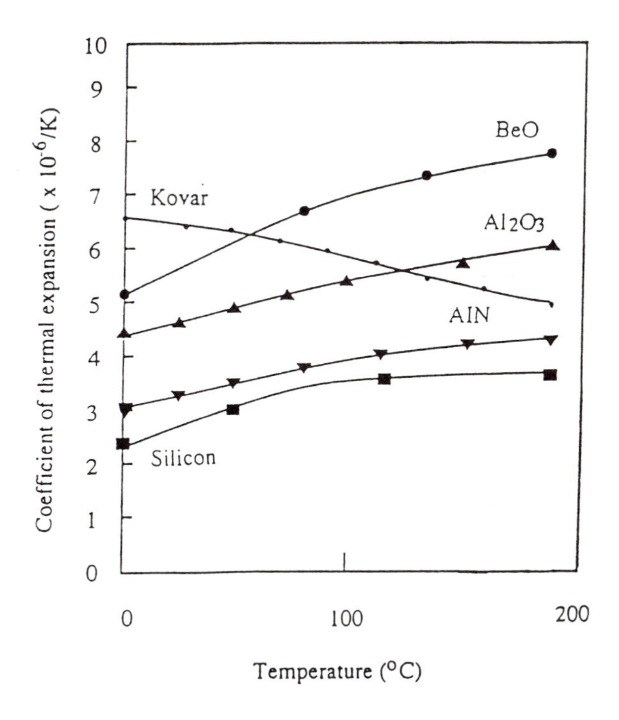

Figure 5.24 The coefficient of thermal expansion of ceramic case materials as a function of temperature [Charles and Clatterbaugh 1994].

5.5.2 Metals

Metal TO cans, the original transistor packages, are still used today for high-power applications because they provide high thermal conductivity at substantially lower cost than BeO and AlN. A listing of the thermal conductivities is given in Table 5.11, and their variation with temperature is pictured in Figure 5.26. The electrical and thermal conductivities decrease with temperature, as expected for metals.

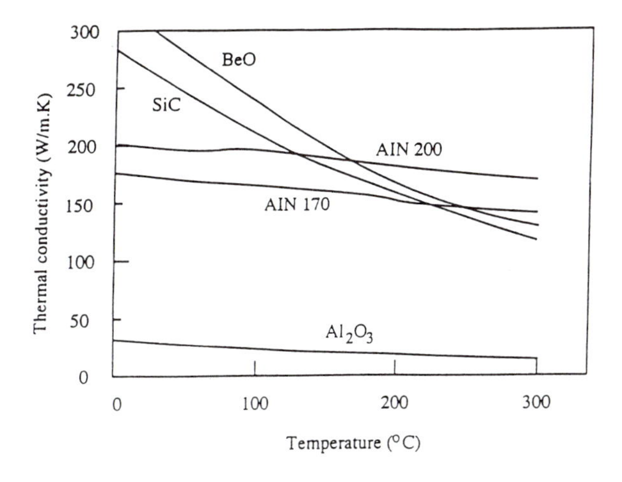

Figure 5.25 The thermal conductivity of ceramic case materials as a function of temperature [Mattox 1994].

Table 5.11 Thermal Properties of Metal Cases

Material	Thermal conductivity (W/mK)	Coeff. of thermal expansion (ppm/K)
Kovar	17	5.8
Cu-W (10:90)	210	6
Cold-rolled steel	78	12
Stainless steel	17	17

The most common metal alloy case material is Kovar, which has a coefficient of thermal expansion closely matched to that of silicon, but has poor thermal conductivity for a metal. It is used in instances where minimizing thermal stress is more important than heat dissipation. An advantage of using Kovar is the ability to oxidize it to produce a surface suitable for glass lead sealing. Power applications use either CuW or cold-rolled steel because of their higher thermal conductivities. Cold-rolled steel is less expensive, but has a higher CTE and a lower thermal conductivity than CuW. The higher CTE creates more thermal stress in the die attach and precludes the use of matched lead seals. In addition, steel is susceptible to corrosion, as is Kovar, and is therefore often plated with gold over nickel. Stainless steel provides corrosion resistance without plating at a cost lower than copper or Kovar, but it has a thermal conductivity as low as Kovar, coupled with a high CTE. Aluminum alloys are available if package weight is a major concern, but they cannot be resistively welded and do not bond well with lead-sealing glasses.

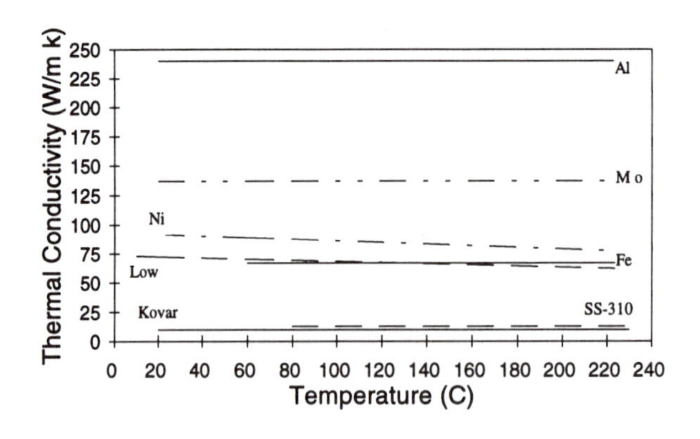

Figure 5.26 The thermal conductivity of metal case materials as a function of temperature [Scott 1974, Holman 1986, Mattox 1994].

Since all the metals anneal and melt at temperatures well in excess of 200°C, with the possible exception of aluminum, the main high-temperature concerns for failure of the metal package are corrosion, whisker growth, and seal failure by fatigue. Corrosion is related not only to temperature but also to the availability of electrolyte, concentration of ionic contaminants, geometry of the metal components (i.e., sites for crevice corrosion), and the local electric field created by the differences in the galvanic potentials of the metals involved. Case metals are typically plated with noble metals to avoid corrosion, but local galvanic cells can be set up at pore sites in the plating. Tin whiskers can be formed in cases plated with pure tin. Whiskers growing inside the case can break under shock and vibration, fall onto the circuit, and produce shorts. The driving force for tin whisker growth is the residual stresses in the plating from initial processing or thermal stresses between the plating and the case, and not from electrical potential [Guttenplan and Violette 1990, Nordwall 1986, Britton 1974]. This should not be a concern for high-temperature operation because (1) it is unlikely that plating with tin is an option, due to its low melt point, high oxidation instability, and low reliability at elevated temperatures > 100°C; (2) the potential for whiskering can be reduced by using a thicker plating, and eliminated by plating 3 to 5 % lead along with the tin and subsequently reflowing the plating. However, again, the low melting point of tin-lead solders precludes their use as case platings at elevated temperatures. A better selection for elevated temperatures is 100 μ in nickel under 50μ in of gold. The gold provides a noble metal surface and the nickel provides a diffusion barrier between the base metal and the gold. Fatigue fracture of the case attach is again an issue involving CTE mismatch, if one metal is used for the header and another for the walls. Unlike ceramic cases, however, metal cases are often supplied as a uniwall instead of modular construction, where the entire case is made by a single stamping or casting [Pecht 1994].

Lids are also made of either metal or ceramic, with the most common lid materials being alumina and Kovar. They are also evaluated on the basis of moisture resistance, corrosion resistance, coefficient of thermal expansion, yield strength, tensile strength, and fatigue strength. In the regime up to 200°C none of these properties decreases to a degree sufficient to warrant alarm in common lid materials.

Corrosion of metallic lid materials, which causes pits and pores to form in the lid, reduces the strength of the lid and can render the package non-hermetic, is controlled by plating the lid with gold, carefully avoiding damage to the plating layer in handling or use. Excessive lid deflection is controlled by choosing a metal which, at the temperature of interest, has sufficient yield strength to avoid plastic deformation, and a sufficiently high modulus to maintain elastic deflection within an allowable range (i.e., one that does not allow the lid to contact the contents of the package). The room temperature values for the strength and modulus given in Table 5.9 can be used for this determination for temperatures up to 200°C.

5.6 Lid Seals

There are three standard processes for lid sealing - welding, solder sealing, and glass sealing. Welding can only be used to join a metal lid to a metal case. Advantages are high yields and no outgassing of moisture or vapors during sealing, which can seal contaminants into the package. Either parallel seam welding or laser welding can be used, with parallel seam welding preferred because of its lower cost; laser welding is an option for welding complex seal geometries or high thermal conductivity metals such as copper-tungsten and aluminum, which are difficult to weld. Because these processes involve, for the most part, directly joining base metal to base metal, there are no high-temperature concerns and no attach material. This method is highly recommended for high-temperature packaging.

Gold-based eutectics, which have a relatively low melting- and thus, sealing-temperature, as shown in Table 5.12, are used to seal metal cases and cases damaged by high-temperature sealing processes. The primary concern with lid sealing materials is resistance to fracture and fatigue from the thermal stresses generated by the CTE mismatch between the lid and case materials. The low melting point, therefore, is a disadvantage for higher temperature use, as it causes these materials to have a poor fatigue life at elevated temperatures. The variation in the modulus and strength of these materials with temperature is given in Figures 5.6 and 5.7 above. Other disadvantages include high electrical conductivity, which restricts their use. For example, they cannot be used in CERDIP packages in which the same material is used to make the lead and the lid seals, as they would short the leads together. Another disadvantage, of course, is their high cost.

Glass sealing is used to seal ceramic packages that require a high-temperature electrically insulating attach, and that can tolerate the sealing temperatures. Advantages of glass seals are their low cost and their good CTE match to ceramic case materials. High temperature concerns with glass lead seals are their softening point, the decrease in modulus and strength near the softening point, and the increase in coefficient of thermal expansion above the softening point. The softening point is the temperature at which a 0.65 mm diameter glass fiber suspended vertically will elongate under its own weight at the rate of 1mm/min, as defined by ASTM C 338. A list of these softening points is given in Table 5.12.

Table 5.12 Maximum Use Temperatures for Lid Seals [Pecht 1994]

Material	Maximum temperature (°C)	Reason
Au97Si3	370	eutectic point
Au80Sn20	280	eutectic point
Au88Ge12	356	eutectic point
7583	385	softening temperature
7572	370	softening temperature
KC-1M	400	softening temperature
KS-1175X1	375	softening temperature
7570	440	softening temperature
KC-405	350	softening temperature
7575	370	softening temperature
KC-810	350	softening temperature
KC-400	342	softening temperature
KC-900	340	softening temperature

5.7 Lead Seals

Lead seals need to perform the following three functions: electrical insulation between the leads, hermetic sealing of the package in the lead region, and mechanical support of the leads. Because of this, lead seals are chosen on the basis of coefficient of thermal expansion, strain point, softening point, and electrical resistivity. The most common lead seals are glass-to-ceramic or glass-to-metal seals, using a glass lead sealing material. The main high-temperature concerns are a loss of mechanical strength with temperature and a loss of seal between the lead and the seal material with temperature as a result of a change in the stress state. The first concern is controlled by keeping the material below its strain point - the temperature at which the viscosity of the glass is equal to 10^{14} Poise, as defined by ASTM C 336. It is more commonly described as the temperature at which internal stresses are relieved and creep phenomena are initiated. A list of the strain and softening points for lead seal materials is given in Table 5.13.

The second concern varies, depending on whether the seal is a matched seal or a compression seal. Matched seals utilize a glass with a coefficient of thermal expansion close to that of the lead and the case materials over the temperature range of interest. Wetting and chemical bonding is the mechanism for seal formation, typically utilizing an oxide that forms on the metal [Partridge 1949]. These seals fail by thermal overstress or fatigue at high

Table 5.13 Glass Lead Seal Materials [Pecht 1994]

Lead-seal materials			
Corning glass code	Type	Strain point (°C) (initial point for seal deformation under load)	Softening point (°C) (seal deformation under light load)
7040	Soda potash	449	702
7052	Alkali Barium Borosilicate	440	712
7056	Alkali Borosilicate	472	718
7070	Lithia Potash Borosilicate	456	--
9010	Alkali Barium	405	646
9013	Alkali Barium (low lead)	423	656
9048	Alkali Strontium	462	688

temperatures or during temperature cycling. They crack or delaminate from the lead as a result of differences in the coefficients of thermal expansion. Corning 7050 and 7052 are often used in this manner to seal Kovar leads and cases.

Compression seals utilize a glass with a coefficient of thermal expansion lower than that of the case but higher than that of the lead, so that when the assembly is cooled after sealing, the case and seal exert a compressive force on the lead, mechanically securing it. In this case, no chemical bonding occurs. These seals begin to degrade as operating temperatures approach sealing temperatures, because the designed-in residual stress decreases. Corning 9010 is used in this manner to seal steel cases to alloy 52 leads. Its critical temperature is in the range of 375°C, which is near but below its strain point of 405°C [Borom and Giddings 1975].

Another type of lead seal is brazing a metal lead to a metallized ceramic ferrule or a metallized port in a ceramic package. These seals are usually made using high-temperature brazes. The temperature concern here is a loss of strength in the braze, which should not be a concern at temperatures less than 200°C. Typical materials are an alumina ceramic with nickel-plated moly manganese metallization, alloy 52 lead, and copper-silver braze.

5.8 Summary

For temperatures up to 135°C, standard high-density devices and packaging technologies can be used. This includes silicon MOSFETs and bipolar devices, aluminum metallization, gold-aluminum wirebonding, epoxy die attach, plastic packages, and copper leadframes.

For temperatures up to 200°C, standard high-density packaging technologies can be used, with some alterations. Aluminum-nickel or aluminum-aluminum wirebonding should be substituted for gold wirebonding to aluminum, and alternatives to eutectic tin-lead solder die attach and conductive epoxy should be used. The high-temperature performance of molding compounds used for plastic encapsulation and housings must be carefully evaluated in this range. In addition, the performance of the silicon devices should be checked to ensure proper design for elevated temperatures.

Above 200°C, it becomes necessary to consider substituting silicon-on-insulator or gallium arsenide devices, together with ceramic packaging technology. This includes aluminum-nickel or aluminum-aluminum wirebonding, chips attached by flip-chip technology with 95Pb5Sn solder, the use of gold-based eutectics or silver-filled glasses for die attach, ceramic cases and substrates, and glass lid and lead seals.

Chapter 6

SECOND AND THIRD LEVEL PACKAGING CONSIDERATIONS FOR THE USE OF ELECTRONIC HARDWARE AT ELEVATED TEMPERATURES

The first five chapters of this book presented the issues involved in developing and packaging components for elevated temperature use. This chapter will focus on the materials and technologies required to assemble these devices into an electronic system that is fully functional at temperatures greater than 125°C. Key areas to be covered include the development of organic and ceramic substrates, solders and conductive adhesives, wires and cables, connectors, and housing materials.

The first section provides information on the ever-growing list of organic substrates for high-temperature use. High glass transition temperatures resins, such as bismaleimide triazine, cyanate ester, and polyimide, are extending the useful range of organic PCBc well above 200 °C. Ceramic substrates are also used for elevated temperature electronics but since they have traditionally performed well at elevated temperatures they are not discussed at length here.

The second section details recent research aimed at the development of high-temperature solders and conductive adhesives. Particular attention is paid to lead-free solders because of growing environmental concerns over the use of lead.

While components form the heart of an electronic system, much of the reliability of the system is determined by the quality of the connections. Thus, the third and fourth sections cover reliability issues related to the use of connector assemblies at elevated temperatures and provide guidelines for the development of high-temperature interconnect system. Wires and cables are discussed in the third section, while separable connectors are d discussed in Section 4.

Finally, the last section discusses engineered polymer housing materials used both as covers for separable connectors and as cases for the entire electronic subsystem. Advances in polymer technology have created materials that can be used reliably at temperatures up to 200°C. Information necessary for the use of these materials in high-temperature designs is provided and discussed.

6.1 Substrates

Most high temperature packaging, including the majority of automotive underhood applications, consists of thick film hybrid cermet and ceramic [Aday et al. 1993]. Thick film hybrid technology has been demonstrated to allow operation to 300^{0}C by Sandia National Laboratories. Conductor/resistor cermet ink systems which have demonstrated function up to 500 ^{0}C include gold, silver-palladium, and, increasingly, plated nickel over copper. The main impediment for wider use of hybrid technology is the high manufacturing cost.

Table 6.1 Properties of Typical Resin Systems [Pecht 1994]

Resin system	T_g (°C)	Elastic modulus (GPa)	CTE (ppm/°C)	Thermal conductivity (W/m-°C)	Dielectric constant at 23°C:1 MHZ
Epoxy	120-140	3.45	69	0.19	3.6
Polyimide	240-300	2.8	50	0.18	3.2
Cyanate ester	260	2.6	55	0.20	3.1
PTFE	-	0.35	99	0.19	2.0
BT	250	-	-	0.20	3.1

Electrically insulated metal substrates offer a low cost alternative to ceramic-cermet circuitry [Johnson et al. 1991]. Metal substrate circuitry has a three layer structure consisting of a metallic base plate, a midlayer of thermally conductive epoxy dielectric, and a top layer of copper conducting traces. Thermal dissipation is the key driver for this technology that utilizes aluminum substrates. These substrates have been demonstrated to have a lower thermal resistance than alumina hybrids as the baseplate [Sakamoto et al. 1993]. The major limitation of this approach is the softening point of the epoxy, which ranges from 150 to 200 ^{0}C.

Advances in organic laminate printed circuit board materials are now permitting polymer-glass laminates to be used in high temperature environments, including automotive underhood [Evans et al. 1993]. In order to perform adequately, these materials must retain their electrical characteristics, such as a high breakdown strength, high insulation resistance, and stable dielectric constant. In addition, they must also provide mechanical support, provide sufficient thermal dissipation, and resist thermal and environmental degradation. Typical room temperature characteristics of resins used in printed wiring boards are given in Table 6.1.

Of particular importance is the coefficient of thermal expansion. Choosing a dielectric material with a lower coefficient of thermal expansion will minimize thermal stresses in the laminate at elevated temperatures. High coefficients of thermal expansion will produce large thermal mismatch stresses, particularly at high temperatures, resulting in void formation and interfacial delamination.

There are significant concerns with the use of polymer laminates at temperatures above their glass transition temperatures (T_g). Above T_g, these materials begin to lose mechanical strength due to resin softening, insulation resistance due to decreases in volume resistivity, and adhesive strength at interfaces. In addition, they exhibit large discontinuous changes in their out-of-plane coefficient of thermal expansion, as shown in Figure 6.1 which together with the weakened adhesive strength can cause in-board delaminations. The relative temperature-related degradation in different materials is often ranked using the thermal index, which is the temperature at which the mechanical and/or dielectric strength of the material drops to 50% of its initial value after 100,000 hours (i.e. 11 years) of continuous use. The loss of strength and modulus for epoxy resins is pictured in Figures 6.2 and 6.3; the loss in insulation resistance is pictured in Figure 6.4. Using this metric, standard FR4 boards are limited to temperatures less than 135 ^{0}C [DuPont 1995]. However, alternate dielectric materials have higher maximum

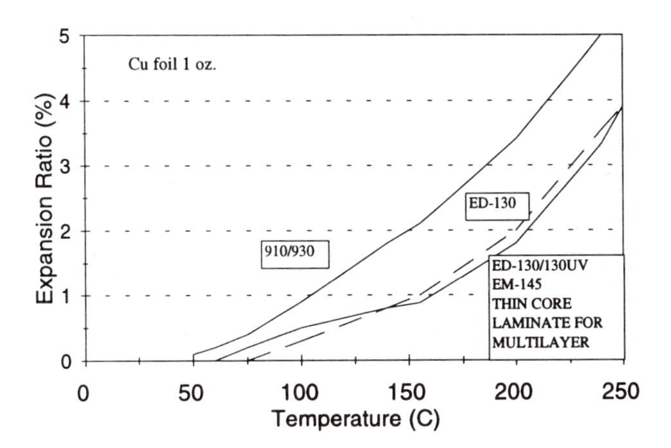

Figure 6.1 The thermal expansion of FR-4 (ED-130, ED-145) and CEM (910,930) printed circuit board laminates [Doosan 1995].

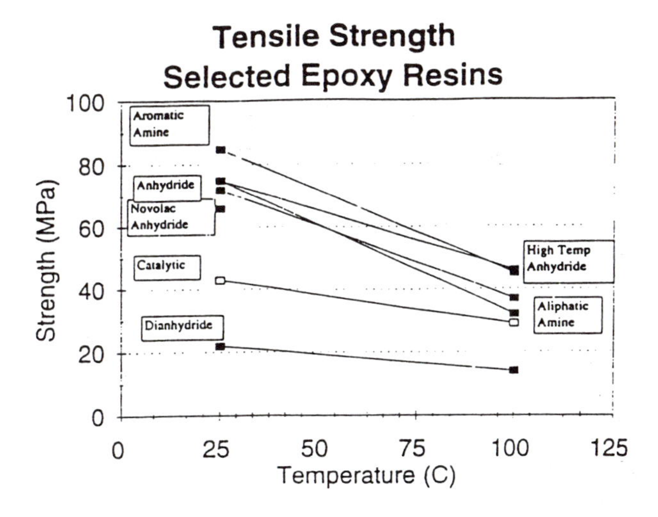

Figure 6.2 The tensile strength of selected epoxy resins as a function of temperature [Pecht et al. 1995].

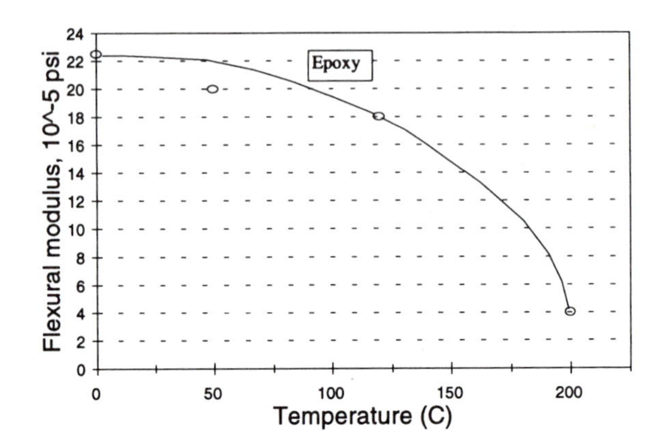

Figure 6.3 The elastic modulus of epoxy resin as a function of temperature [Harper 1970].

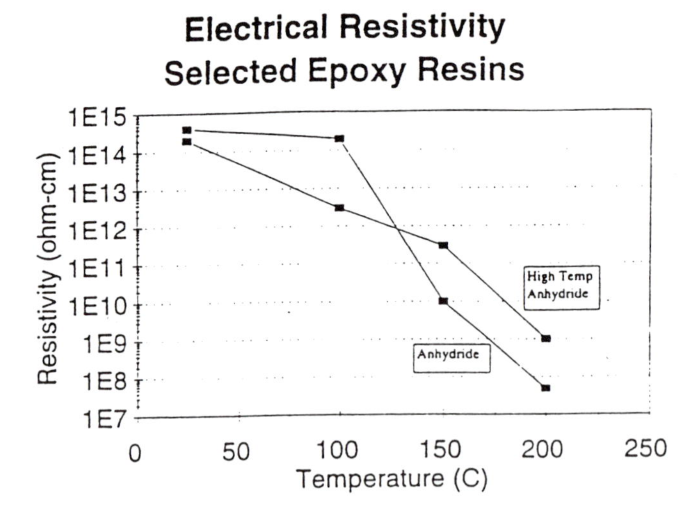

Figure 6.4 The volume resistivity of selected epoxy resins as a function of temperature [Pecht et al. 1995].

materials have higher maximum continuous use temperatures. Boards made of bismaleimide triazine (BT) can be used to 190^{0}C, and boards made of cyanate ester or polyimide can be used up to 260^{0}C, with polyimide-kevlar boards being restricted to 200 C and below, but polyimide-quartz boards being used to 260^{0}C.

E-glass/PTFE laminates do not exhibit degradation in their mechanical or electrical characteristics at temperatures up to 300^{0}C, the highest temperature for all laminates. However, they are not recommended for use above 120^{0}C because of the weakened adhesion of the copper layer. The effect of temperature on copper adhesion is present but is much less pronounced in FR4 and other standard boards as shown in Figure 6.5 [Doosan 1995]. Adhesion at the polymer to metal conductor interfaces is crucial for long term performance. The adhesion is often degraded not only by elevated temperature, but also by exposure to water and chemical agents. However, since adhesion is primarily an interface rather than a bulk property, a wide variety of coupling agents, primers, and surface treatments can be used to enhance it. For example, typical polyimide - copper interfacial adhesion may be increased more than threefold at elevated temperatures by incorporating adhesion promoters [Geffken 1991]. Similar technologies are being used to extend the useful operating temperature range of PTFE-based laminates.

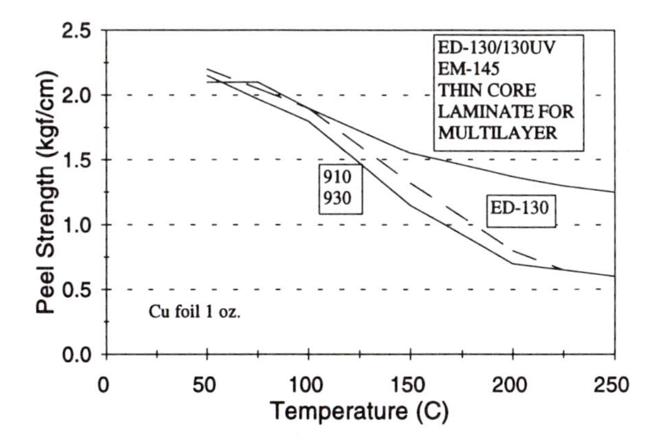

Figure 6.5 The peel strength of the copper cladding on FR-4 (ED-130, ED-145) and CEM (910, 930) printed circuit board laminates as a function of temperature [Doosan 1995].

Adhesion is also important at other interfaces as well. Because of the low thermal conductivity of most polymers it is often necessary to bond them to a heat sink. Metal matrix composites used as heat sinks can provide a low coefficient of thermal expansion and a high thermal conductivity. As a result these lightweight materials have been used as cores for multilayer printed circuit boards in order to remove heat and to constrain the laminates [Heller 1993]. Adhesion of these different material layers at elevated temperatures is essential to long term reliability.

While polyimide printed circuit boards are capable of withstanding temperatures in excess of 200^{0}C [Elsby 1989], drawbacks include the complicated processing required to produce them, especially in multilayer constructions; the adhesives required to attach them to heat sinks; and most importantly their high moisture absorption. Moisture adsorbed into a dielectric polymer can plasticize the polymer dielectric, considerably reducing the glass transition temperature and increasing the dielectric constant [Seraphim et al. 1985]. Furthermore moisture trapped within a multilayer structure may rapidly volatilize during processing or use resulting in catastrophic delamination or blistering. Other electrical properties such as impedance and dissipation factor are also affected by the moisture content [Hsu 1991].

Along with the loss of mechanical and dielectric strength at elevated temperatures, organic boards are subject to the formation of conductive filaments. This is a function not only of temperature but of moisture content and voltage as well, as shown by the formula for filament formation below [Rudra 1992]:

$$t_f = \frac{af(1000L_{eff})^n}{V^m(M-M_t)} \tag{6.1}$$

where t_f is the average time to failure, a is an acceleration factor that varies for different board materials and features, f is a correction factor for multilayer constructions, L_{eff} is the effective length between conductors, V is the voltage, M is the moisture content, M_t is the threshold moisture content, n = 1.6, and m = 0.9. The effect of temperature and humidity are hidden in the value for the moisture content which is a function of both and can be determined from the moisture isotherms for the board material.

A novel technique for developing high temperature substrates is to combine conductors made of new, highly temperature stable, polymer-based conductive inks (Ormet® inks) with high temperature polymer dielectric materials and metallic heat sinks. The majority of the electrically conductive polymer thick film inks currently in use are in the form of thermosetting or thermoplastic resin pastes with metal or metal coated powder as the conductor. The metal filled pastes suffer from degradation of electrical conductivity with aging after curing, and they respond poorly to temperature and humidity fluctuations.

The new inks were developed to overcome these shortcomings and are based on transient liquid phase sintering technology. These inks contain metal and alloy powder mixtures with a permanent polymeric adhesive binder. The metal powder mixtures are designed to form an electrically continuous, high melting point alloy through transient liquid phase sintering at relatively low temperatures. During heating, the action of the adhesive flux mixture helps flux the metals, enabling rapid transient liquid phase sintering to occur between the high melting metal powder and the low melting alloy powder which results in a high melting point alloy. For this reason, the resulting conductor can provide stability with little degradation due to oxidation or corrosion.

These inks are called Ormet® inks, which stands for ORganic-METallic inks. They are analogous to the ceramic-metallic (cermet) inks that have been used in hybrid circuit manufacturing for many years. They provide high conductivity (bulk electrical conductivity is $\sigma = 4 \times 10^4$ S/m), comparable to cermet materials as well as good adhesive strengths.

In transient liquid phase (TLP) bonding two or more different metals are selected which exhibit a eutectic point or melt at some composition and at some temperature T_l. A mixture of the two metals is prepared in the proportion of the desired final alloy composition, the final composition being a solid at T_l. Heating of the mixture to T_l results in the formation of a liquid phase. The liquid phase has a very high solubility in the surrounding solid phase, thus diffusing rapidly into the solid and eventually solidifying at temperature T_l. Diffusional homogenization creates the alloy of the final composition without the need to heat the mixture above its equilibrium melting point, T_m.

In a TLPS ink, the above phenomenon is used in partial sintering of metal powders. A high melting point metal or alloy powder and a low melting point metal or alloy powder are mixed together with the polymer flux/binder system. The metals undergo transient liquid phase sintering during curing.

In practice, the formation of transient liquid phase alloy is characterized by a powder mixture which melts once and only once at some low temperature T_l. This has been observed in numerous metal mixtures. As the temperature is raised for the first time, the mixture undergoes a melt endotherm which is shortly thereafter followed by an alloying reaction exotherm. After cooling, subsequent temperature excursions beyond the original melt temperature do not produce melting anew. This is the signature of a typical transient liquid phase sintered metal mixture.

The manufacturing process for these inks uses established additive ink processes (MCM-C technology) with the well established printed circuit technology (MCM-L technology). A multilayer printed circuit can be easily and inexpensively fabricated by screen printing alternating layers of TLPS conductive inks and dielectric inks. Interconnection between layers can be easily accomplished by printing openings for vias in the dielectric layers. The circuits can be constructed on both conventional substrate materials as well as non-standard ones.

Inks that can withstand elevated temperatures are currently being developed with a focus on addressing the following concerns:

- Long term high temperature stability of the materials and the interfaces, including immunity to fatigue, distortion, electrical degradation, electromigration, oxidation, corrosion and outgassing.
- Low CTE which is stable with temperature. Significant changes in CTE over the projected range of service temperature extremes can cause high stresses to be developed. This when combined with the decreasing thermal conductivity which occurs at elevated temperatures and the resultant localized heating due to device operation, can cause stresses to exceed the design strength of the board.
- Low density of materials for acceptance in weight-sensitive avionics and automotive applications. Low cost of materials and manufacturing processes to present a cost-effective alternative to hybrid ceramic technology.

This new high temperature ink has a sintering temperature of 255 ^{0}C which is compatible with the polymer-flux binder curing temperature to allow the curing of the polymer and the sintering of the alloy to occur simultaneously. The polymer flux binder has a T_g in excess of 250^{0}C and good adhesion to polyimide. This binder was paired with a metallurgical system capable of withstanding temperatures greater than 350^{0}C. The combination produced a high temperature conductive ink which was demonstrated to have a bulk conductivity less than

60 μΩ-cm at room temperature and adhesive strengths of over 1 kg/mm^2.

The metallurgical and polymer systems employed did not exhibit any physical or energy transitions below 280^0C once processed. This indicates that the original low melting point alloy was completely absorbed into a new alloy composition with the high melting point metal. The newly formed metal composition melts at a temperature above 350^0C. SEM analysis revealed a homogeneous structure of the metal matrix which would produce reliable, consistent, and predictable conductivity after a period of "burn-in" in which the metals fully interdiffuse. Electrical performance behaved linearly up to approximately 280^0C. At 280^0C, the polyimide dielectric began to flow and interfere with resistance measurements. Substantial changes in resistivity were noted at 300^0C. The underlying cause of these effects is probably the degradation of the underlying dielectric rather than the ink. Upon cooling to room temperature, the resistivity decreased to a level 26% higher than the original room temperature measurement.

This conductive ink was used in conjunction with high temperature polymer insulative dielectric layers to form conductive traces, vias, and interconnects on metallic substrate materials. The polymer dielectric employed is a thermosetting polyimide precursor which was formulated for laminating copper foil onto polyimide film for flexible circuitry. Testing at Sandia indicated that the polymer material, once fully cured had a glass transition temperature in excess of 235^0C and a CTE of 60-70 ppm/K below the glass transition temperature. These new materials and the additive processes used to make them are designed to facilitate a cost-effective means of producing electronic packages capable of extended high temperature operation.

A final source of concern for organic boards is the use of conformal coatings. While silicone and polyimide based coatings can survive continuous use above 200^0C, urethanes, acrylics, and epoxy coatings can not be used [Pecht 1994].

In summary, at present the state of high temperature substrate circuitry is limited to high cost, high performance ceramics, lower cost polyimide systems with temperature limitations of 200^0C, and metal substrates with epoxy-copper laminates limited to 150 -200 ^{0}C. However, the new Ormet® ink processes and the newer PTFE-based laminates still hold out promise for a low cost, high temperature organic substrate solution.

6.2 High-Temperature Solder Materials and Conductive Adhesives

The purpose of a second-level package interconnect is to provide an electrical and mechanical joint between a chip carrier and a printed wiring board. The joint must conduct current from the package to the board. In surface-mount applications, the joint has the additional task of mechanically holding the package in place.

In high-temperature electronics applications, the second level interconnect must operate under severe environmental conditions with temperatures that can cyclically vary from -55 up to 200°C (a temperature profile for under-hood electronics for automobiles). The combination of this temperature cycle and the difference in coefficients of thermal expansion of the materials joined by the interconnect (e.g., a ceramic component [6x10-6mm/mm°C] and a polyimide board [19x10-6 mm/mm°C]) results in cyclic strain on the joints. The cyclical temperature/strain is termed thermomechanical fatigue. The interconnects could also be exposed to high-humidity conditions (>85% relative humidity). Mechanical shock (such as hitting a pothole with a car) and vibration (such as engine noise in a car or aircraft) are also environments that could degrade the reliability of interconnects.

A number of materials have been proposed to satisfy these requirements for interconnects for second-level packaging of high-temperature electronics. The two most common families of materials for this application are solders and electrically conductive adhesives (ECAs).

6.2.1 Currently used solder for second level interconnects

For electronic packaging applications, the most commonly used interconnect material is near-eutectic tin-lead (Sn-Pb) solder (60Sn-40Pb by weight percent). An outline of the characteristics of this alloy follows. Further details can be found elsewhere [Frear 1990, 1994; Yost et al. 1993]. The near-eutectic Sn-Pb solder is the standard by which all other interconnect materials are compared. Furthermore, other materials and processes used in electronic packages were developed, for the most part, to be compatible with this solder.

The Sn-Pb near-eutectic alloy melts at approximately 183°C and solder processing (joining the package to the board) is normally performed around 220°C. This solder has a two-phase structure: a tin-rich phase and a lead-rich phase. A Sn-Pb phase diagram showing the melting temperature as a function of composition is found in Figure 6.6 This solder wets metallized surfaces well provided the surfaces are clean. Flux is used to remove oxides on the metallized surfaces and assist in the spreading of the solder. As the solder wets and spreads over the surface, the tin reacts with the metallization to form an intermetallic compound that is indicative of a good bond. After solidification, near-eutectic Sn-Pb solder has excellent electrical conductivity. The solder also has adequate time-dependent deformation characteristics allowing it to take up strain caused by the difference in coefficients of thermal expansion of joined materials without damaging the component or board. The mechanical and thermomechanical stability of the alloy is sufficient for electronic packaging applications.

The alloy does evolve to a heterogeneous coarsened structure that accelerates failure under thermomechanical fatigue conditions. The heterogeneous coarsened structure forms parallel to the direction of imposed strain and consists of coarsened lead- and tin-rich phases. In the low-temperature portion of the cycle, damage in the form of defects is imparted to localized regions of the microstructure. In the high-temperature portion of the cycle, diffusion mass flow recovers this damage and heterogeneously coarsens the structure. The individual grains inside the coarsened structure eventually grow to the point at which they can no longer accommodate the deformation, and intergranular cracks form and propagate through the coarsened regions. This failure behavior is accentuated when temperatures exceed 150°C, because the mass diffusion that coarsens the microstructure is accelerated. For high-temperature applications, the melting temperature of near-eutectic Sn-Pb solder is too low. Environments that exceed 183°C cause the solder to melt and application temperatures that do not exceed the melting point, but do approach it, cause the solder to lose structural integrity. Therefore, other interconnect materials that operate at higher temperatures are needed.

6.2.2 High-temperature solders

Solder alloys used for electronic applications must have a small two-phase (liquid+solid) region for ease of processing. A eutectic alloy has the optimum condition (a direct transformation from liquid to solid). If the two-phase region is too large, the joint members have the opportunity to move with respect to one another during solidification. This results in an irregular, coarse, and sometimes cracked joint surface and often indicates poor mechanical properties. Therefore, most solder alloys for interconnects are near-eutectic alloys.

High lead-content tin-containing solders offer a high operating temperature interconnect alternative. Although the composition is outside the eutectic, the two-phase temperature band is small. The high lead-content 95Pb-5Sn solder alloy has a melting point of

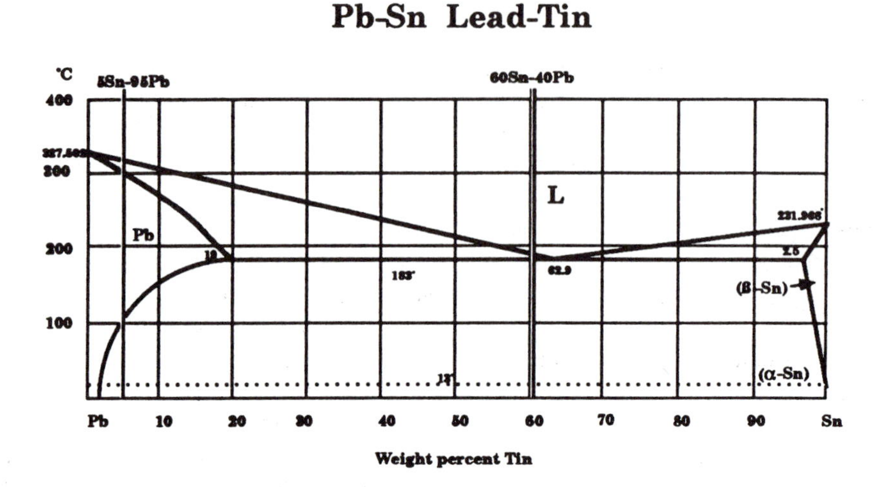

Figure 6.6 A Sn-Pb phase diagram showing the melting temperature as a function of composition.

300°C. This alloy has found extensive use as a first-level solder interconnect for flip-chip applications joining chips to chip carriers. Other alloys similar to the 95Pb-5Sn alloy have similar properties, 95.5Pb-1Sn-1.5Ag was developed to minimize intermetallic growth on silver metallizations, also has a high melting point of 309°C [Bader 1975]. These alloys consist of a matrix of large lead grains with small precipitates of tin in the bulk of the solder joint.

With a melting temperature above 300°C, the high lead-content solder alloys offer the ability to make interconnects that will not melt for almost all high-temperature electronics applications. Unfortunately, there are a few drawbacks to these alloys. The low tin-content of the alloys raises the melting temperature and also decreases wetting behavior. The reaction of tin at the metallization interface enhances wetting, and as the tin-content decreases, the wetting behavior suffers [Yost et al. 1993]. The processing temperature for the high lead-content solders must also exceed 340°C to get good solder flow. The processing equipment and other electronic packaging materials for the second-level package must be redesigned to withstand these temperatures. Finally, the lifetime of high lead-content solder, under conditions of thermomechanical fatigue for second-level interconnects, is less than half of that of 60Sn-40Pb solder [Frear et al. 1987]. These solders have large lead-grains that form the matrix of the solder; these are so large that they are unable to accommodate the strain and fail intergranularly during thermomechanical deformation.

6.2.3 Lead-free high-temperature solders

The family of lead-free alloys that is best suited for high temperature applications is based upon the Sn-3.5Ag eutectic and the Sn-5Sb eutectic alloys. Potential ternary additions of antimony, silver, bismuth, or copper can be made to the tin-silver solder. Table 6.2 lists a few alloy compositions that are eutectic, or near-eutectic, and their related melting temperatures. The eutectic Sn-Ag alloy melts at 221°C. This set of alloys melts in the range of 210°C to 234°C. At these melting temperatures, the operating temperature of the electronic systems can be raised to above 200°C.

Environmental and public health concerns are driving a move toward tougher regulation of lead. These legislative initiatives reflect concerns that even small concentrations

Table 6.2 Melting Temperatures of Some High Temperature Solder Alloys

Alloy Composition (wt.%)	Melting Temperature (°C)
60Sn-40Pb	183
95Pb-5Sn	300
95Sn-5Sb	234
96.5Sn-3.5Ag	221
Sn-4.7Ag-1.7Cu	217
Sn-3.4Ag-4.8Bi	210

of lead in children's blood can seriously affect mental development. This concern has led to the ban on lead in plumbing solder, paint, and gasoline from which over 95% of ingestion occurs. Concerns also exist that the continued use of lead, and its disposal in waste streams may further release lead into the environment. From an environmental perspective, the implementation of a lead-free high temperature solder alloy would obviously be beneficial.

The wetting behavior of the high tin-content solders is not as good as that of lead-containing alloys, but results in the formation of good solder joints by reflow or wave-soldering methods. The presence of lead improves wetting by lowering the surface tension of molten tin. High tin-content solders have a higher surface tension, so they do not wet as well as 60Sn-40Pb, but they react with the substrate metallization to form a good metallurgical bond. This is evidenced by the interfacial intermetallic layer that forms.

A drawback to Sn-Ag and Sn-Sb alloys is that their response to time-dependent deformation is much slower than Sn-Pb (roughly an order of magnitude slower at the same stress level), so damage can be imposed on the joined components. These solders also exhibit a somewhat brittle fracture mode when deformed in tension. The failure path is along the interface between the intermetallic and the solder [Frear and Vianco 1994]. The energy required for fracture is about half that for 60Sn-40Pb which fails through the solder. This failure mode could be important when tensile stresses are imposed at room temperature and lower. Tensile stresses arise in second-level solder joints when boards or components deform out of plane.

The thermomechanical fatigue life of the high tin-content, lead-free solder is better than that of lead-based alloys. The Sn-3.5Ag and Sn-Ag-Cu alloys have lives about 10% longer than Sn-Pb; Sn-Ag-Bi has a life over twice that of Sn-Pb. This is due to the limited microstructural evolution that occurs in the tin-matrix of lead-free alloys and the limited time-dependent deformation of the alloys. The tin-grain size in Sn-Ag solders is much finer than in high lead-content alloys, so the deformation mechanism of grain boundary sliding and rotation does not result in the rapid failure found with 95Pb-5Sn. Thus, one of the benefits in utilizing a high-temperature lead-free solder is improved reliability under thermomechanical fatigue conditions.

6.2.4 Fluxes for high-temperature solders

The purpose of flux in a soldering operation is to clean and prepare the metallization so that it can be wetted by molten solder. The flux consists of a solvent carrier (e.g., alcohol), surfactant (to assist wetting of the flux), and the acid that is bonded into a solid form. Upon heating, the solvent evaporates, and the solids of the flux become molten and begin to remove the surface oxide. The active flux forms a continuous film over the surface that inhibits access of the oxygen to the hot surfaces until it is displaced by the spreading molten solder.

Almost all fluxes currently in use were formulated for the near-eutectic Sn-Pb solder alloy. The process temperature range of these fluxes is 220° to 240°C. Flux chemistry and formulation (solvent, acid, surfactant, etc.) have not yet been developed for process temperature between 240°C and 340°C. New fluxes must be developed for high-temperature electronic solder applications.

6.2.5 Metallizations

The metallizations used to form joints at process temperatures in excess of 240°C must also be evaluated. Traditional metallizations, such as copper or nickel heavily oxidize at these temperatures and fluxes may not be able to adequately remove sufficient oxide to enable wetting. A potential solution to this problem is to use noble metals. Gold is an obvious choice, but its expense may limit its use. Yeh and Strickman [1992] propose using palladium in either a plated or evaporated form as a solderable metallization. They found that even without flux, using a forming gas cover, good wetting by tin-based solders is possible. Wetting was enhanced as process temperatures increased over the range of 215° to 350°C. As with the solder type and flux, the metallization must be part of the total solution to high-temperature electronics requirements.

6.2.6 Electrically conductive adhesives

Electrically conductive adhesives (ECAs) are composite materials consisting of a dielectric curable polymer and metallic conductive particles. ECAs are low processing temperature alternatives to solder alloys. The polymer is an adhesive material that chemically reacts with metals to form a bond. The metallic particles in the adhesive form a network in the cured joint that forms a conduction path from the package to the board.

The polymer portion of the adhesive can come in two forms: thermoplastic and thermosetting. A schematic illustration of these two types of polymers is shown in Figure 6.7. The thermoplastic polymer is a linear chain of linked mers (nominally, hydrogen-carbon bonds). After curing, the thermoplastic polymer becomes more rigid, but remains flexible. As the temperature is increased, the polymer essentially becomes molten. This allows for easy repair of a bonded thermoplastic joint. The thermosetting material is similar to the thermoplastic, but the chains of mers are cross-linked in three dimensions during the curing stage. The number of cross-links determines the glass transition temperature (Tg) of the polymer. In general, the more cross-links, the higher the glass transition temperature. A curing agent (e.g., a hardener) is added to a thermoplastic material to cause the cross-linking reaction. Curing can be performed using UV light, heat, or catalysts. Thermosetting materials are harder than thermoplastic, but can be processed and cured below Tg and do not become "molten" above Tg. The curing operation is not reversible, making repair difficult. Examples of thermosetting materials include epoxies and acrylics. ECAs are typically made using either thermosetting or thermoplastic polymers, or a combination of the two.

A number of metals have been used for the conductive portion of the ECA. The metals typically are in the form of flakes, plates, rods, fibers, or spheres. The size of the metal particles ranges between 5 and 20μm. For the ECA to have suitable strength and electrical conduction requires a trade-off in metal filler content. The greater the metal content, the better the electrical conduction, but the poorer the strength of the adhesive. The metal content depends upon whether the adhesive is an isotropic or anisotropic conductor (described below), but does not usually exceed 40% by volume.

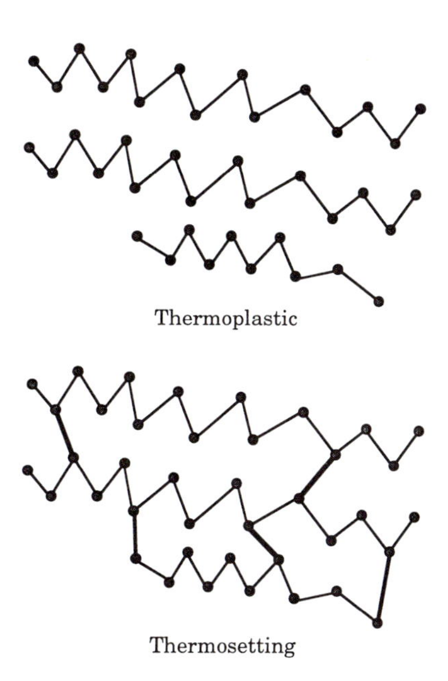

Figure 6.7 A schematic illustration of thermoplastic and thermosetting polymers.

The types of metal used in ECAs are silver, gold, nickel, or copper or variations of these (such as gold on nickel, silver on glass, or nickel on polymers). The silver-filled ECAs are the most common, because the cost of the metal is moderate and it has good electrical conductivity and low reactivity with oxygen. Gold has better physical properties than silver but the cost is prohibitive for most applications. Nickel is less expensive, but also has lower conductivity and has been found to corrode when aged in a humid environment resulting in poor adhesion between the nickel and the epoxy matrix. Copper fillers oxidize rapidly and delaminate from the polymer matrix.

The mechanism of conduction in ECAs is hypothesized to occur via percolation where current travels through a three-dimensional matrix of contacting particles. There must be enough particles present in the adhesive so that a conduction path can form. The bulk resistivity of 60Sn-40Pb solder is 15x10-6 Ω-cm, while that of ECAs ranges from 7x10-5 to 5x10-4 Ω-cm [Hvims 1995]. Therefore, the conduction of the solder is 5 to 33 times greater than of ECAs. This loss in electrical conductivity does not severely affect second-level interconnects and is due to the percolation path required for ECAs, whereas the entire solder joint is a conductor. For the current to pass from one metal particle to another, it must pass through, at a minimum, the oxide that forms on the metals or, at worst, some distance through the polymer matrix. It has been proposed that contact resistance is overcome by tunneling or bulk conduction through the semiconducting metal oxide layer. However, work by Li et al. [1995] indicates that the conduction is better than if any of the above mechanisms occurred. It has also been proposed that, in silver-loaded ECAs, silver diffuses out of the particles into microcracks in the polymer matrix, thereby creating additional conduction paths. To date, the details of the conduction mechanisms are not resolved.

ECAs are available in two types of conductors, isotropic and anisotropic. In an isotropic adhesive, the path of conduction is uniform in all directions and the filler metal content must be high to create a percolation path for the conductor. An anisotropic ECA (or AECA, also called a z-axis conductor) has only one direction of conduction and is useful for fine-pitched applications. In general, anisotropic ECAs are prepared by dispensing metal particles in the polymer matrix at a level far below the percolation threshold. Pressure is imposed on the contact so that the particles come into contact with one another in one direction (parallel to the direction of imposed pressure). In the areas where no pressure is imposed, there is no conduction path and the adhesive is electrically insulated. This is illustrated in Figure 6.8. In fine-pitch applications, a line of anisotropic ECA can be applied on all the pads on the board. Applying pressure on the package leads results in fine-pitch conduction paths with insulating adhesive between each lead.

Thermosetting ECAs can be applied in a variety of forms. The adhesive can be purchased in two parts (polymer loaded with metal, and a separate curing agent) that are mixed immediately prior to use. In this form, it is extremely difficult to eliminate the air bubbles that detrimentally affect the strength and conduction of the adhesive. The most commonly used ECAs come as a manufacturers premix. In this form, the air bubbles that can be incurred in mixing the metal particles, the polymer, and the curing agent are significantly reduced through the use of a centrifuge. However, because the curing agent is present, the materials must be stored at –40°C prior to use to slow the curing reaction. The shelf life, even at low storage temperatures, is short. The premix comes as a paste or a film. A third option is to use modified acrylics, placing part A of the premix on one side of the joint, part B on the other side. When assembled, the acrylic undergoes rapid free-radical polymerization.

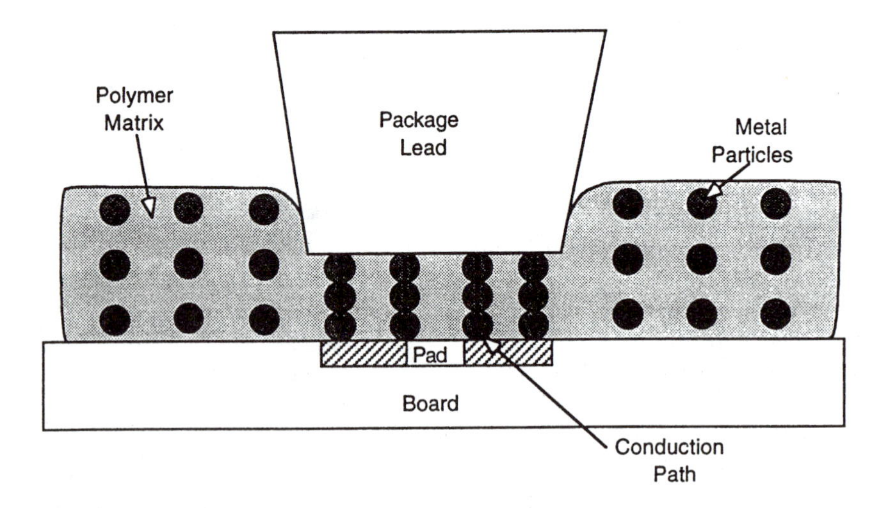

Figure 6.8 In the areas where no pressure is imposed, there is no conduction path and the adhesive is electrically insulated.

The bonding surface of the metallizations to be joined must be oxide-free for the adhesive to form a cohesive joint. Bare copper gives very poor adhesion [Chang et al. 1993]. The adhesives do not wet or react well with metal oxides, and thus are very similar to solders. To overcome this problem, the copper can be plated with nickel and a thin layer of gold. The gold is an excellent bonding surface because its surface oxide is negligible and it provides a diffusion barrier for the nickel. An additional difficulty with ECAs is that they do not allow the luxury of using flux to remove oxides during processing. Therefore, all surface treatments for the removal of oxides must occur prior to dispensing the adhesive. The stability of ECA joints is directly related to the quality of the bond interface with the metallization. The less surface oxide present, the better the electrical stability. If surface oxide is present, there can be de-adhesion at the interface that leads to a loss in the ability to carry electrical signals. The bonding of ECAs to other polymers, such as printed wiring board surfaces (FR-4 or polyimide) is excellent, which creates problems in defining the ECA joint location during processing. A solder will only wet metal surfaces, so special processing steps are not needed to define the solder joint location. For ECAs, the method of depositing the ECA must be precise (or use an anisotropic ECA) so that electrical shorts do not occur.

ECAs offer a number of interesting possibilities for high-temperature electronics because a number of formulations have operating temperatures that approach 250°C. Table 6.3 lists some of these materials. For thermosetting materials, the Tg extends up to 250°C. Furthermore, due to extensive cross-linking, the behavior of the thermosetting polymers above Tg tends to be more like a solid than a viscous fluid. Some thermoplastic materials can operate at higher temperatures (for example, the silicone in Table 6.3 can go up to 200°C).

The ability to perform rework on second-level interconnects may limit the ability to use ECAs in high-temperature electronics. Thermoplastic adhesives can be easily reworked simply by heating and "melting" the joint. Thermosetting bonds are not reversible but can be softened by heating well above Tg and then removed. Some acrylic adhesives can be cleaved at room temperature and physically removed, although the brittle nature of the acrylics that allows for easy removal may also represent too great a reliability risk.

Table 6.3 High Temperature Conductive Adhesives from Hvims [1995]

Polymer	Metal	T_g (°C)	Operating Temperature (°C)
Epoxy	Ag-flakes	130	up to 150
Epoxy	Ag-flakes	>200	-55 ° to 250
Epoxy	Ag-flakes	80	-50 to 150
Polyimide	Ag-flakes	249	20 ° to 250
Epoxy	Ag-flakes	85	up to 225
Epoxy	Ag-flakes	150	20 ° to 175
Silicone	Ag-flakes	N/A	-50 ° to 200

6.2.7 Reliability of electrically conducting adhesives

In order for ECAs to be useful in high-temperature electronic applications, they must have the same, or better, reliability than that of solders. The ECAs must be able to survive, without serious degradation of physical or electrical properties, high temperatures for extended periods, mechanical shock, humidity, and thermal cycles (thermomechanical fatigue).

6.2.8 Strength and aging

The effect of aging on the ability of the joint to withstand mechanical shock of is important in ECAs. There is some evidence that ECAs are susceptible to even mild CTE mismatches between boards and components which can result in failure under mechanical shock conditions flexing (e.g., PCMCIA cards) [Gaynes et al. 1995].

Generally, aging at high temperatures increases the strength of ECAs because of the increased curing of the adhesive. However, that the standard deviation of the strength increases with aging [Gaynes et al. 1995]. The strength and adhesion of the joint can be severely decreased at high temperatures (>85°C) in humid conditions, due to oxidation of the metal particles and the joined metal surfaces. As adhesion is lost between the polymer and the metal (both particles in the ECA and the metallized interfaces), electrical continuity also decreases because the intimate metal/metal conduction path is disrupted. With high Tg materials, the polar groups of the molecular chains enhance water diffusion through the polymer, accelerating the oxidation process [Rusanen and Lenkkeri 1995].

Trade-offs must be made in order to achieve the proper strength in ECA joints. High-strength adhesives tend to be brittle and are susceptible to water diffusion and oxidation of metal surfaces, which promotes loss of adhesion. More flexible ECA materials have a lower Tg and thus, lower operating temperatures. Conformal coating of the ECA joint with a thermosetting polymer is an option that enhances joint strength, but it makes rework virtually impossible. Moreover, this does not address the potential loss of conduction due to metal oxidation, because the conformal coating is also permeable to water.

6.2.9 Silver migration

One of the greatest concerns with the reliability of silver-loaded conductive adhesives is silver-migration. In this mechanism, the silver is ionized and migrates in a humid environment, forming electrically conductive dendrites that can result in electrical shorts between interconnects. Silver migration can only occur if the silver particles are directly exposed to a corrosive, humid environment. The silver in ECA joints is encapsulated inside the polymer, and it is difficult to promote dendrite growth from within the joint if corrosion does occur. Work by Hvims [1995] and Rörgren and Liu [1995] indicated that silver migration was never observed in any of the accelerated corrosion tests performed.

6.2.10 Thermomechanical fatigue

ECA joints can be exposed to thermomechanical fatigue if they join materials with different coefficients of thermal expansion under temperature cycling conditions. These conditions are typical for almost all ECA applications, particularly for high-temperature electronics.

Under conditions of thermal shock (very rapid changes in temperature), ECA joints are susceptible to brittle failure in much the same way they are for isothermal mechanical shock. Liu et al. [1995] found ECA-joined J-leads to be particularly susceptible to thermal shock. The problem of thermal shock conditions is that the polymer cannot plastically respond to the deformation quickly to relieve the strain so the joint fractures. The best method to avert thermal shock failures is to either enhance adhesive flexibility by decreasing Tg or stiffen the board with a conformal coating.

Under the more common slow thermal cycles, the failure behavior of ECAs improves. It was thought that the deformation of the soft polymer adhesive with hard inclusions at high temperatures would result in interfacial separation between the metal particles and the matrix but this was not observed; the metal particles retained adhesion with the polymer matrix during thermomechanical aging [Rörgren and Liu, 1995].

Figure 6.9 shows an optical micrograph of a high-temperature epoxy-based silver-loaded ECA after 300 cycles of -55° to 125°C at 10% shear strain. Even after 500 cycles, the adhesive still had not failed electrically. Under the same test conditions, 60Sn-40Pb failed after 220 cycles. Cracks appear in the ECA joint, but they run perpendicular to the direction of deformation and electrical conduction paths are retained. The ECA performs well under thermomechanical fatigue conditions. Further material development will enhance this performance.

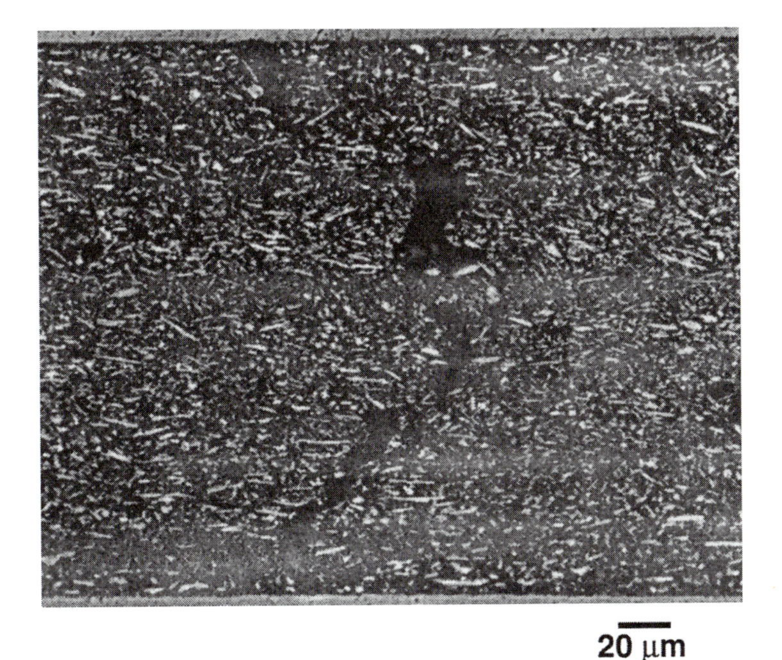

Figure 6.9 Optical micrograph of a high-temperature epoxy-based silver-loaded ECA after 300 cycles of -55 to 125°C at 10% shear strain.

6.2.11 Comparison of ECAs and solder

Neither solder nor ECAs offer the penultimate solution for all high-temperature electronic applications. Each material type offers its own relative merits, and selection of the appropriate material should be made on a case-by-case basis. In order to attach a second-level package, nine processing steps are required for solder, while only five are required for ECA. The decreased number of processing steps results from the fluxing and cleaning operations needed for solder. The decreased number of processing steps also translates into shorter processing times and decreased handling. However, the ECAs material cost is roughly three times as much as Sn-Pb solder.

ECA joining can be performed at lower processing temperatures, but can still be used for high-temperature applications (up to 250°C). With solder, the processing temperature always exceeds the maximum operating temperature.

One of the major problems with ECAs is that adhesive is not selective, as it is with solder. Solder only wets cleaned metal surfaces; ECAs will wet almost any surface. The ECAs needed for high-temperature applications are thermosetting, and are therefore much more difficult to rework than solder. The cure times for ECAs are twice as long as the soldering operation, but time is saved by avoiding fluxing or post-solder cleaning. Adhesives are also difficult to process because they have less density than solder, and tend to stick to the squeegees used to spread materials. The amount of ECA paste required for the operation is less than for solder for similar applications.

ECAs are brand new materials compared to solder, a material that has been in use for thousands of years. More characterization work and alloy development is necessary. The current electrical performance is satisfactory, though not equal to solder. The reliability aspects, in some respects (e.g., thermomechanical fatigue), often exceed those of Sn-Pb solder alloys. This offers further promise for the use of electrically conductive adhesives for high-temperature applications.

6.3 Wire and Cable

Wires and cables are used to conduct signals and power from one subassembly to another. They are composed of a metallic conductor, a polymeric insulating or dielectric material, and a polymer jacketing. Coaxial cables also have a shield or braid that may be made of metal or a carbon or metallized film. In this section, the use of each of these elements at elevated temperatures will be discussed. Since no coaxial cables are used in any of the modules under study, the effects of temperature on signal transmission performance will not be discussed.

6.3.1 Conductors

Conductors are chosen on the basis of electrical conductivity, tensile strength, density, and continuous use temperature. Conductors are most often made of silver-plated annealed copper for maximum electrical conductivity. Silver plating increases the dc conductivity 2%, and increases the conductivity more at higher frequencies, where a greater portion of the signal is carried by the plating layers, a phenomenon known as the skin effect. Tin plating does not alter the dc conductivity of the copper wire; nickel plating decreases it 4%. Table 6.4 lists some of the properties of common conductor materials, along with their maximum continuous use temperatures [Sclater 1991].

Table 6.4 Properties of Conductor Materials [Sclater 1991]

Conductor	Conductivity (% IACS)	Tensile Strength (annealed) (ksi)	Continuous Use Temp (°C)	Density Factor (Cu = 1.00)
Bare copper	100	36	150	1.00
Tin-plated copper	100	36	150	1.00
Silver-plated copper	102	36	200	1.00
Nickel-plated copper	96	36	260	1.00
Bare copper-clad	40	55	200	0.93
Silver-plated Cu-clad steel	40	55	200	0.93
Cd-Cr-copper	85	58	260	0.98
Cd-copper	80	55	175	0.98
Cr-copper	85	58	200	0.98
Zr-copper	85	55	200	0.98

In most instances, stresses on the wire are small, so loss of strength at elevated temperatures is not a serious concern. The loss of strength would be minimal anyway, because conductors are supplied in the annealed state to maximize conductivity. Cables requiring more strength use wire made of a copper alloy or a copper-clad steel, in which the strength comes from materials selection, not from cold working, and thus is not subject to annealing.

Pure copper is limited to use at 150°C. At 180°C, oxidation can blacken the copper surface in minutes. Plating the copper allows the conductor to be used at higher temperatures. Nevertheless, there are temperature limitations to the use of certain platings. Because of their lower melting temperatures and their tendency to oxidize, tin and tin-lead platings are limited to use at temperatures less than 150°C. Silver plating can be used to 200°C, at which time it can begin to form sulfide and oxide films. Silver-plated copper must also be kept away from highly humid environments, in which it undergoes a galvanic corrosion mechanism commonly known as red plague. Applications above 200°C require the use of nickel-plated copper. This is not an option for high-frequency applications, because a sizeable fraction of the signal is carried in the plating, but it is an option for lower frequency signals, which are carried by the bulk copper.

Aluminum wire can also be used as a conductor. It has 60% of the conductivity of copper, so larger diameters are required to produce the same resistance, a factor offset by its lower density. Other disadvantages of aluminum are the inability to use conventional eutectic tin-lead solder for joining and the threat of galvanic corrosion.

The electrical and thermal conductivity of both aluminum and copper wires decreases at elevated temperatures, as shown in Figures 6.10 and 6.11. High ambient temperatures also limit the current-carrying capability of the wire by limiting the amount of self-heating that can be accommodated before the maximum temperature rating of the insulation is reached. The

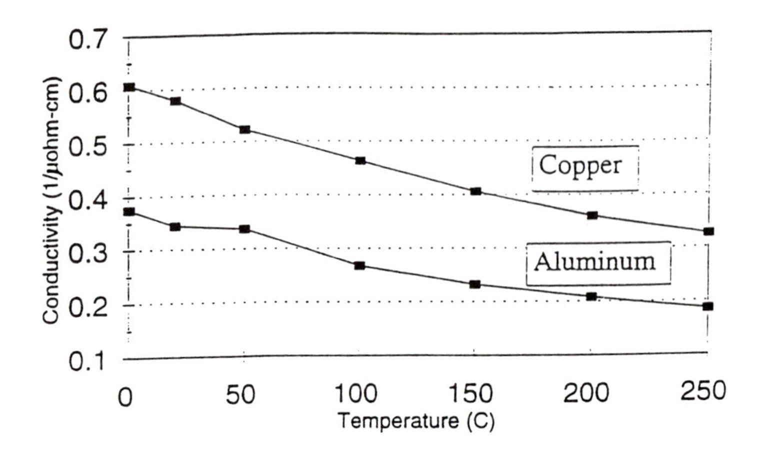

Figure 6.10 Electrical conductivity of conductors as a function of temperature [Charles 1994]

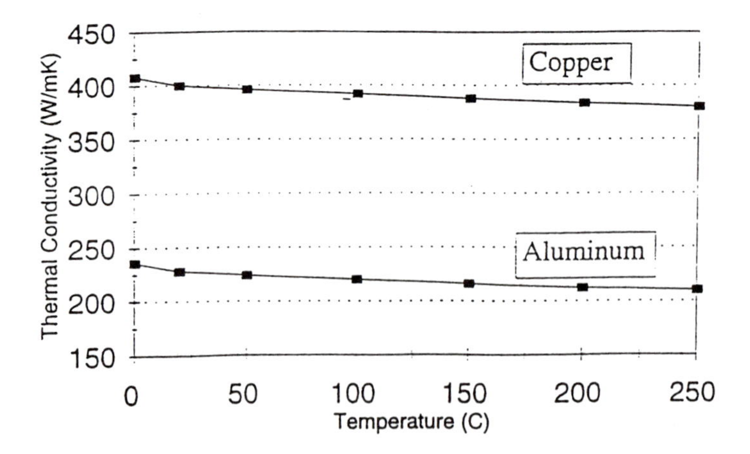

Figure 6.11 Thermal conductivity of conductors as a function of temperature [Charles 1994]

temperature at which the insulation degrades is often more limiting than any effects of temperature on the conductor.

6.3.2 Insulating materials, shields, braids, and jackets

Wire insulations consist of plastics extruded directly onto the wire material. Common insulation materials include polyethylene, polypropylene, polyester, polyvinyl chloride and extrudable variants of teflon, such as teflon PFA and teflon FEP. These materials need to be tough, flexible, and waterproof, and offer a high dielectric strength and insulation resistance. Maximum use temperatures for these materials are given in Table 6.5; room temperature properties are given in Table 6.6. The use of these materials at elevated temperatures is limited by softening, melting, depolymerization, and a loss of electrical resistivity. Variations with temperature in the tensile strength, electrical resistivity, dielectric strength, and dielectric constant of polymer insulating films are given in Figures 6.12 to 6.16. The lowered strength and softening that occurs at elevated temperatures can result in mechanical tears and breaks in the insulation and subsequent shorts, either to other wires or to the shield in coaxial cables. The loss of dielectric strength can cause dielectric breakdown in the insulation and shorting between wires and between the conductor and the shield in a coaxial cable. The loss of electrical resistivity can lead to current leakage to the drain in coaxial cables. Decreases in the dielectric constant with temperature will speed up signal propagation, while increases will slow it down. Both behaviors are observed, depending on the insulation material selected.

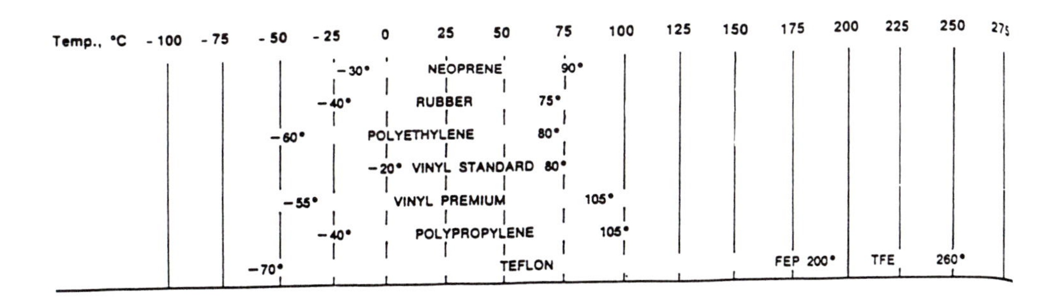

Figure 6.12 Nominal ranges of operation for wire insulation materials [Matisoff 1987]

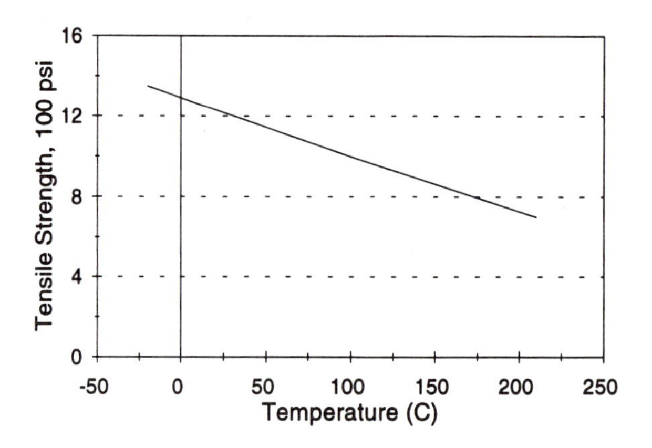

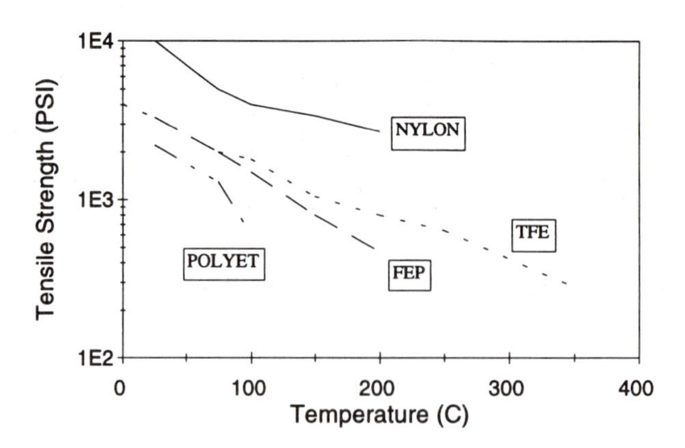

Figure 6.13 Tensile strength of polymer insulating films as a function of temperature a) polyimide, b) nylon, polyethylene, teflon TFE, teflon FEP [Harper 1970].

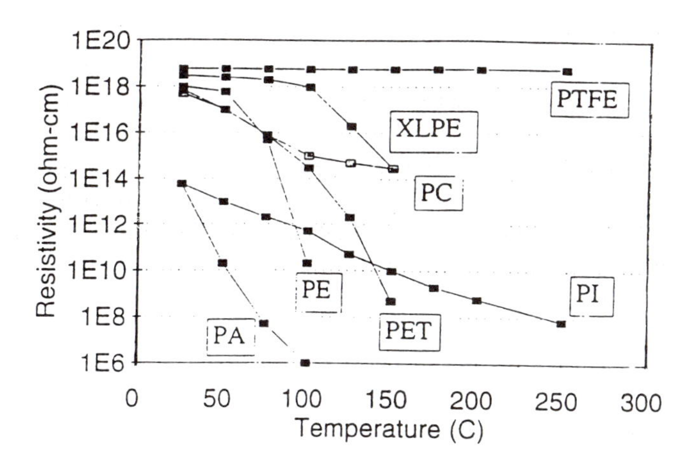

Figure 6.14 Volume resistivity of polymer insulating films as a function of temperature [Mark 1985].

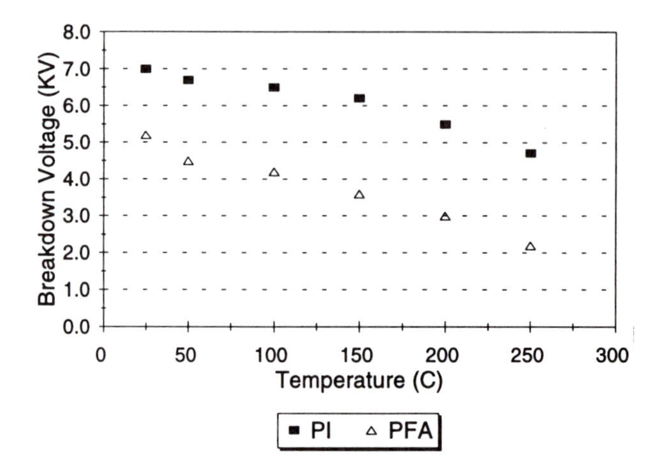

Figure 6.15 Dielectric breakdown strength of polymer insulating films as a function of temperature.

Another high-temperature effect is outgassing of the polymers, which can leave a contaminant coating on the wire and on other surfaces throughout the assembly, causing leakage on insulating components and reducing the conductivity of wires and contact pads. In addition, elevated temperatures and humidity can lead to the growth of silver filaments through the insulation when a silver-plated copper conductor is used in coaxial cables. This phenomenon is a function not only of temperature but also of bias and humidity, together with the thickness and moisture absorptivity of the insulation [Krumbein 1988].

Shields for coaxial cables can be made of metal wire or metallized polymer films. Metal wire shields and braids are subject to the same concerns as conductors, while metallized polymer films are limited by the same factors as the insulating polymer films. Aluminized mylar, one of the most common metallized polymer shields, has a maximum use temperature of 150°C because of the film [Sclater 1991].

Jacketing materials are also insulating polymer films. However, they group together several insulated wires or overcoat an entire coax, so they tend to be thicker and more elastomeric. A list of typical jacketing materials, their room temperature properties, and their maximum use temperatures is given in Table 6.7.

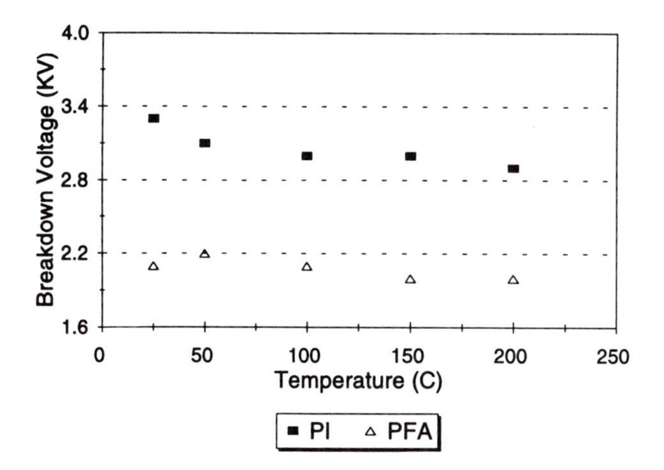

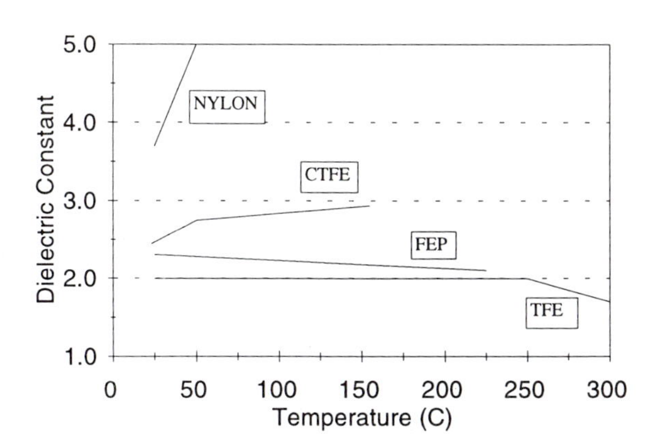

Figure 6.16 Dielectric constant of polymer insulating films as a function of temperature a) polyimide and teflon PFA b) nylon, CTFE, FEP, TFE [Harper 1970].

Table 6.5 Maximum Use Temperatures for Wire Insulations [Matisoff 1987, Modern Plastics Encyclopedia 1995, Lau 1991, Dally 1993]

Material	Established Maximum Temperature (°C)	Absolute Maximum Temperature	Reason
PVC	105	105	Glass transition temperature
Polyester (mylar)	150	180	Melting temperature
Polysulfone	130	190	Glass transition temperature
Polyethylene	80	121	Melting temperature
Nylon (polyamide)	115	260	Melting temperature
Teflon TFE	260	327	Melting temperature
Teflon FEP	200	275	Melting temperature
Polypropylene	105	168	Melting temperature
Silicone rubber	200	260	Softening/ depolymerization
Kapton HF	200	388	Melting temperature
Tefzel 280	180	270	Melting temperature
Teflon PFA	250	300	Melting temperature

6.4 Connectors

The function of a connector is to provide a separable connection between a module or subassembly and the rest of the electronic system. In order to perform this function, it must transmit signals without introducing unacceptable levels of distortion, and it must maintain mechanical contact between separable interfaces. Connectors are composed of a base metal spring to maintain the mechanical contact and carry the signals, a noble metal plating to provide and maintain low contact resistance at the separable interfaces, and an insulating housing [Mroczkowski 1994].

There are two sets of properties by which the base material of a connector is specified and for which temperature dependencies must be identified: those related to its spring characteristics and those related to its signal-carrying abilities. The first set includes the elastic modulus, the yield strength, the tensile strength, and the stress relaxation behavior. The second set includes the electrical and thermal conductivity. Values for these properties for common connector spring alloys are given in Table 6.8.

Table 6.6 Room Temperature Properties of Plastic Films [Pecht 1994, Sclater 1991]

Plastic Film	Moisture Absorption (%)	Tensile Strength (MPa)	Volume Resistivity (ohm-cm)	Dielectric Constant (at 1 kHz)	Dielectric Strength (MV/m)
Polyimide (Kapton)	3.0	230	------	3.4-3.5	150-300
Polyester (Mylar)	< 0.8	170	------	3.25	300
TFE (Teflon)	< 0.01	28	10^{18}	2.0	17
FEP (Teflon)	< 0.01	21	10^{18}	2.0	255
PFA (Teflon)	-------------	24	10^{18}	2.1	40
ETFE (Tefzel)	-------------	41	10^{16}	2.6	40
Polyamide (Nomex)	3.0	75	10^{12}	2.0	18
Polyvinyl chloride	< 0.5	35	5×10^{13}	3.0	40
Polyethylene	< 0.01	21	10^{16}	2.2	20
Poly-propylene	< 0.005	170	10^{17}	2.1	160
Polysulfone	0.22	68	-----	3.1	295

Table 6.7 Room Temperature Properties and Maximum Use Temperatures of Jacketing Materials [Sclater 1991]

Material	Max Use Temp (°C)	Tensile Strength (MPa)	Volume Resistivity (ohm-cm)	Dielectric Constant	Dielectric Strength (MV/m)
Thermoplastic elastomer	125	16	2×10^{16}	2.2	20
Nylon polyamide	105	68	10^{12}	4.0	16
Neoprene	90	12	-----	-----	-----
Polyurethane	75	27	2×10^{11}	-----	16
Glass braid	260	------	-------	------	-----
Silicone rubber	200	-------	-----	3.1	------

Table 6.8 Properties of Common Connector Spring Materials [Loewenthal 1984, Mroczkowski 1994]

Material	Tensile Strength (ksi)	0.2% Yield Strength (ksi)	Elastic Modulus (ksi)	Electrical Conductivity (% IACS)	Thermal Conductivity (W/mK)
Beryllium Copper 172 290 - M	140	111	19,000	18.1	121
Beryllium Copper 172 190 - ½HM	125	100	19,000	19.6	121
Brass C260	-------	75	16,000	27	-----
Phosphor Bronze C510 Spring temper	99	92	16,000	19.8	69
Phosphor Bronze C521 Spring temper	109	102	16,000	13.2	62
Copper-nickel C725 Spring temper	103	102	19,000	10.9	54

Normal force (the force holding the separable interfaces in contact) is used to measure the spring properties. It is a function of the modulus of the base metal alloy given in Equation

$$F_N = Ed\left(\frac{CI}{L^3}\right) \tag{6.2}$$

For a cantilever beam leaf spring, where F_N = the normal force (in g), E is the elastic modulus, C is a geometric constant, I is the moment of inertia, L is the beam length, and d is the beam deflection. I is defined as

$$I = \left(\frac{bh^3}{12}\right) \tag{6.3}$$

where b is the beam width and h is the beam thickness. As is obvious from this formula, decreases in the modulus, which occur at elevated temperatures as shown in Figure 6.17, must be accounted for during design by using larger beams or allowing for more beam deflection, both of which reduce the I/O density. Otherwise, these changes will result in a drop in the normal force and a less reliable mechanical contact. However, in beryllium copper, the decrease in modulus is minimal for temperatures up to 200°C.

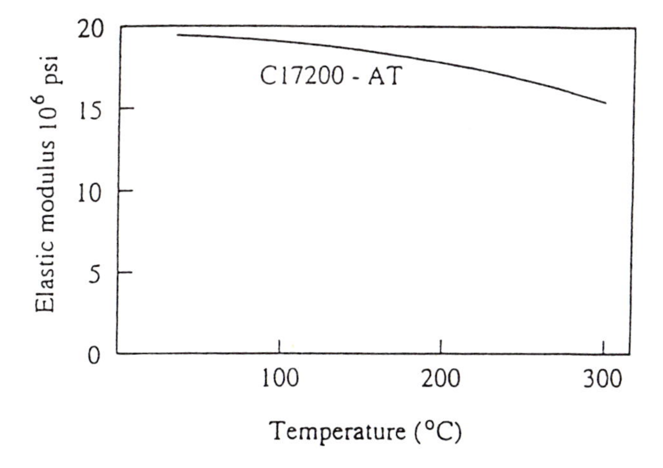

Figure 6.17 Elastic modulus of beryllium copper as a function of temperature [Guide 1991]

Normal force is necessary during mating in order to wipe contaminant films from the surface, while maintaining the normal force after mating ensures that contacts do not become separated under vibrational loading. The contact resistance is a function of the normal force, which asymptotically approaches a constant low level at forces greater than 100g per contact. The effects of a drop in normal force during use on contact resistance and signal integrity are moderated somewhat by a phenomenon known as sticking. This is an adhesive force between the separable interfaces of a mated contact which helps maintain contact between the surfaces, even if the force applied by the spring is reduced [Mroczkowski 1995].

In designing contacts, higher yield strengths are preferred because they allow a greater amount of stress to be transferred to the beam before the onset of plastic deformation. This allows the design of smaller, more densely packed contacts. Higher tensile strengths are also preferred, because they allow this higher yield strength to be achieved without increasing the yield/tensile strength ratio. This ratio, which is related to the toughness of the material, is kept low to avoid permanent set. Thus, decreases in the yield strength and tensile strength at higher temperatures, as shown in Figure 6.18, can cause the spring to be plastically deformed or to take a permanent set. However, as neither beryllium copper, phosphor bronze, or brass lose an appreciable fraction of their strength until temperatures above 350°C, this is not a concern for operation at elevated temperatures up to 200°C [Loewenthal 1984].

Common spring alloys are subject to stress relaxation, however, at temperatures up to 200°C. Stress relaxation refers to the decrease in the force applied by a spring over time at a fixed elevated temperature and fixed beam deflection. It occurs because of viscoelastic creep of the spring alloy, and is greater in alloys with lower moduli and lower melting points. Figure 6.19 shows the amount of stress relaxation that occurs after 1000 hours at different temperatures for springs initially stressed at 50 to 75% of the yield strength. These data were used to generate a table of maximum use temperatures for the base metal alloys, based on the criterion of 75% as the maximum allowable stress relaxation [Harkness 1990]. Table 6.9 indicates that beryllium copper and beryllium nickel connector systems can be used at temperatures of 200°C. However, some of the lower cost spring materials, such as brasses and phosphor bronzes, should not be considered for use at these temperatures. As mentioned above, however, the effect of these losses in spring force on contact resistance and signal integrity are moderated by the sticking phenomenon.

Base metal alloys must also possess high electrical and thermal conductivity. Electrical conductivity is necessary for fast and accurate signal transmission, while both electrical and thermal conductivity are needed to limit the temperature rise of the contact. The temperature dependence of the electrical and thermal conductivities of common spring alloys are given in Figures 6.20 and 6.21.

Temperature rise is a function of the heat generated by the resistance to current flow in the contact. The conduction of that heat away from the contact is given by

$$\Delta T = \frac{J^2 L^2}{2\sigma\kappa A^2} \tag{6.5}$$

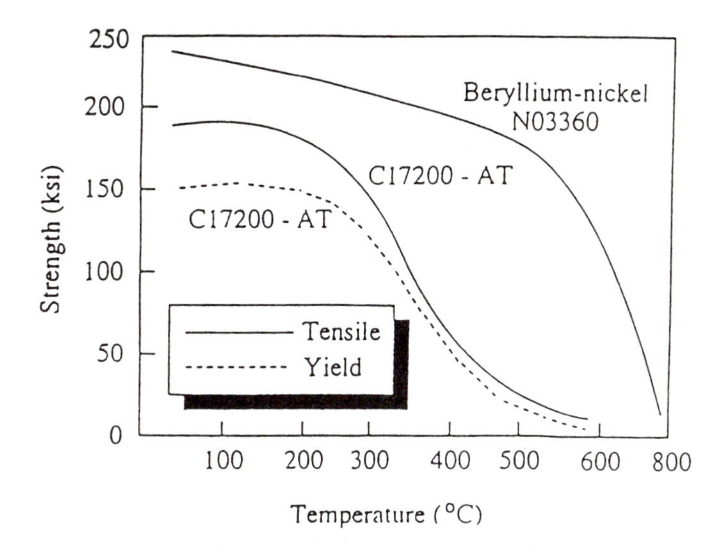

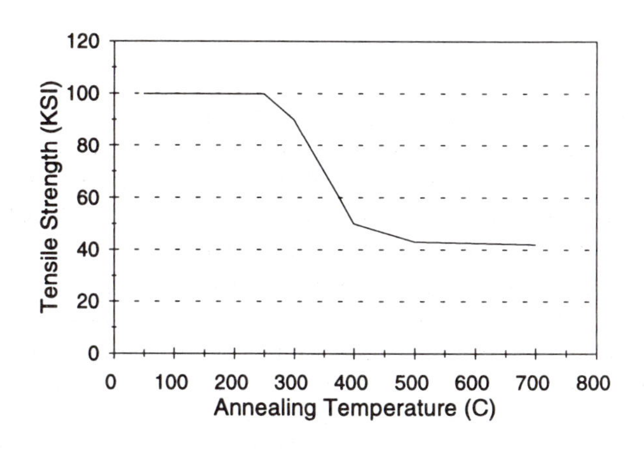

Figure 6.18 Strength of connector spring alloys as a function of temperature a) BeCu and BeNi [Guide 1991] b) Brass [ASM 1990].

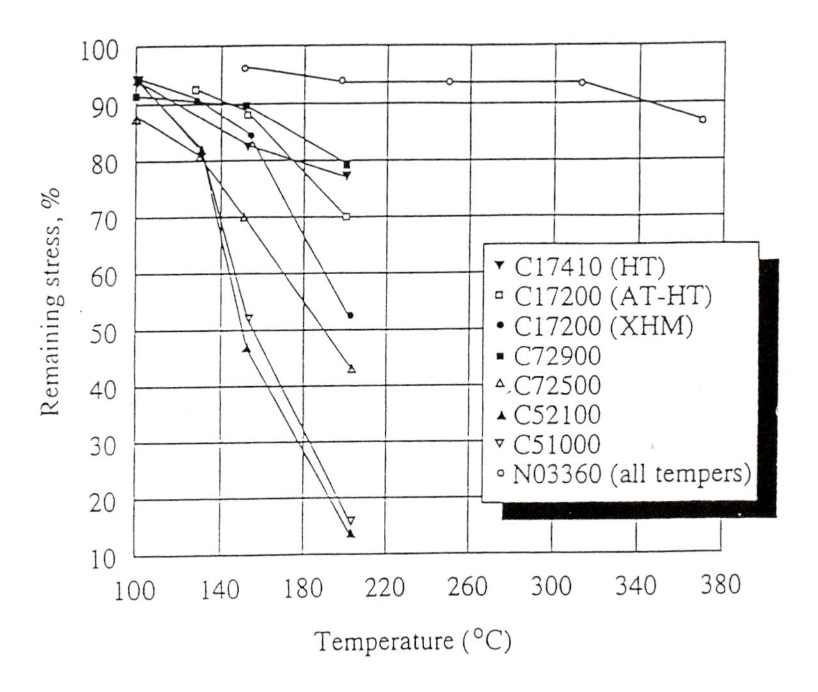

Figure 6.19 Stress relaxation of connector spring alloys as a function of temperature [Harkness et al. 1991]

where ΔT is the temperature rise of the contact, J is the current, L is the beam length, A is the cross-sectional area of the beam, σ is the electrical conductivity, and κ is the thermal conductivity. The temperature rise should be minimized since the effects of temperature on the mechanical and electrical properties of the contact are governed by the contact temperature, which is the sum of the ambient temperature and the temperature rise.

In summary, a high modulus will allow a higher I/O density without lowering the normal force on each contact. High yield strength and tensile strength allow the production of small, higher density connectors with sufficient strength to avoid plastic deformation and permanent set. Good stress relaxation properties allow contacts to be designed closer to the yield strength limit without introducing concerns about strength and spring force at elevated temperatures. Finally, electrical and thermal conductivity ensure that the connector is able to transmit signals with high fidelity, low heat generation, and efficient heat dissipation.

Table 6.9 Maximum Use Temperatures for Connector Spring Alloys [Harkness 1990, Brush-Wellman 1991]

Material	Maximum Temp	Reason
Beryllium copper-C174HT	220	75% stress relaxation/1000 hrs
Beryllium copper-C172XM	170	75% stress relaxation/1000 hrs
Beryllium copper-C172HT	175	75% stress relaxation/1000 hrs
Copper nickel (Tin) - C725	140	75% stress relaxation/1000 hrs
Spinodal Cu Ni - C729	220	75% stress relaxation/1000 hrs
Phosphor bronze - C521	130	75% stress relaxation/1000 hrs
Phosphor bronze - C510	130	75% stress relaxation/1000 hrs
Beryllium nickel - N03360	> 400	75% stress relaxation/1000 hrs

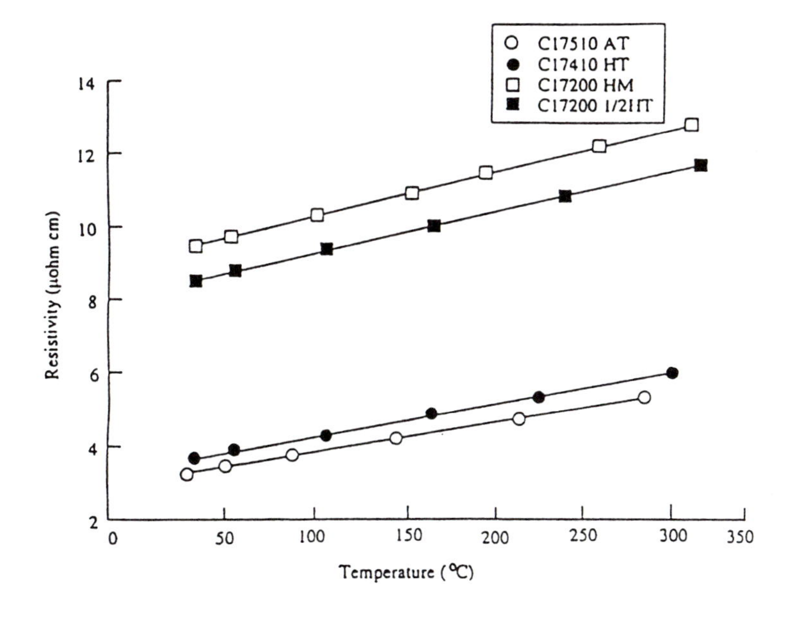

Figure 6.20 Electrical resistivity of connector spring alloys as a function of temperature [Brush-Wellman 1991].

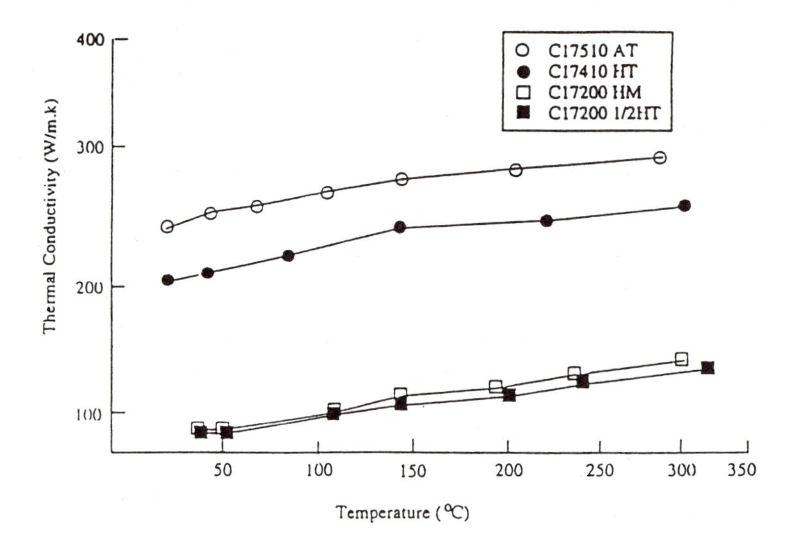

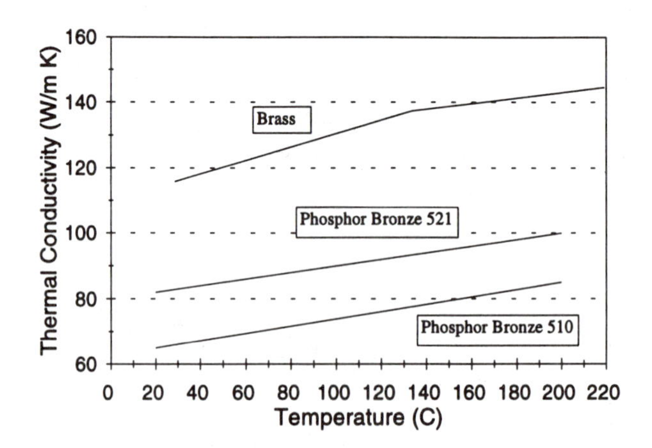

Figure 6.21 Thermal conductivity of connector spring alloys as a function of temperature a) BeCu [Brush-Wellman 1995], b) brass and phosphor bronze [Holman 1986, Touloukian et al. 1970].

The factors that determine the high-temperature reliability of the contact are a function not only of the spring, but also of the surface plating and the housing. Properties that affect the plating are corrosion resistance, hardness, wear resistance, and electrical resistivity [Pecht 1994]. High-temperature concerns include corrosion, intermetallic formation, and diffusion of impurities to the contact surface, all of which are accelerated at elevated temperatures. These effects can cause concerns at temperatures below those at which stress relaxation limits the base metal. Effects due to the degradation of the plastic housing at elevated temperature, such as motion and misregistration of the pins or electrical leakage through the housing, will be addressed in the next section.

For tin and tin-lead plated contacts, the maximum use temperature is set at 100°C because of intermetallic formation between the plating and the underlying copper. These intermetallics are resistive, increasing the contact resistance to an unacceptable level. In addition, they are brittle and can flake off, leaving sites for corrosion and possibly obstructing the mating of contact surfaces. Using a nickel underplate elevates the maximum temperature to 110°C by eliminating the formation of copper-tin intermetallics, but replaces them with nickel-tin intermetallics that grow at one-third the rate.

Gold-plated contacts are not recommended for use above 125°C. Figure 6.22 indicates that the limitation is due to the diffusion of copper through the gold to the surface and its subsequent oxidation, which increases the contact resistance [Antler 1970]. This problem can be alleviated by using a nickel underplating (100 μin) and a thicker overplate of gold (50 μin vs. 20 μin). However, the limitation is also due to the diffusion and segregation to the contact surfaces of small quantities of cobalt and nickel added to the plating as hardening agents [Mroczkowski 1995]. Use of palladium is an alternative, but palladium alloys are limited by the formation of frictional polymers. Temperatures of 200°C should be achievable by using a thick, soft gold plating (100 μin) over a 100 μin nickel underplate, although this increases cost and decreases wear resistance.

6.5 Connector Housings and Box Materials

While there are few temperature-related concerns over the use of metal or ceramic box materials at temperatures up to 200°C, care must be exercised when choosing a molded plastic for use as a connector housing or box material at those temperatures. Several available polymer compounds available are able to withstand continuous use at temperatures in excess of 200°C, as shown in Table 6.10. The continuous use temperature (thermal index) of liquid crystal polymers is 220°C; that of polyamide-imide is 220°C; and that of polyetherimide is 170°C. The thermal index is the temperature at which the mechanical and/or dielectric strength of the material drops to 50% of its initial value in 100,000 hours, or 11 years, of continuous use. In addition, polyethersulfone, polyphenylene sulfide, polyetheretherketone, polyester (PET, PBT), polyamide (nylon), and polyimide have heat deflection temperatures @ 264 psi greater than 200°C, as shown in Figure 6.23. Of these, the liquid crystal polymers (LCP) have the highest strength at 35,000 psi, the highest modulus at > 2,000 ksi, and the best flow characteristics [Modern Plastic Encyclopedia 1994].

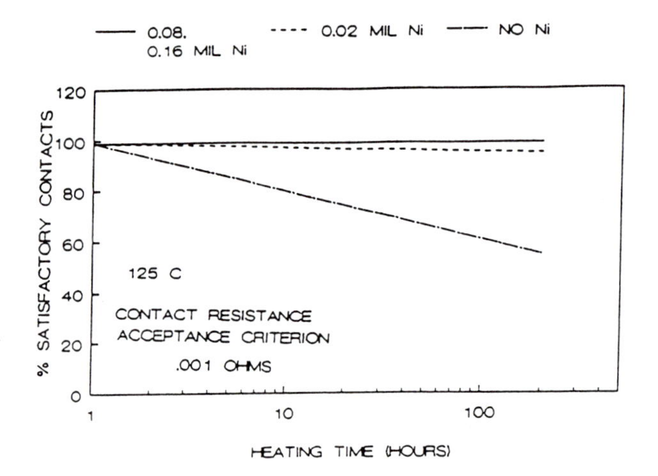

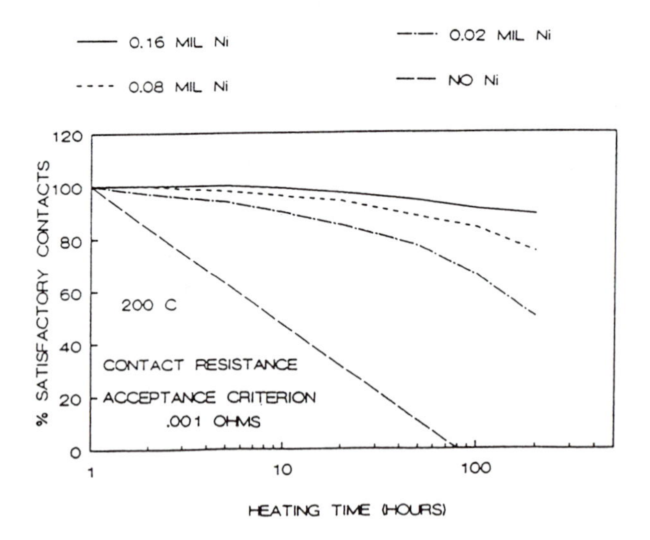

Figure 6.22 Effect of temperature on the resistance of plated contacts [Antler 1970].

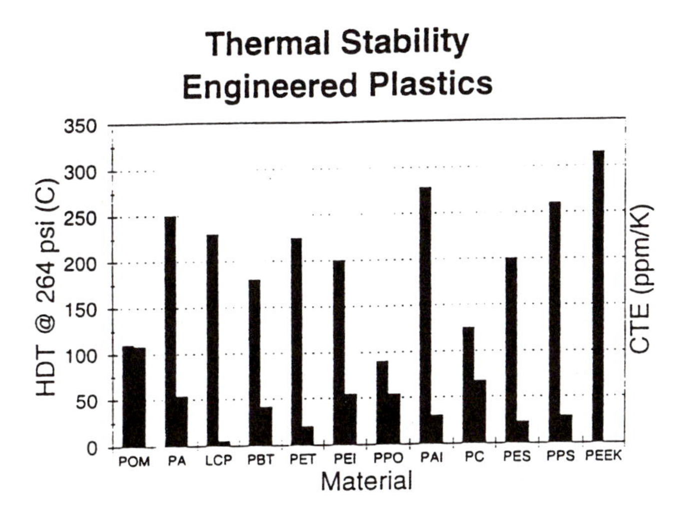

Figure 6.23 Thermal stability of engineered plastic housings as illustrated by the heat deflection temperature and the coefficient of thermal expansion [Modern Plastic Encyclopedia 1994, Amoco 1991, GE 1990, Hoechst 1990, 1991].

Table 6.10 Room Temperature Properties and Maximum Use Temperatures of Engineered Plastic Housings [Amoco 1991; GE 1990; Hoechst 1990,1991; Mroczkowski 1994; Modern Plastics Encyclopedia 1994]

Material (Properties for 25-40% Reinforced)	Continuous Use Temp (Thermal Index) (°C)	Heat Deflection Temp @ 264 psi (°C)	Tensile Strength (ksi)	Coeff. of Thermal Expansion (ppm/K)
Acetal (POM)		110	10	108
Polyamide	130	250	19	54
LCP (Polyester)	220 (mech) 240 (elec)	230	35	5
PBT	130 (elec) 140 (mech)	180	17	41
PET	150	225	22	20
PEI	170	200	24	55
PAI	220	278	28	30
PC		125	17	68
PES		200	20	23
PPS	210	260	18	29
PEEK		315	23	------

At elevated temperatures, particularly above the glass transition temperature, the strength and modulus of the engineered plastics decreases. Figures 6.24 to 6.26 illustrate the temperature dependencies of the strength and modulus for several of the higher use temperature materials. The electrical resistance of the materials also decreases with temperature, as pictured in Figure 6.27, and the dielectric properties change, as pictured in Figure 6.28. All of these changes can adversely affect the ability of the housing to provide the needed mechanical support and insulation. Additionally, the linear expansion of the housing with temperature can cause misalignment of the pins and sockets at the connector level, and stresses between the case and the components at the box level. The coefficient of thermal expansion of the plastics is generally greater than that for the metal or ceramic housings, as shown in Figure 6.23 and it increases with temperature, as shown by the curvature of the graphs in Figure 6.29.

The housing materials can be joined with a series of adhesives, including epoxies, urethanes, silicones, and acrylics. The maximum use temperatures of these adhesives and the materials they join are given in Table 6.11. Silicone rubbers, which are stable at temperatures up to 260°C, and fluoroelastomers can be used in those areas of the housing requiring an elastomer for strain relief.

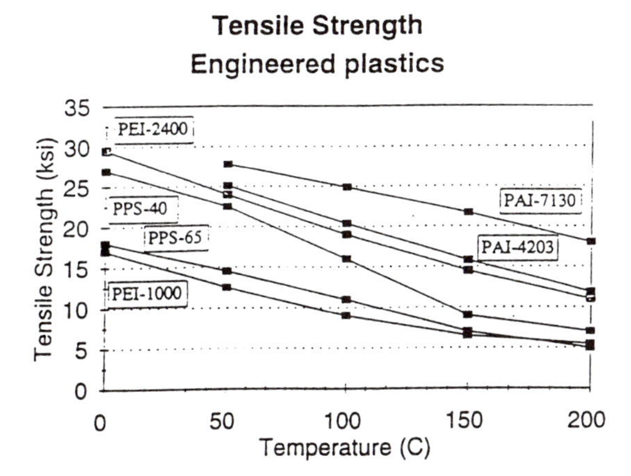

Figure 6.24 Tensile strength of engineered plastic housings as a function of temperature [GE 1990, Amoco 1991, Hoechst 1991].

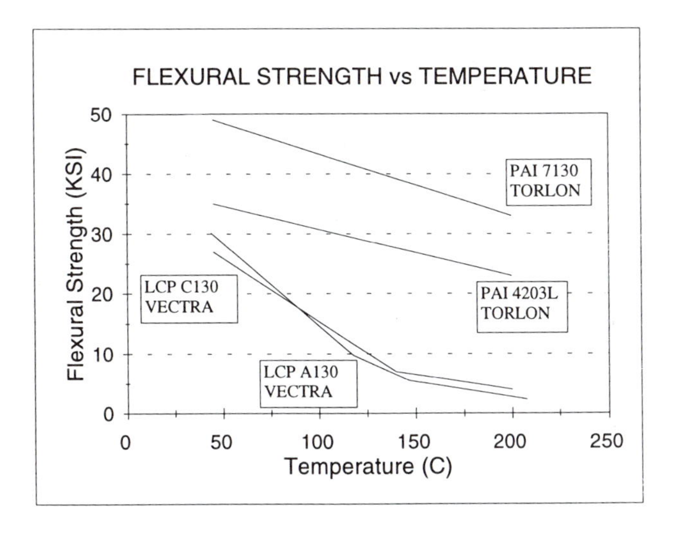

Figure 6.25 Flexural strength of engineered plastic housings as a function of temperature [Amoco 1991, Hoechst 1990, 1991].

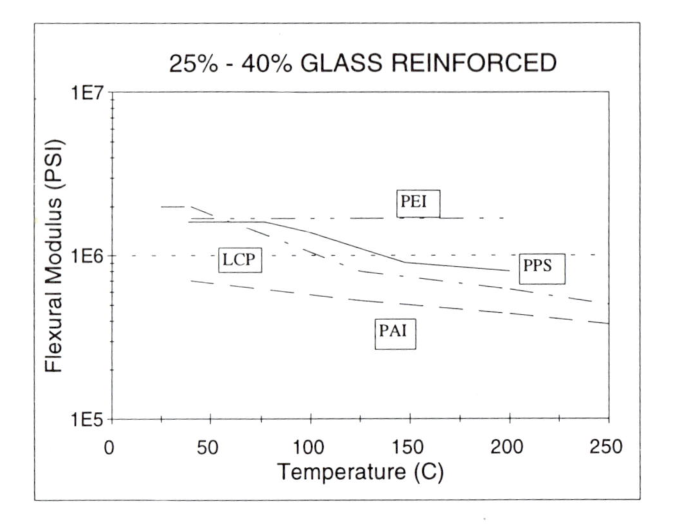

Figure 6.26 Flexural strength of engineered plastic housings as a function of temperature [Amoco 1991, Hoechst 1990, 1991].

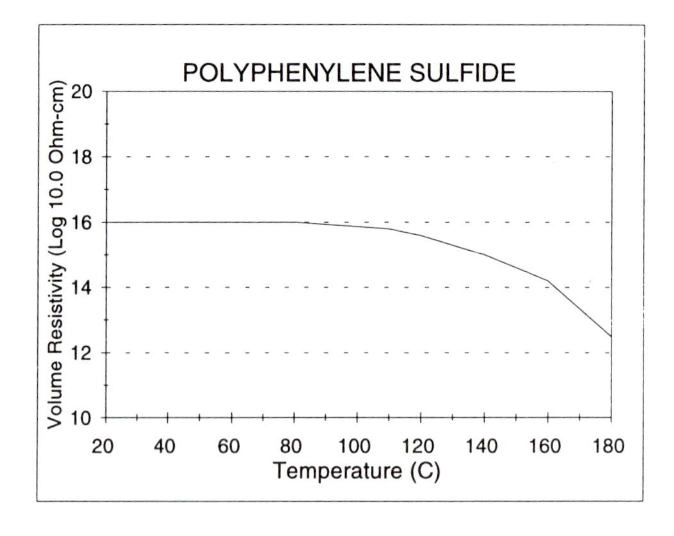

Figure 6.27 Volume resistivity of engineered plastic housings as a function of temperature [Hoechst 1990, 1991].

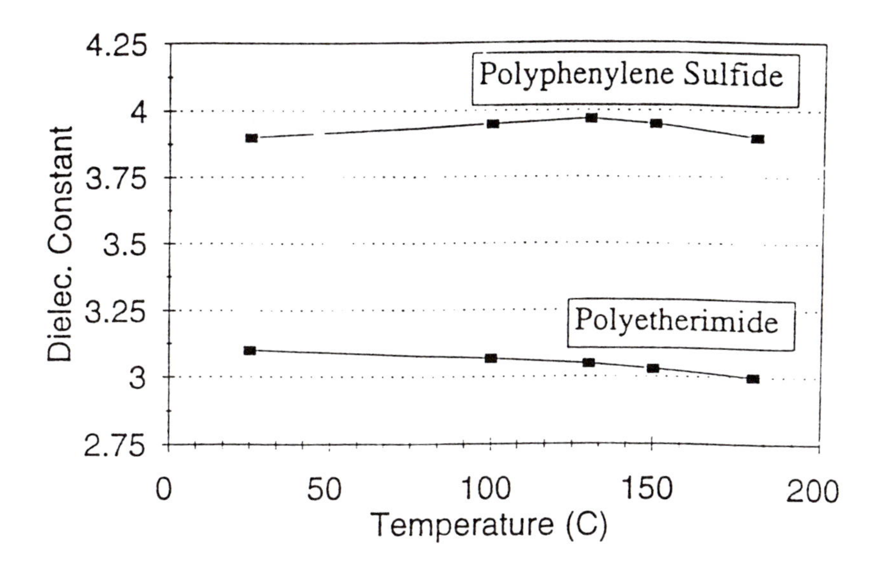

Figure 6.28 Dielectric constant of engineered plastic housings as a function of temperature [GE 1990, Hoechst 1991].

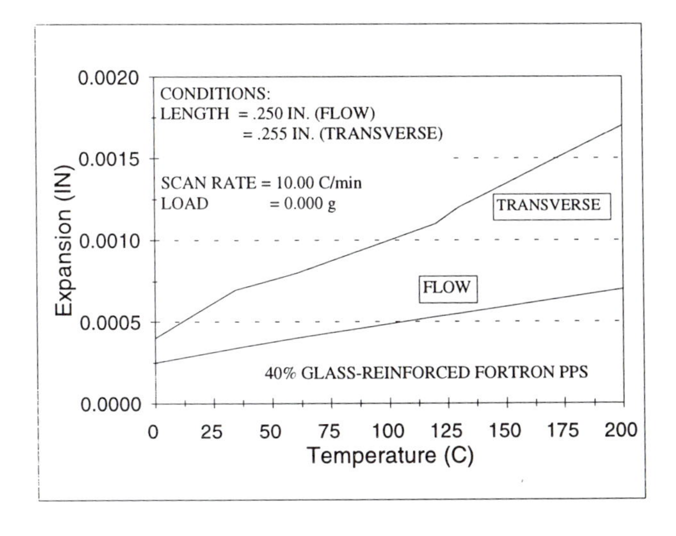

Figure 6.29 Linear thermal expansion of polyphenylene sulfide as a function of temperature [Hoechst 1991].

Table 6.11 Maximum Use Temperatures of Adhesives [Amoco 1991, GE 1990]

Material	Material Joined	Max. Use Temp. (°C)
Scotch-Weld Epoxy	Polyetherimide	175
Furane Epibond Epoxy	Polyetherimide	95
Emerson & Cuming Ecobond Epoxy	Polyetherimide	230
Urethanes	Polyetherimide	70-115
GE RTV Silicone	Polyetherimide	205
Lord Versilok Acrylic	Polyetherimide	120
Dexter Hysol Epoxy	Polyamide-imide	70
Lord Cyanoacrylate	Polyamide-imide	100
Amoco Amide-Imide	Polyamide-imide	260

Chapter 7

THERMAL MANAGEMENT FOR HIGH TEMPERATURE ELECTRONICS

A well-designed electronic packaging scheme must ensure acceptable operating and non-operating temperatures for the various levels of packaging. These levels most commonly are the chip, package, board, and system, as illustrated in Figure 7.1 [Nakayama 1988]. The large size differential involved is immediately apparent by comparing the typical die size range of 2 to 5 mm on each side, with the typical system size of tens of centimeters on each side. The surface heat flux (heat transfer rate per unit area) at the chip level for a 5 mm by 5 mm chip dissipating 10 W is 4 x 10^{5} W/m^2, which is only one order of magnitude below that on the surface of the sun [Antonetti et al. 1989]. In fact, the power dissipation within the chip is localized to even smaller regions, resulting in extremely large heat generation rates per unit volume. Clearly, implementing appropriate thermal management schemes at each packaging level is a challenging task, even for conventional electronics designed for near-room-temperature ambient environments.

Conventional electronics are often designed to operate within a maximum chip (die) temperature for a specified ambient operating environment. In practice, limits on the device and ambient temperatures determine the available thermal options. Current limits and future projections for various electronic products are presented in Table 7.1 [Semiconductor Industries Association 1994]. While both the operating and ambient temperatures for most electronic equipment categories in Table 7.1 are expected to remain unchanged in the foreseeable future, automotive electronics projections are considerably different. They show higher levels and increases with time for both junction and environment temperatures, compared to conventional products. From a thermal design perspective, higher allowed temperature operation may permit the use of simpler and lower cost thermal management schemes. In some cases, passive cooling may be adequate, allowing the elimination of fans and other forced convection hardware.

Determining the passive cooling options requires a detailed thermal characterization at the various packaging levels during the design phase. For a given set of packaging materials and component layout, it is crucial to assess whether the temperatures at various locations are being maintained at design levels by the selected thermal management strategy.

Many near-term, high-volume elevated temperature electronics applications, including those for automotive components, will probably use silicon technology [Dreike et al. 1994]. Design modifications to allow high-temperature use will be affected largely through changes in packaging materials and component layout. At higher junction temperatures (above about 250 °C), where conventional silicon technology is no longer a feasible option, considerable

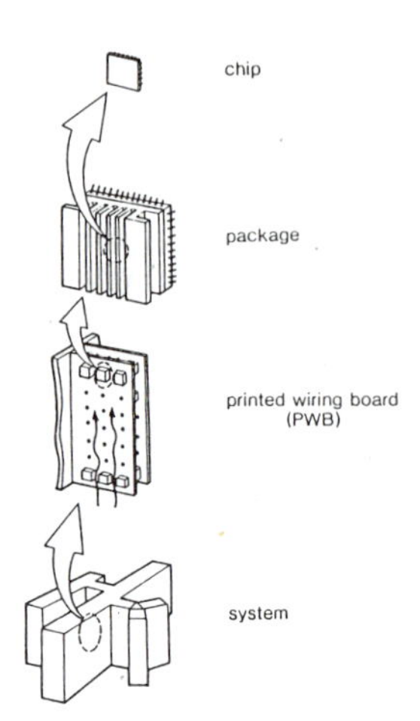

Figure 7.1 The packaging hierarchy (from Nakayama 1988). Thermal management issues must be addressed at each level.

Table 7.1. Semiconductor Industry Association Roadmap, 1994 Projections from 1995 to 2010

Power (single chip packages; W)	
Commodity products (<$300)	1
Handheld products (<$1,000)	2
Cost/performance products (<$3,000)	16-55
High-performance products (>$3,000)	80-180
Automotive	3
Junction temperature maximum (°C)	
Commodity products (<$300)	125
Handheld products (<$1,000)	115
Cost/performance products (<$3,000)	100
High-performance products (>$3,000)	100
Automotive	165-200
Ambient maximum temperature (°C)	
Commodity products (<$300)	55
Handheld products (<$1,000)	55
Cost/performance products (<$3,000)	45
High-performance products (>$3,000)	45
Automotive	140-175
Chip size (mm^2)	
Commodity products (<$300)	55-90
Handheld products (<$1,000)	240-530
Cost/performance products (<$3,000)	250-620
High-performance products (>$3,000)	450-1400
Automotive	55-90

changes must be made in the device material and/or technology, as well as in the packaging.

This chapter deals with thermal management and thermal characterization, with emphasis on high-temperature electronics applications. The former refers to defining the cooling schemes available for a given application, while the latter involves determining operating temperatures, either through measurements or, increasingly, through experimentally validated computational modeling. Successful thermal management requires an acceptable "micro-climate" at all levels of electronic packaging in order to meet targeted performance and life specifications.

The thermal challenges associated with high-temperature electronics arise from a combination of elevated environment temperature and device power dissipation. Depending upon conditions, the appropriate thermal management techniques may be passive, active, or a combination.

7.1 Thermal Management Considerations for Elevated Temperature Operation

Identifying the appropriate thermal management strategy for a high-temperature application requires knowing the operating and storage environment temperature ranges, the power dissipation requirements, and the maximum allowable junction temperature. Two possible sets of design conditions can be identified for steady state operation. If the environment temperature is above the maximum operating junction temperature, $T_{env} > T_{J,\,max}$, an ultimate heat sink with temperature below the maximum junction temperature must be provided for heat removal from the electronic system [Mahefkey 1994] (see Figure 7.2). Heat removal can be achieved through passive, active, or combined passive/active techniques. Such a scheme will require expending external power to eventually reject heat to the environment from the heat sink, since $T_{env}>T_{sink}$. The second possibility is that $T_{J,max} > T_{env}$. This case is similar to conventional electronics cooling applications, and allows direct heat rejection to the environment by a variety of heat dissipation techniques.

The first law of thermodynamics can be applied on a rate basis to the closed system contained within the boundaries of the electronics box in Figure 7.2.

$$\dot{E}_{out} = \dot{E}_{in} + \dot{Q}_{gen} \tag{7.1}$$

where $\dot{E}_{out}$ is the rate of heat rejected by the system, $\dot{E}_{in}$ is the rate of heat transferred into the system from the environment, and $\dot{E}_{gen}$ is the net heat dissipation rate of the various components. When $T_{env} > T_J$, all the terms in Equation 7.1 are present. When $T_{env} > T_J$, is absent, since the environment temperature is below the maximum junction temperature.

This brings out the unique issues in the thermal management of high-temperature electronics. The case of a higher environment temperature than the junction temperature needs special attention. For both this case and the reverse, the selection of coolants for convective cooling also deserves special consideration. In order to remove the high heat fluxes in modern packages, liquid cooling is often used in conventional electronics thermal management [Joshi and Kelleher 1992]. Liquids may be used for direct immersion cooling or can be circulated through cold plates for indirect heat removal. Conventional liquid coolants include de-ionized water, fluorocarbon liquids, and silicone oils. The high-temperature stability of these liquids and their thermophysical properties at elevated temperatures must be examined prior to their use.

The importance of radiation heat transfer in high temperature applications must also be

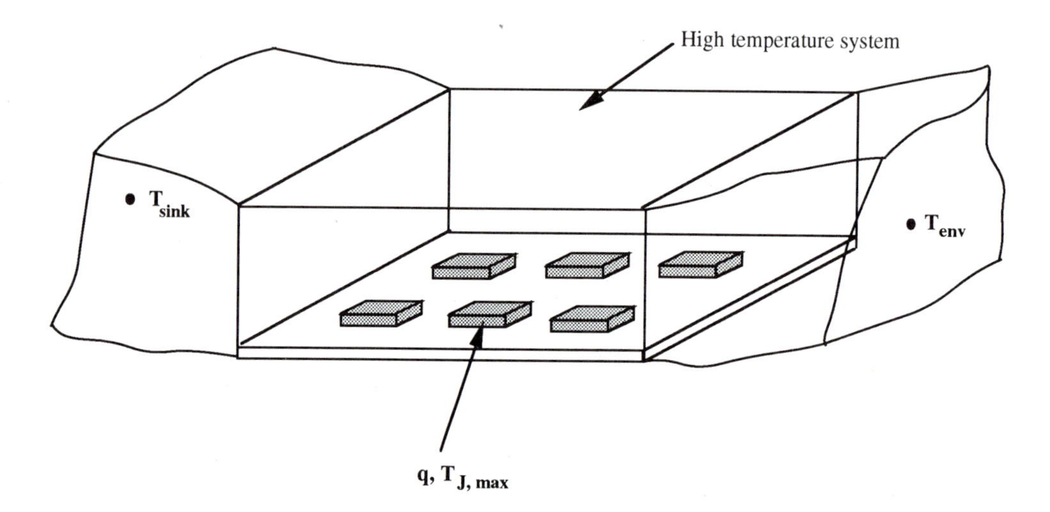

Figure 7.2 The general thermal management problem for elevated temperatures electronics. The three temperatures involved are shown. The system may be exposed to the high-temperature environment through one or more of its boundaries. For steady-state operation, if this temperature is higher than the maximum allowable junction temperature, heat rejection to a heat sink at a temperature below the maximum allowable junction temperature, heat rejection to a heat sink at a temperature below the maximum allowable junction temperature must take place across some region of the system. Remaining boundaries of the system may be insulated.

considered. For an isolated gray diffuse surface at an absolute temperature of T, losing heat to an environment at temperature T_{env}, the heat loss rate per unit surface area by radiation is $q'' = \sigma\varepsilon(T^4 - T_{env}^4)$, where σ is the Stefan-Boltzmann constant and ε is the surface emissivity. At T=800 K, T_{env}=700 K, and for ε=0.5, q"= 4805 W/m^2. To assess the importance of radiation, this value must be compared with the net surface heat flux, which includes the internal heating and convection effects. Radiation will be an important mode if the radiative surface flux component is a significant fraction of the net surface heat flux.

Candidate thermal management schemes for high-temperature applications can be selected once the heat flux and the maximum temperature difference available for heat transfer have been identified. As Figure 7.1 indicates, a complete thermal management strategy involves selecting an appropriate heat-removal scheme for each level of packaging. For low heat fluxes, passive thermal management techniques can be used that do not require expending external energy for the heat removal. Interest in such techniques is currently very strong, due to their design simplicity, low cost, and high reliability. For a maximum temperature rise of about 85°C above the environment (assuming $T_{J, max} < T_{env}$), heat fluxes of up to about 5 W/cm^2 can be removed by passive techniques.

For higher heat fluxes, a combination of passive and active techniques can be used. These may include, for example, using a high thermal conductivity substrate at the board level, along with forced-air cooling at the box level. For this range of heat fluxes, the use of cold plate technology, thermoelectrics, and flow-through cooling are also possible. For even higher heat fluxes, a variety of active cooling techniques are available that require the use of external power for cooling. This could operate a coolant circulator (forced convection), or a thermoelectric cooler.

For the highest heat fluxes, direct-contact liquid cooling techniques may be considered. A number of direct liquid-immersion cooling techniques have recently been examined for removal of very high heat fluxes. Considerable improvement in thermal performance can be achieved through many of these techniques. For example, heat flux handling capabilities on the order of 200 W/cm^2 have been demonstrated with micro channel heat sinks under flow boiling conditions with temperature rises below 100 K [Bowers and Mudawar 1993]. These immersion cooling methods require an intimate contact between the coolant liquid and electronic components. Design complexity, cost, and reliability issues for such techniques must be carefully investigated.

7.2 Passive Techniques for Low (< 5 W/cm^2) Power Dissipation Applications

7.2.1 Natural convection with direct immersion in dielectric liquids

This cooling option offers a heat transfer capability comparable to high-velocity forced-air cooling, and at the same time results in considerable design simplicity and high reliability. Consider the relative cooling capabilities of three fluids- air, water, and Fluorinert FC- 75 (a dielectric liquid manufactured by 3M)- during laminar natural convection along a vertical flat plate of height L, dissipating a uniform surface heat flux of q" in terrestrial gravity field of intensity g. For given values of q" and L, the temperature rise of the surface at its trailing edge above the ambient, $[T_s(L)-T_\infty]$, depends only on the thermophysical properties of the fluid. This dependence involves the volumetric expansion coefficient, β , the thermal conductivity, k , the kinematic viscosity, ν, and the Prandtl number of the fluid (ratio of kinematic viscosity to thermal diffusivity), Pr. Both Fluorinert FC-75 and water result in a much smaller temperature rise than air, as seen in Figure 7.3 [Joshi and Kelleher 1992].

7.2.2 Thermally conductive substrates and heat sinks

Heat removal from electronic components depends heavily on the thermal conductivity of the substrate. For a glass epoxy substrate (FR4) cooled in air, the ratio of substrate to fluid thermal conductivity is about 60. This makes substrate conduction an important heat removal mechanism. Over the past decade, a number of high thermal conductivity organic and metal-matrix composite substrates have been developed. The thermal conductivities of some of these new materials are presented in Figure 7.4. In the organic composites, use is made of fibers, such as graphite, with a very large unidirectional thermal conductivity. With a 50% fiber volume fraction, effective thermal conductivities of over 200 W/mK have been reported. With metal-matrix composites such as graphite/aluminum, overall thermal conductivities of 600 W/mK have been reported. These may possibly be used as thermal backplanes within circuit cards.

Graphite composites impregnated with unidirectional carbon fibers have also been developed. Thermal conductivities of up to 800 W/mK have been reported for such carbon brick composites. Heat sinks fabricated from such materials can be attached directly to the package for efficient thermal spreading.

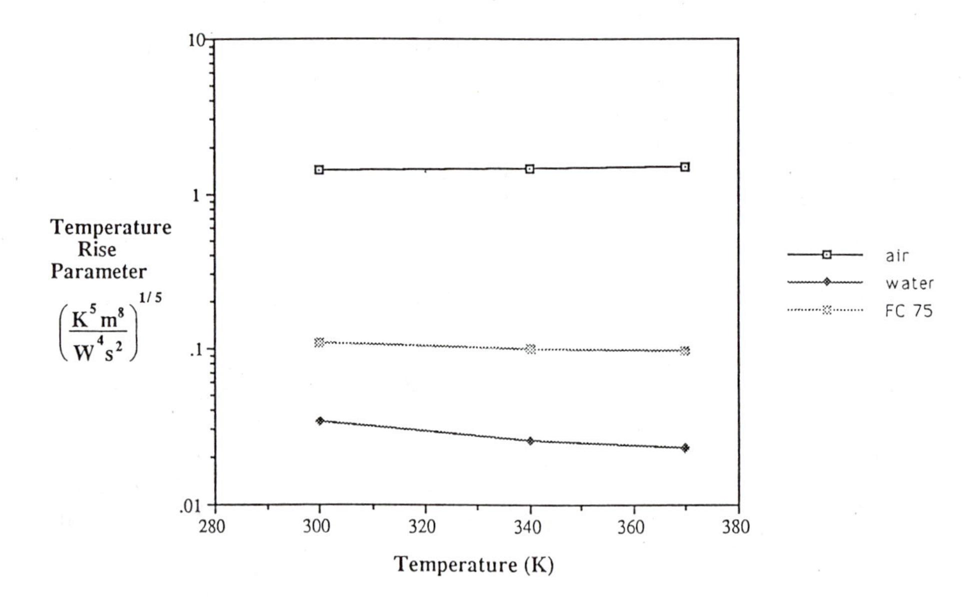

$$\text{Temperature Rise Parameter} = \left(\frac{g}{q''^{4} L}\right)^{1/5} [T_s(L) - T_\infty] = \frac{1}{F(Pr)}\left[\frac{\nu^2}{\beta k^4 Pr}\right]^{1/5}$$

$$\text{where } F(Pr) = \left[\frac{Pr}{4 + 9\sqrt{Pr} + 10\,Pr}\right]^{1/5}$$

Figure 7.3 Comparison of cooling capabilities of liquids compared to air during laminar natural convection over a vertical flat plate.

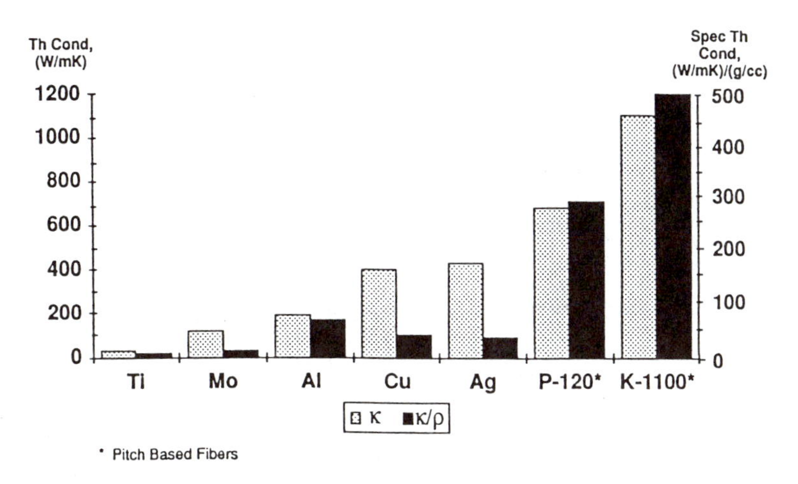

Figure 7.4 Thermal conductivities of several advanced materials.

7.2.3 Diamond-film heat spreaders

Use of thin-film polycrystalline diamond, deposited over ceramic substrates, has been demonstrated to be highly effective in heat removal from electronic packages. Gersch [1994] examined diamond films deposited over both Al_2O_3 and AlN and found an over 65% reduction in the junction-to-case resistance for some cases. The increased use of this technology in electronics packaging depends on a number of factors, including the ability to deposit films with acceptable adhesion on a variety of substrate materials, to handle thermomechanical stresses due to CTE mismatches, and to get high enough deposition rates to make the process economically viable. Potential applications include cooling very high power standard electronic modules (SEM), high-temperature electronics, and passively cooled systems.

7.2.4 Thermosyphon loops

Thermosyphon loops are gravity-assisted devices, in which the coolant liquid is contained in a closed loop (see Figure 7.5) [Hamburgen and Fitch 1992]. The coolant is heated (and sometimes evaporated) by the heat sources (electronic components). The heated, lighter fluid rises due to buoyancy and is cooled at the top. The cooled liquid returns by gravity along the walls, to be heated again by the heat source. The successful use of such systems requires a vertical or near-vertical orientation, which may only be feasible for certain applications.

7.2.5 Dielectric liquid-filled bags

To avoid direct contact between a liquid coolant and electronic components, the use of dielectric liquid-filled bags has been demonstrated by 3M Corporation. These bags, containing a Fluorinert-type liquid (manufactured by 3M), are placed in contact with the electronic components. Natural convection in the liquid in the bag assists in heat removal from the packages. The removed heat is transferred out from the fluorinert bag to a cold plate or rejected directly to the environment.

7.2.6 Miniature heat pipes

Heat pipes have been used extensively for thermal control applications in spacecraft. Their successful use allows an effective thermal conductivity several times larger than that of

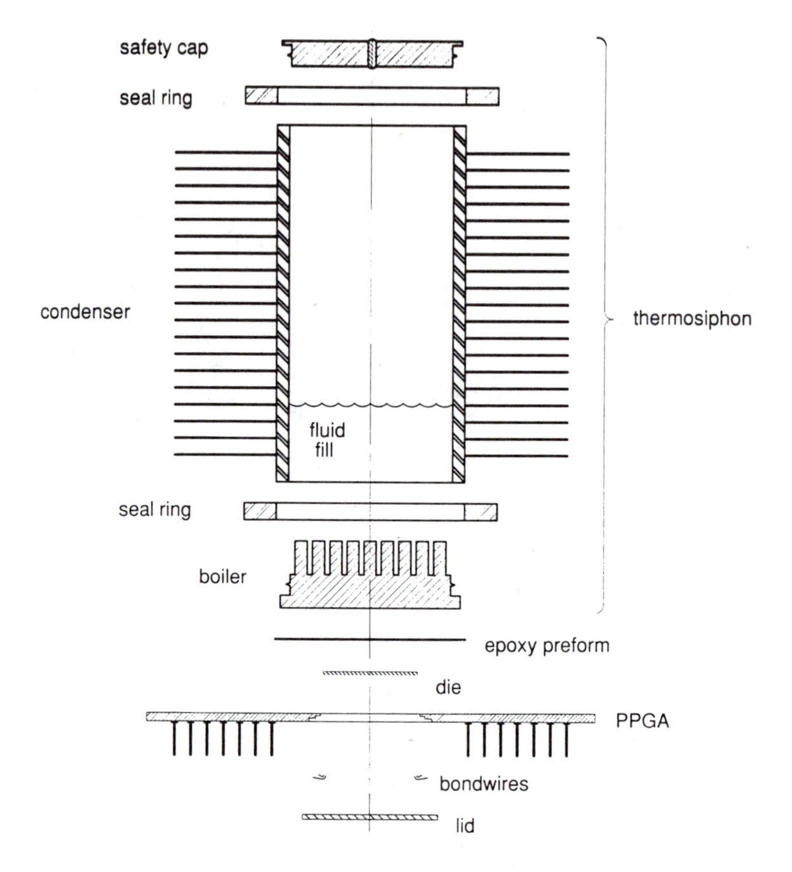

Figure 7.5 Schematic sketch of a thermosiphon used for cooling a 150 W bipolar emitter coupled logic (ECL) chip prepackaged in a plastic pin-grid array (PPGA) package [Hamburgen and Fitch 1992].

metallic materials, such as copper. A heat pipe is a closed device, employing a working fluid (e.g., ammonia or water) and an internal wick structure. The evaporator end of the heat pipe is placed in contact with the heat-generating device (e.g., electronic components). The evaporated working fluid is driven to the condenser end by an established pressure gradient. Upon heat rejection, the condensed working fluid is returned to the evaporator section by capillary action through the wick. These devices differ from thermosyphons only in the mechanism for the return of the condensed working fluid. Since they do not rely on gravity for this purpose, heat pipes can be used in any orientation. An application of a heat pipe to an edge-cooled circuit card is illustrated in Figure 7.6 [Peterson 1994].

7.2.7 PCM cooling

Phase-change materials (PCMs) have been used for a number of years for solar energy storage and for space applications involving pulsed power loads. Such materials have a large latent heat, allowing absorption of large thermal loads during transient periods. The thermal energy can later be released to a heat sink or heat transfer environment over extended periods. A number of different organic and metallic materials have been employed for such applications, depending upon the phase-change temperature range of interest.

Many electronics applications involve high-power dissipation only for relatively short durations. PCMs appear quite promising for such transient high-power dissipation applications

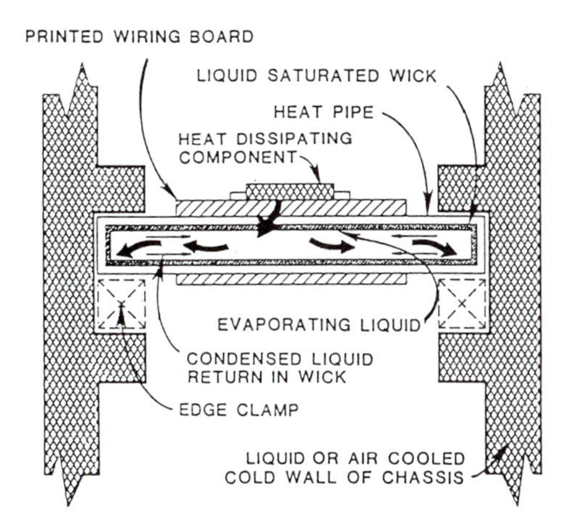

Figure 7.6 Use of a heat pipe for cooling an edge-cooled circuit card [Peterson 1994].

[Pal and Joshi 1995]. The energy dissipated during the active phase could be stored within the PCM by melting. The phase change can be reversed during the power-off phase by natural convection cooling to the environment, or by incorporating a backup cooling technique during the inactive phase.

7.3 Active Cooling Techniques for High-Power (>10 W/cm^2) Applications

A number of active cooling schemes have been demonstrated to provide high heat-flux capabilities. All of these employ direct liquid immersion of electronic components in order to achieve very high heat transfer coefficients. In some techniques, boiling heat transfer is employed for achieving even higher convection coefficients. Typically, the complexity in design is greater than in passive schemes. Bergles and Bar-Cohen [1993] have presented a technique for selecting the liquid coolant for these schemes by evaluating a figure of merit (FOM). This incorporates all the important thermo-physical properties of the liquid influencing the heat transfer coefficient into a single parameter.

7.3.1 Single-phase forced-liquid cooling

Direct forced-liquid cooling may be employed in applications where the electronic components are mounted within a channel or enclosure through which a dielectric liquid is circulated. Studies to determine the thermal characteristics of single-phase forced-convection channel flows are summarized by Incropera and Ramadhyani [1993] for electronics cooling applications. This technique offers an increase of one order of magnitude in the allowable heat fluxes over forced-convection air cooling.

7.3.2 Enhanced pool boiling

Pool boiling results in the continuous formation and detachment of small vapor bubbles at the hot surfaces. The detachment of the bubbles from the surface causes a vigorous stirring of the surrounding liquid and an associated enhancement of heat transfer. Considerable research has been done over the last decade to determine the thermal characteristics of electronic modules under pool boiling conditions, and a large body of literature on this topic is available, as summarized recently by Bar-Cohen [1993]. Dielectric liquids used to cool electronics typically are highly wetting. This requires large superheats (increase in heater surface temperature above the boiling temperature of the liquid at prevailing pressure) to initiate boiling, especially for smooth surfaces. This problem has been handled by using enhanced boiling surfaces with engineered roughness features, which provide predictable boiling response and smaller superheat.

7.3.3 Flow boiling

In flow boiling, the coolant is typically circulated over the heated components in a closed loop. The vapor formed by evaporating liquid must be condensed, prior to its return to the component area. The location and size of the condenser is a significant issue in the design of these systems. The coolant can enter the module region either in a saturated state (at the boiling temperature) or in a subcooled state. Thermal performance correlations for various cases are available in the literature.

7.3.4 Jet impingement

Jet impingement has been employed in the cooling of electronic components because of the extremely high local heat transfer coefficients attainable. Both single-phase (Incropera and Ramadhyani [1993]) and phase-change liquid jet impingement cooling schemes have been investigated. While superior thermal performance has been demonstrated in such schemes, their incorporation in a high-reliability, low-cost, compact design needs careful evaluation.

7.3.5 Micro-channels

The use of micro-machined channels on the back of silicon chips has been demonstrated for dissipating extremely large heat fluxes (>100 W/cm^2). Increased convective heat transfer results due to the very large fluid exposed surface area. Many of the concerns regarding compactness, reliability, and costs are also applicable to this scheme.

The high and low heat flux ranges have been discussed above. For the intermediate range of heat fluxes (5-10 W/cm^2) combinations of passive/passive and passive/active techniques can be used. An example of that first combination is a diamond-film heat spreader at the package level, along with a finned natural convection heat-removal scheme at the box level. An example of the second category is the use of a PCM at the module level and a forced-air cooling scheme at the box level.

Another example in the latter category is cold plate technology. Use of cold plates has also been made in a variety of avionics and other high-power electronics applications. In avionics, SEMs (standard electronic modules) have been demonstrated to have heat removal capabilities in excess of several hundred watts for a module approximately 25 cm by 25 cm. In these modules, the electronic components are mounted on internally finned hollow cards, through which liquid coolant (or air) is circulated, as seen in Figure 7.7. Heat is transferred from the electronic components by conduction across the module to the flowing fluid. Significant recent advances in this technology have used cards made from composite materials, with large thermal conductivity along the card plane for better heat spreading.

7.4 Computational Simulations for Thermal Design of High-Temperature Electronics

In the thermal design of electronic modules for elevated temperature operation, the possible use of passive cooling must be examined first. Only when this is inadequate should some of the more complex techniques described above be used. In order to rapidly assess various cooling designs, computational modeling can be used. Two classes of approach are possible for such analyses. Traditionally, thermal modeling has been performed at the printed wiring board (PWB) level, using conduction-based solvers. The effects of convection are included in these analyses in the form of boundary conditions on the component and board surfaces. Due to the large uncertainties in prescribing the convection coefficients, such analyses often can be significantly in error.

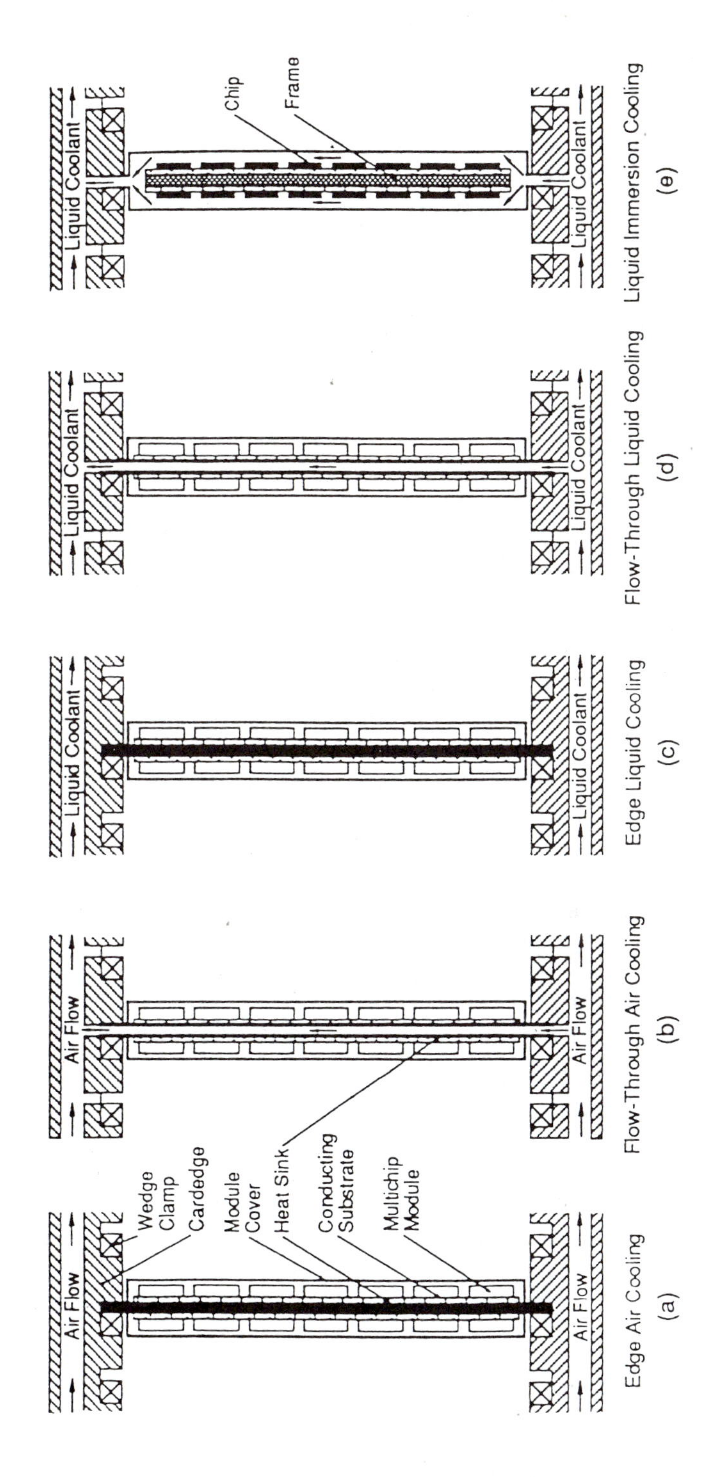

Figure 7.7 Cooling options for SEM-E module [Bowers and Mudawar 1993].

Thermal simulations of entire electronic systems can be performed by using the computational fluid dynamics (CFD) technique. In this approach, the balance statements for mass, momentum, and energy are considered throughout the entire system. The differential equations describing these, along with the boundary and initial conditions, are discretized into a set of coupled algebraic equations. These are invariably non-linear and are solved by iteration. The major advantage of CFD-based techniques over conduction-type PWB-level simulations is the ability to solve for temperatures throughout an entire system, without having to prescribe heat transfer coefficients on the PWB surfaces. Air flow patterns and air temperatures within enclosures are also calculated, which can provide useful information during the design phase. A variety of commercial codes for carrying out system-level thermal analyses are currently available. The limitations of the CFD-based technique include long computational time and high memory requirements. Also, while solutions for laminar flow can be obtained easily, transitional or turbulent flow simulations must still be actively researched.

In designing electronics for elevated temperature use, computational modeling can allow the rapid evaluation of several options. Two examples follow to illustrate this idea. The first is an automotive power train controller module, the second is a module used in avionics applications as a lamp dimmer.

7.4.1 The automotive power train controller card [Romanczuk et al. 1993]

Electronic modules for controlling engine and power train functions are examples of products finding increasing usage in adverse operating environments. The trend is towards moving these modules closer to the engine and decreasing their size, while increasing functionality and reducing cost at the same time. Modern vehicle aerodynamic styling trends, such as the cab-forward design, have resulted in a more crowded engine compartment, reduced space for air flow, and increased ambient temperatures.

The high-temperature operating environment is created in these applications by a combination of increased air temperatures in the engine area and higher device power dissipations. Based on the techniques discussed in Section 7.3, one proposed solution is to change the substrate material. This has been implemented in a proposed design for a future Chrysler Motors Corporation Jeep Truck Engine Controller (JTEC) by Romanczuk et al. [1993]. The proposed design, illustrated in Figure 7.8, uses a high thermal conductivity substrate with an aluminum base. This is covered with a thermally conductive dielectric layer and then 1 oz. copper circuitry. The electronic components are attached to the substrate by reflow soldering.

The design goal- to keep the maximum junction temperature below 150 ^{o}C- was achieved, as verified by both a board-level thermal analysis and experiments. In the analysis, nine components with the highest power dissipation were considered. Both natural convection and forced convection air-cooled operating conditions were simulated by prescribing heat transfer coefficients on the board surface. Measurements were taken of the junction and substrate temperatures for a single transistor, cooled by natural convection.

7.4.2 A lamp dimmer for avionics applications [Hwang et al. 1995]

This assembly is a compact enclosure containing the circuitry to dim cabin lights in aircraft. The high-temperature operation in this application is caused primarily by the high power dissipation in some of the components, not elevated ambient air temperature. A schematic of the assembly is provided in Figure 7.9. It consists of a control board and rectifier board with electronic components, a header assembly, and a transformer in an enclosure. Heat from the components is primarily conducted to and from the case, and finally transferred out by convection and radiation.

In the baseline case, the base plate is made of stainless steel, with a relatively low thermal conductivity of 16 W/mK. Heat transfer is also restricted due to the small base-plate thickness.

Figure 7.8 Proposed design for a Chrysler Motors Corporation Jeep truck engine controller (JTEC) card [Romanczuk et al. 1993].

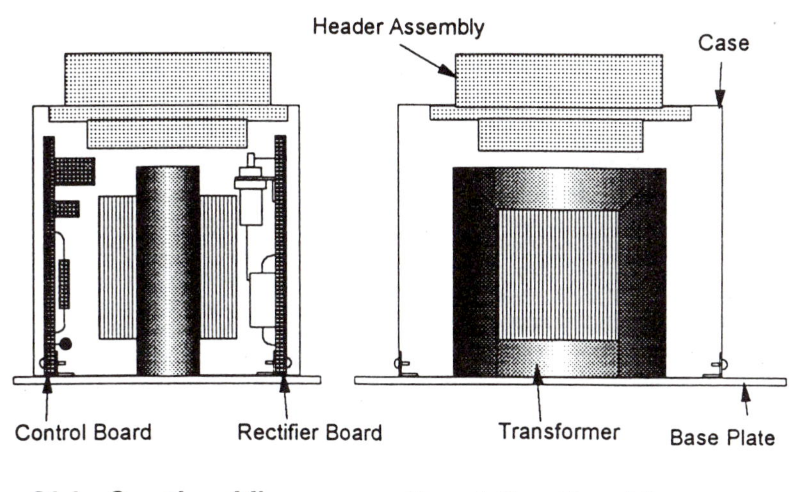

Figure 7.9 Schematic of lamp dimmer assembly [Hwang et al. 1995].

The space between the transformer and rectifier board is very small, resulting in a strong heat transfer interaction between the two. Components on the bottom side of the header assembly can dissipate heat only by conduction through the relatively low thermal conductivity encapsulant. The rectifier and control boards contact the base plate only through a few screw holders, resulting in a large thermal contact resistance between these and the base plate.

Many modifications can be made to the baseline design to make it suitable for elevated temperature operation. First, a careful study needs to be made of the temperature limitations imposed by various materials used in the packaging as illustrated in earlier chapters. Alternate materials may be picked during the preliminary design phase to allow the system to function properly at the maximum junction temperature. Also, a number of changes in assembly configuration can be made for improved thermal performance. Two simple changes are illustrated in Figures 7.10-7.11. The control board is placed upside down in Figure 7.10, in order to allow a shorter conduction path from the hot components to the base plate. This also improves the natural convection heat transfer, since the highest convection coefficients arise near the bottom of the board, where the highest power dissipating components are located. In Figure 7.11, heat conducting pads are used on the back of the two boards to allow better thermal spreading. This effect is also achieved by using a higher thermal conductivity substrate, such as a ceramic, instead of FR4.

To quantify the improvement in thermal performance, these solutions must be evaluated through thermal characterization. In the design phase this may be done by extensive computational simulations supported by selected experiments. For this example, a system-level computational fluid dynamics model is created. For the simulations in this problem, the input data are collected in Table 7.2. To reduce the computational effort, thermal contact resistances at various interfaces are neglected. Similar parts, including resistors, inductors, or capacitors, are sometimes combined as one larger part.

The rectifier board is simplified to have seven components: one equivalent inductor (combination of L1 and L2), two thyristors (CR1, CR10), and two equivalent capacitors (C1, C2). The control board is also simplified to have seven components: two resistors (R2, R3), two transistors (Q1, Q2), two thyristors (Q3, Q4), and one large equivalent resistor. The header assembly is simplified to be a plate with the properties of polyester encapsulant. The base plate is assumed to be insulated at the bottom. Forced convection boundary conditions are applied on the various external enclosure walls, with a heat transfer coefficient of 30 W/m^2K. The ambient temperature is fixed at 40 °C. Calculations are carried using a 25x23x24 grid. The model configuration is illustrated in Figure 7.12.

Results of the baseline case for a transformer dissipation of 5 W are illustrated as temperature fields in a vertical plane through the control board in Figure 7.13 and a horizontal plane across the middle of the transformer in Figure 7.14. The maximum temperature is approximately 212 °C, found on resistors on the control board. Comparisons were made between model predictions and measurements at selected locations; the two agreed within about 10%.

Improvement in the thermal performance was studied for these two cases. The upside-down control board configuration provided a large drop in the temperatures of hot components near the top of the control board. In the second modified design, the substrate thermal conductivity was increased 2.5 times and provided component temperature reductions of approximately 20%.

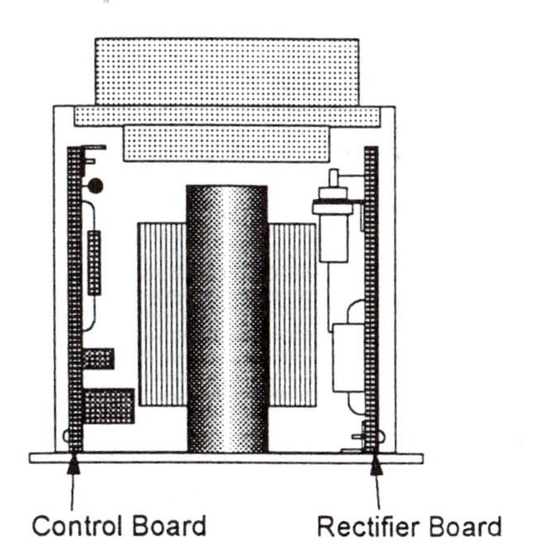

Figure 7.10 Proposed design modification, with an upside-down orientation of the control board [Hwang et al. 1995].

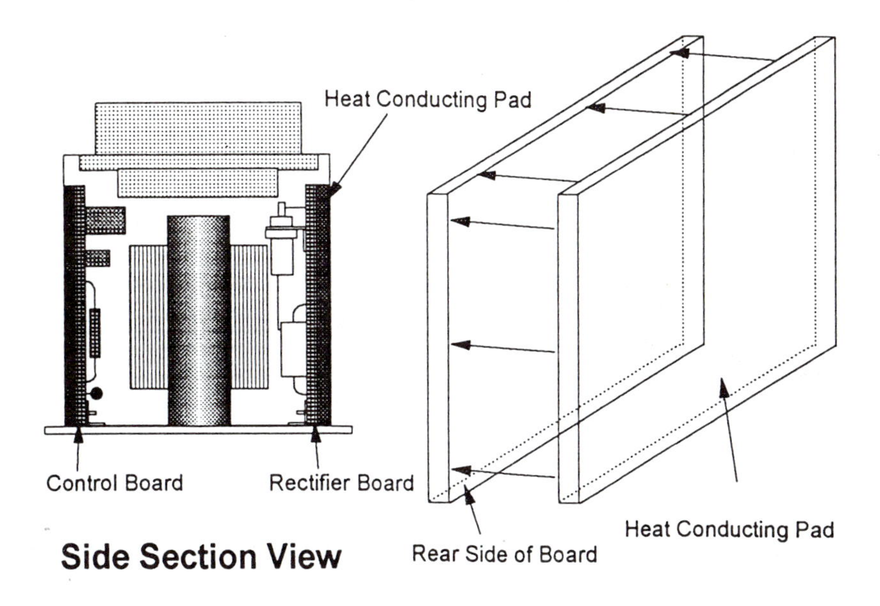

Figure 7.11 Proposed design modification, with the use of heat conducting pads [Hwang et al. 1995].

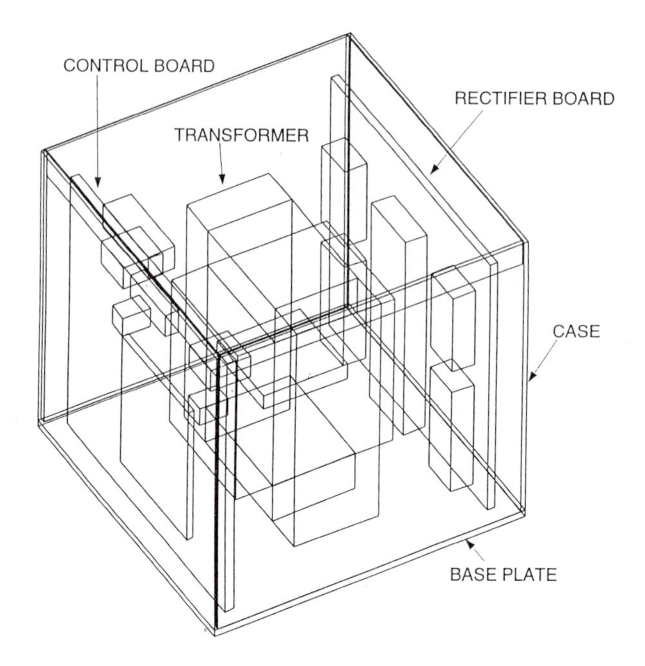

Figure 7.12 Computational model of the lamp dimmer assembly [Hwang et al. 1995].

Table 7.2 Input data

Element	Component	Thermal Conductivity [W/mK]	Heat Dissipation [W]
Rectifier Board	Board Inductor (L_1&L_2) Thyristor (CR1, CR10) Capacitor (C1,C2)	0.35 122 50 30	0 1.0 0.3 0.25
Control Board	Board Resistor (R2) Resistor (R3) Transistor (Q1,Q2) Thyristor (Q3,Q4) Resistors	2.4 40 40 1.0 20 0.82	0 1.57 0.818 0.16 0.63 0.203
Transformer	Center Core Frame	258 0.6 54	5/10 0 0
Head Assembly	-	0.3	0
Base Plate	-	16.3	0
Case	-	180	0

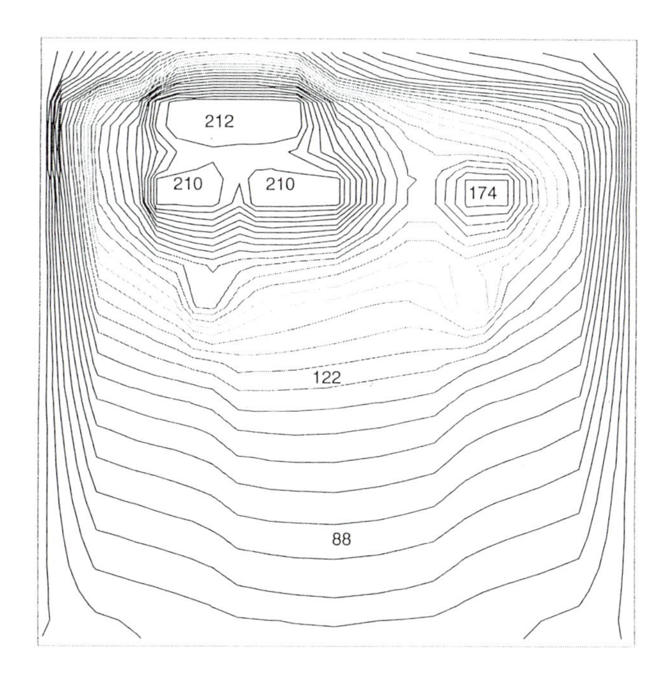

Figure 7.13 Temperature contours for the baseline case. A vertical plane through the control board is displayed.

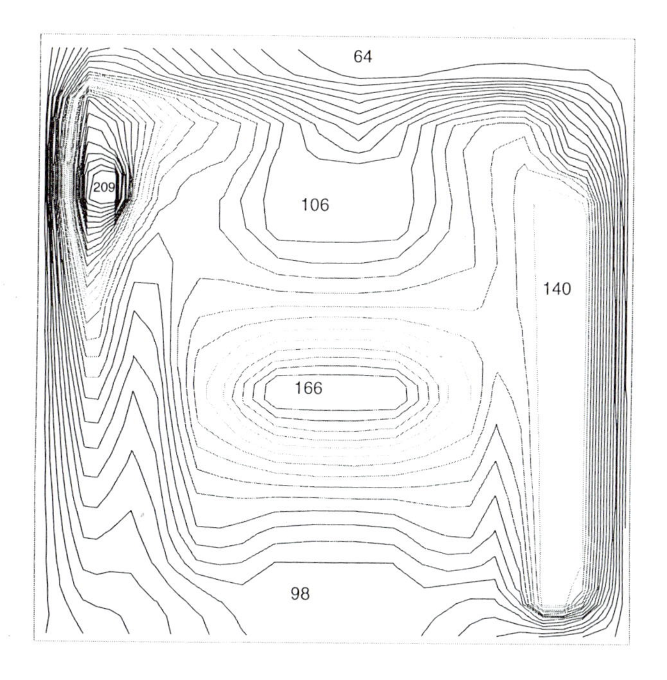

Figure 7.14 Temperature contours for the baseline case. A horizontal plane through the middle of the transformer is displayed.

7.5 Summary

This chapter addresses the unique problems arising in the thermal management and characterization of high-temperature electronics. Possible passive, active, and combined passive/active schemes are presented for such applications. For thermal characterization, current modeling approaches are described and illustrated by considering two examples of high-temperature applications.

Chapter 8

APPLICATIONS

8.1 Commercial Avionics - High Speed Civil Transport (HSCT)

High-temperature electronics can contribute significantly to the performance of the HSCT primary control system, and thus to its economic viability, by reducing wiring weight, electromagnetic and radio frequency interference, connector count, and measurement noise. This section identifies the main environmental and performance requirements for high-temperature electronics to be used in HSCT propulsion and flight control subsystems, and presents a proposed HSCT development schedule to provide insight into the program time available for the development of high-temperature electronics.

8.1.1 Character of the U.S. HSCT

The proposed U.S. HSCT, illustrated in Figure 8.1, will cruise at Mach 2.4 at altitudes up to 70,000 feet. The mainly composite structure airplane will be over 300 ft long, have a 130 ft wingspan, carry up to 300 passengers in a tri-class configuration, and fly 5000 nautical miles non-stop. U.S. Industry and NASA are developing the technology foundation for a High Speed Civil Transport (HSCT) through the NASA High Speed Research (HSR) program as well as independent industrial research and development efforts. An international working group is also studying the feasibility of such an aircraft and laying the ground work for an international agreement on critical regulatory issues such as emissions, noise, and sonic boom. Several studies to date indicate that a potential market of well over 600 airplanes will exist over the period from 2005 to 2015 if the seat cost is relatively close to subsonic airplanes. High productivity, roughly twice that of current aircraft in long overseas routes, is a major factor in compensating for the higher development and manufacturing costs for the HSCT. This paper discusses the control system architectural requirements for the HSCT airplane with emphasis on the potential of high-temperature electronics to improve the operating economics of the system. Most of the discussion centers on the actuation system electronic architecture because it is this portion of the control system that is inherently exposed to the most severe HSCT thermal environment.

8.1.2 Baseline control system architecture

A baseline architecture (Figure 8.2) was developed based on anticipated system components available circa the year 2000, and experience with fly-by-wire subsonic transport airplanes. This architecture is characterized as a consolidated/federated control system, utilizing subsystem control units operated within environmentally controlled electrical equipment bays. The various elements of the system are integrated via a fiber optic data bus.

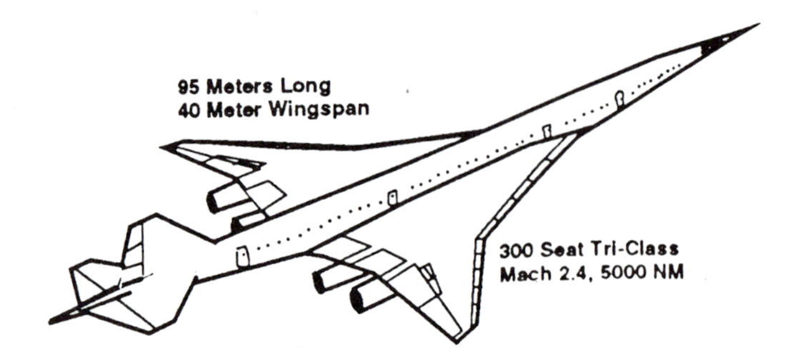

Figure 8.1 Boeing high speed civil transport environment.

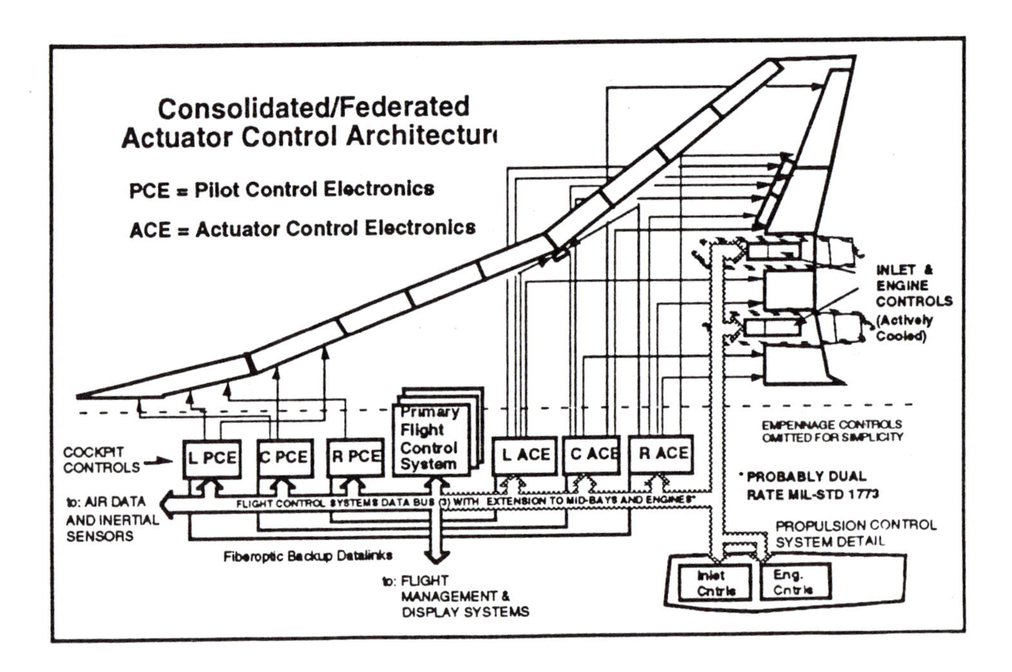

Figure 8.2 Current baseline control system uses electronics only in environmentally controlled areas.

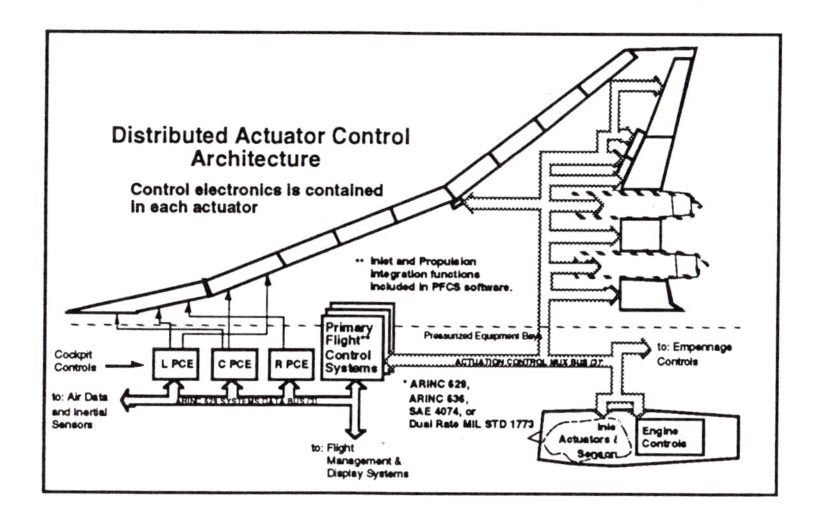

Figure 8.3 Distributed HSCT flight control actuation system reduces wiring complexity.

To avoid placing electronics in hot areas on the wings and empennage, all the flight control actuators are "dumb" actuators. The control electronics for the actuators are contained in centralized actuator control electronics (ACE) units, installed in the pressurized, air-conditioned fuselage.

Separate ai- cooled inlet and engine control units are provided in the nacelles because both the engine and inlet utilize a relatively large number of transducers and actuators in a constrained area, and because these devices are interrelated by fairly complex control laws. The air cooling for the propulsion system electronics is provided by ducting cabin cooling discharge air out to the nacelles.

Using separate inlet and engine control units minimizes pneumatic line lengths to pressure transducers, minimizes wiring and pneumatic connections crossing the inlet engine interface, and permits independent development and test of the two systems. This arrangement also facilitates independent maintenance of the propulsion system elements, a basic design feature of the nacelle. Although inlet and engine control systems are physically contained in separate boxes, the propulsion system control laws are functionally highly integrated both internal to the propulsion system and with the airframe. This integration is achieved through the control system design process and the communication capabilities of the flight control data bus which can make the boundaries between the various controllers virtually transparent, if desired.

The major drawbacks of this architecture are: (1) the long wiring runs to flight control actuators, (2) the large number of connector pins between control electronics and sensors and actuators, and (3) the cooling system required for propulsion electronics. Although none of these drawbacks are prohibitive, each provides an opportunity for improving the economics of the system.

Each actuator requires a wiring bundle containing typically 17 to 26 wires. The wire runs for the flight control (wing and empennage) actuators are on average nearly 100 feet long, while those for the propulsion actuators are 15 feet long. The weight of these wire runs has been estimated at as much as 600 pounds. The wire runs are also expensive to design and fabricate, difficult to install, difficult to maintain, and subject (if copper wire is used) to electrical interference. Service studies of both subsonic commercial aircraft and Concorde [Ganley 1991] have shown that wiring and connector problems are the most frequent cause of propulsion system maintenance action in regular service.

The long wire runs also necessitate connectors at each airframe production break. We have estimated that the total number of connector pins in the proposed system is close to 10,000.

Electronic system reliability (MIL-HDBK-217F) is estimated, in part, as a function of connector pin count, where the connectors are subjected to high-temperatures, vibration, or chemical contamination. Although we are improving connector reliability, and high-temperature will be addressed as an engineering design problem, connectors, especially those exposed to handling during maintenance, will tend to degrade the operational reliability of the airplane. Another way to address the situation which is more beneficial to us, in a new design such as HCST, is to eliminate signaling wires altogether using multiplex signaling (such as, ARINC [Aeronautical Radio, Inc] 429, ARINC 629, MIL-STD 1553B) methods.

The cooling required for the inlet and engine control units, although it does not impose serious weight or performance penalties, does make the propulsion system controls dependent on the operation of the environmental control system. Eliminating or simplifying the cooling system is also highly desirable.

8.1.3 Proposed distributed control system configuration

The modification of the control system shown in Figure 8.3 is proposed to correct the deficiencies of the baseline. In this system the flight and inlet control actuators will use high-temperature control and signal interface electronics mounted on each actuator. The flight and inlet control actuators are connected via data bus to the flight control computer system thus eliminating much of the dedicated actuator and sensor wiring required by the baseline and the cooling requirement for the inlet controllers.

It is assumed that the engine control unit will be engine mounted. It will be fuel or ram air-cooled to completely eliminate the air cooling duct to the nacelle. This assumption will be reviewed in coming years as the high-temperature electronics and fiberoptic technologies mature and as HSCT control system architecture studies progress.

High-temperature electronics are critical to implementing this architecture which distributes electronics (without active cooling), reduces the system weight, and simplifies the physical and functional complexity of flight and propulsion controls.

8.1.4 Temperature environment

The temperature distribution for HSCT actuators varies significantly depending on the location of the device. A conservative assumption is that the highest ambient temperature to be encountered will be 193°C (380° F). since this is the stagnation temperature at the airplane speed placard limit. The engine bays will be hotter but electronics in them can be supplied with ram air via ducting from the inlet if fuel cooling is not used. More detailed analyses may show that a reduction in the maximum temperature is possible, but for the present, this conservative assumption (193°C) is preferred to define the electronics challenge.

8.1.5 High-temperature electronics requirements

In order to implement the distributed architecture, the elements of a smart actuator electronics module, Figure 8.4, must be implemented in a high-temperature chip or multi-chip module. The minimum requirements for this chip set indicated in the figure include: a 16-bit microprocessor with on-chip ROM, RAM; and I/O; a high speed serial bus interface; 3 analog to digital (A/D) and 3 digital to analog (D/A) signal converters; and other support components which are briefly described in this report. Although a more elaborate design may be desirable, the 16-bit electronics module could satisfy both the signal multiplexing and control functions required. The functions to be performed by this module are fairly limited and should be virtually identical in form for every actuator on the airplane. Thus, in order to satisfy the high-

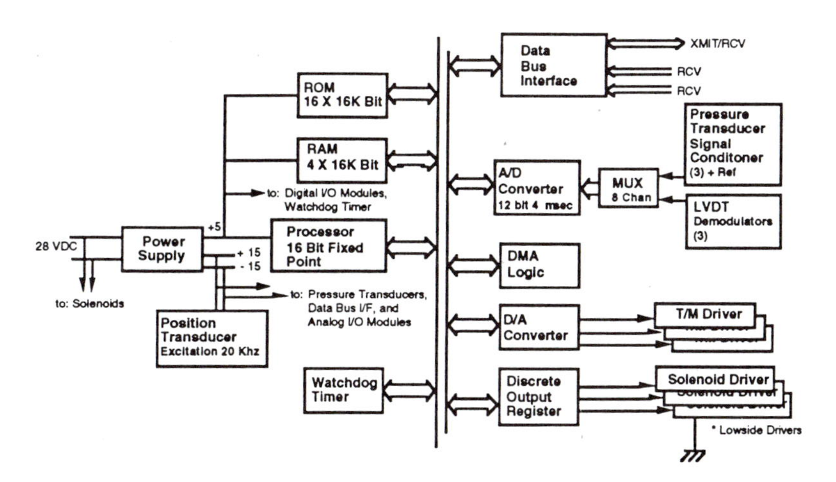

Figure 8.4 High temperature actuator electronics module.

temperature requirement at minimum cost, we should not develop components more elegant than necessary to meet common actuation control, sensing, and multiplex requirements.

In the following paragraphs we discuss the rationale for the functional requirements and offer some thoughts on alternative approaches to satisfying the requirements.

Processor. We know from experience that a fast 16-bit processor with double precision arithmetic capability is adequate to perform bus interface, actuator position, and rate loop closures and fault detection/ accommodation for a single flight control actuator. The processor should have an interrupt structure tailored to real time context switching (for example, Texas Instruments 9900/9989 series processors). The properties of the actuation control loops do not require floating point and other complex instruction features. A hardware floating point instruction set may simplify software development and facilitate the use of higher order languages, but these features increase the complexity and reduce the reliability of the module and ultimately the unit cost of a large production run. A standard software package, tailored by adjusting coefficients and data within a fixed logical structure, with a fixed point assembly language development environment will be adequate for every installation of the module. The issues of processor specification should be analyzed thoroughly with both life cycle cost and technology trade studies prior to starting development of the processor.

Data Bus Interface. A high speed fiber optics data bus is incorporated in the architecture because the various electronic subsystems are widely separated around the HSCT airplane and we are confident that a suitable optic bus can be developed. The fiberoptic data bus provides a low weight, high data rate means to communicate among the various subsystems thus reducing vehicle weight while providing electrical isolation among subsystems, important in controlling fault propagation. Copperwire ARINC 629 (as configured on the Boeing 777), is not satisfactory in the distributed actuator control application because its bus interfaces are relatively large, it introduces difficult-to-manage timing uncertainties, it must be shielded against high energy RF signals, and it is relatively slow (for example, 1-2 megahertz). A simple fiber optic bus system, satisfactory for primary flight control, is essential to implement the flight/propulsion control integration concept. This bus should provide a minimum 10 to 20 megahertz throughput depending on the protocol, and provide high reliability and message integrity.

Presently, the three most promising candidate bus concepts being considered within Boeing for Advanced Vehicle Management System and actuator control applications interconnection are: The Boeing Dual Rate 1773 (1 or 20 megahertz), the SAE 4074 Linear Token Passing Bus (50 megahertz), and ARINC 636 or Fiber Digital Data Interface (FDDI) (100 megahertz) protocols (Bauer-1993). However, development work is required to evolve and certify a suitable bus from any of the above candidates.

A high data rate bus is required to interconnect the actuator mounted electronics modules with the various flight and propulsion control computers mounted in the fuselage. However the difficulty with a fiberoptic data bus in the 193°C temperature environment is that the cheapest and most reliable optoelectronic interface components (Light Emitting Diodes) degrade seriously in optical performance and reliability at temperatures above 125°C [Bauer et al. 1993]. A number of multiplex data bus options may be selected as development options for inclusion as the bus interface element in the actuator control electronics chip set. The selection must be based on a balance between airframe performance requirements and semiconductor constraints. Some of the possible options are as follows:

- Option 1, Figure 8.5. Use a MIL-STD-1553B (copper wire) data bus in the severe environment areas with a translator/optical isolator gateway unit installed in the fuselage to connect to the main optical flight control data bus.
- Option 2, Figure 8.6. Use relatively low data rate individual optical datalinks emanating from a fuselage mounted interface unit to communicate to the actuators.
- Option 3, Figure 8.7. Use the high data rate fiberoptic bus and Peltier coolers to cool the critical optical/electrical interface components at the actuator.
- Option 4. Use the fiberoptic concept of Option 3, with high data rate optical interfaces designed to operate at the necessary elevated temperatures.

Option 1 is appealing because MIL-STD-1553B is well understood, has a wide market base, and the necessary parts are compact, capable of operating at elevated temperatures (200°C) and are available at reasonable prices. Unfortunately it is a wire bus and thus is vulnerable to common mode faults resulting from HIRF and lightning. It also adds electronic units to the architecture which have no other function beside message handling, and it is a command-response, master-oriented bus which introduces common mode failures that ARINC 629 was specifically designed to eliminate. On the other hand an architecture similar to this was flight tested very successfully on the Condor Research Aircraft [Lambregts et al. 1991].

Option 2 is an optical digital variation of the consolidated/federated architecture in the current baseline. This may still be attractive because it eliminates most of the individual command and sensor lines (16+ per actuator). By using a dedicated optical link to the actuator rather than a bus, both the protocol and data rate requirements are greatly reduced. This, in turn, reduces the cost and risk of building the high-temperature optical electronic interface. The optical links and interface boxes to the main bus do, however, partly defeat the purposes of pursuing the distributed architecture (for example, reducing wiring complexity and connector count).

Option 3 is probably the best alternative if practical Peltier or other cooling mechanisms can be developed with a 193°C heat sink and if the high data rate optical bus interface terminal can be reduced to a form consistent with actuator mounting real estate.

Option 4 is highly desirable but risky because of the low probability of successfully developing the high-temperature optical/electronic interface in the HSCT developmental time frame, and at reasonable cost.

To summarize, selection of a bus protocol is required before serious chip development can commence. Further study of life cycle cost and high-temperature technology is required to select the bus protocol.

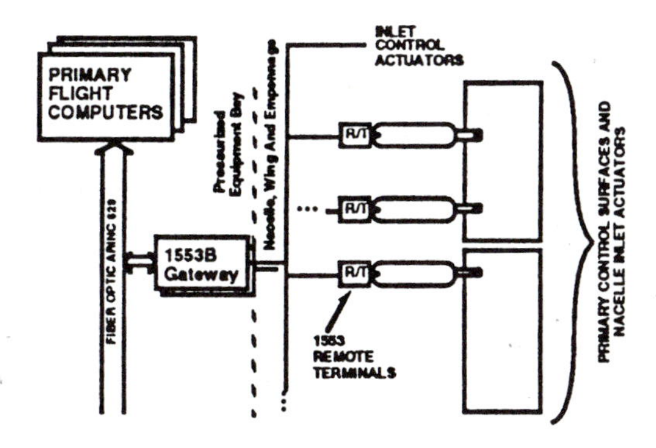

Figure 8.5 Copper wire Mil-Std 1553 gateway is feasible (Option 1).

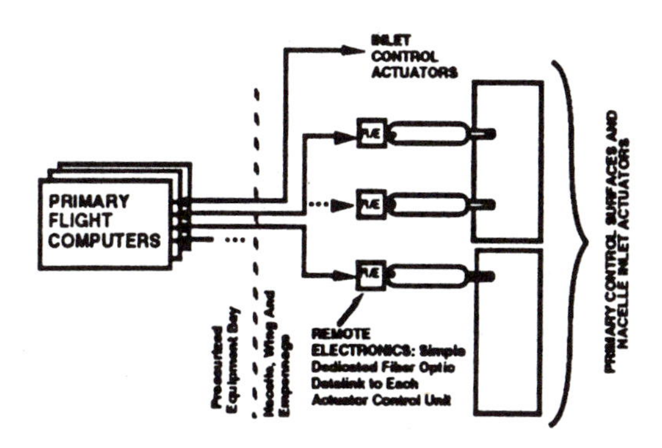

Figure 8.6 Dedicated fiber optic link simplifies the actuator electronics (Option 2).

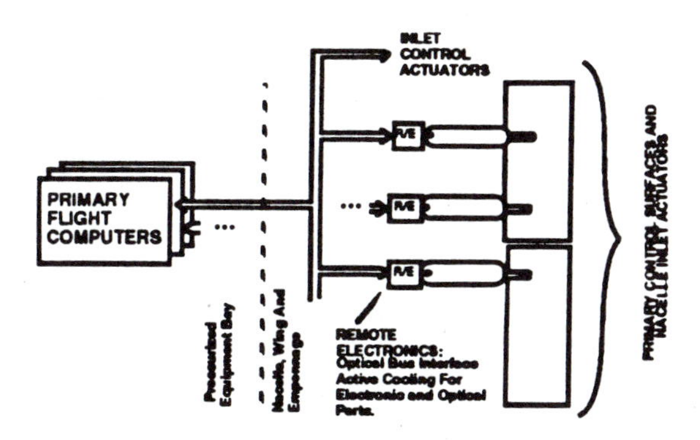

Figure 8.7 Some form of cooling is required for fiberoptic databus interfaces (Option 3).

Position interface. A typical flight control or inlet actuator will have one or two position feedback sensors. One measures the actuator output position, and in some actuator applications a second position feedback device is used to measure the position of the second stage spool in the hydraulic servovalve. Usually in aerospace applications these position feedback devices are LVDTs; however, in some applications resolvers are used. In general, since we can expect to use two position measurements, three are specified to provide a spare.

An interesting point is that the output of LVDTs is usually converted to digital format by analog demodulation followed by a conventional analog to digital (A/D) conversion. Resolvers may be converted to digital format by phase shift demodulation and measuring time between zero crossings. Thus, one could develop the analog input conversion process for the chip using either a multiplexed analog to digital converter or an array of direct device to digital converters. This may be significant, since the direct device to digital converters could be easier to implement in the high-temperature environment than a conventional 12 bit A/D converter.

Pressure transducer. Four pressure transducer input ports are provided on the actuator control electronics circuit. Two are for use in actuator monitoring, force fight reduction, and stabilization. The other two are provided to process the output of pressure transducers used to measure inlet aerodynamic pressures. At the moment it is not clear whether all the pressure transducer requirements can be satisfied with strain gage type transducers or if the inlets may require high accuracy pressure transducers using frequency encoding. These distinctions may affect the issue of whether to use a multiplexed A/D converter or direct data to digital conversions.

Sensor excitation. Pressure transducers and position sensors require excitation. Based on the sensor compliment envisioned at this time, an oscillator and driver will be required for the LVDTs. The pressure transducers will probably require a DC power source. Depending on the type of transducer, this may be a high precision source.

Servovalve driver. A current feedback amplifier will be required to drive the servovalve. The maximum current requirement is approximately a 100 ma. This could be pulse width modulated or a linear device. The choice depends largely on the compatibility of the approach with the thermal environment and the actuator control electronics packaging constraints.

Solenoid drivers. The solenoid drivers are required to drive "mode" valves and pressure reducers used to implement actuator bypass and blocking logic. Typically these devices operate at 1 amp and a nominal 28 volts DC. Inductive spike suppression diodes will be required either on the solenoids or in the actuator control electronics

Watch dog timer. The smart actuator chip set will provide a watch dog timer function to detect and declare a fault if the processor should stop for hardware reasons (for example, loss of clock) or fail to complete any of its processing functions within a reasonable period due to software or memory fault. When a timing fault is indicated, the watch dog timer will place the actuator in the bypass condition by de-energizing the appropriate solenoid driver. The functions involved shall be performed independent of the processor circuitry.

Built in test**.** Sufficient built in test (BIT) features will be incorporated to correctly detect and identify failures to a specific actuator controlled by the specific actuator control circuit. Since the maintenance plan for this system is to remove the actuator from the aircraft for either hydraulic or electrical failures, the critical function of the BIT system is to isolate faults between the actuator and the airframe wiring. Fault isolation internal to the actuator will be dealt with on the bench. Bench test interfaces will be required in the actuator control electronics but need not be accessible in the airplane on-line BIT environment.

Power supply. One of the more significant challenges in the system may be the power supply. While all the other functions required for the system could conceivably be placed on a single chip, the power supply will be implemented using a variety of discrete components. The input power will be nominal 28 volts DC per MIL-STD 704. The power will be structured with battery backup and an uninterruptable switching system. However, to protect against transient disturbances, the controller processor will be required to operate without interruption during a 5 ms power drop out. This requirement will probably not extend to the power driver elements of the system since to do so would result in excessively large storage devices at the actuator. The possibility of superimposing the serial communications protocol on the 28 volt DC power should be considered.

8.1.6 Market for actuator electronics module

If we assume sales of at least 600 airplanes and perhaps 50-80 applications of a standard smart actuator module per airplane, we will have roughly a 30,000 module market. Furthermore, if we assume a price of $1000 per chip we will have a fundamental market worth about $30 million dollars.

This estimate is probably conservative for a number of reasons. First, the airplane studies have shown that a much larger market will exist if ticket prices converge on the subsonic fare structure. Second, the number of chips used per airplane may be significantly larger since this paper only addresses the flight control actuation and inlet actuation. If available, a chip of this nature might well be used in various subsystems such as landing gear, brakes, and ECS. Finally, if economy of scale reduces the price, it would probably be applicable to various other aerospace, automotive, and industrial high-temperature servo actuator control applications.

8.1.7 Schedule

The nominal HSCT technology development schedule is depicted in Figure 8.8. Given a program go-ahead in year 2001, and technology demonstration efforts whose design phase will start in 1996, the window for HSCT oriented high-temperature electronics technology development is rapidly closing. In other words, major technology level validation tests must be completed before the program will commit in its design phase to use technology at a component level. Thus, to find their way onto the HSCT, fundamental high-temperature technologies must be demonstrated in the laboratory in 1996 or 1997 and be available as prototype chips in late 1997 for proof of concept demonstrations in the laboratory and perhaps in flight test before the firm avionics system configuration is selected at the end of 2000 (see Figure 8.8).

8.2 Military Avionics

For over 40 years, developments in solid-state electronics have provided the U. S. Air Force with the most sophisticated and capable avionics systems in the world. Steady advancements in solid-state devices and integrated circuits have enabled modern electronic warfare, navigation, and flight and propulsion control electronics. While conventional silicon (Si)-based solid-state devices have continually been the workhorse of microprocessors, digital signal processing, large memory, analog I/O, and power control electronics, compound semiconductors such as gallium arsenide (GaAs) and HgCdTe are generally used in radar and sensor electronics, respectively. Recently, interest has also been generated in wide-bandgap semiconductors such as silicon carbide (SiC) for advantageous use in high-temperature and high-power device applications. Wide-bandgap electronic devices are capable of operating at

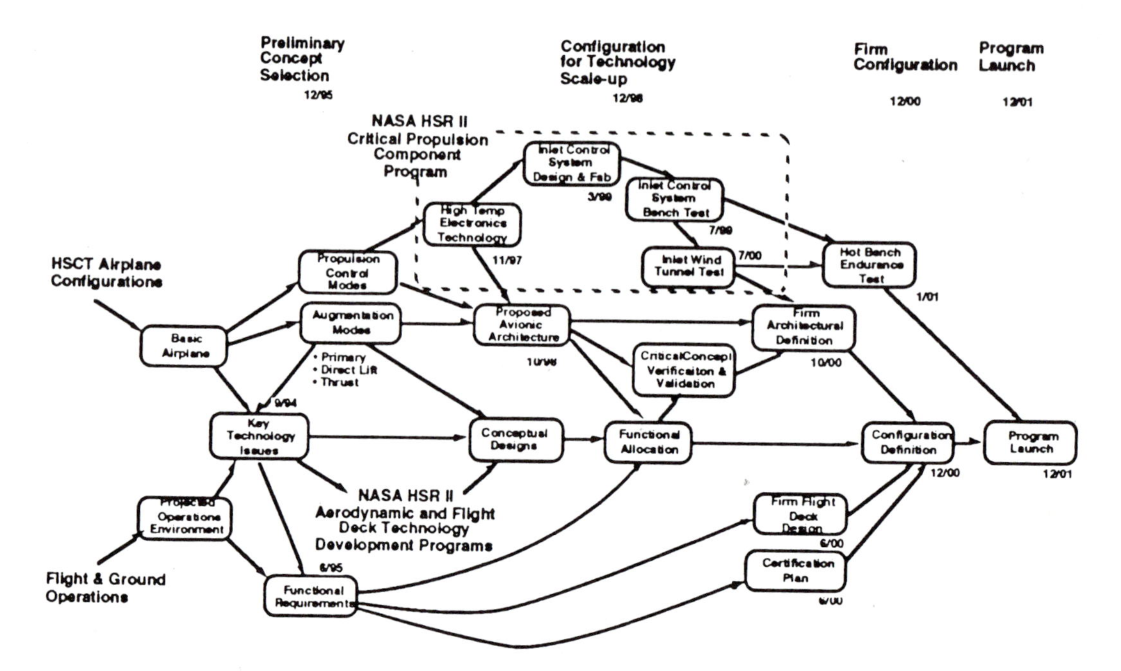

Figure 8.8 Time for high temperature technology development is limited if 2001 go-ahead is to be observed.

both higher temperatures and higher efficiencies compared to Si-based devices. The higher efficiency will reduce the amount of heat dissipated by the electronics, thus enabling a reduction in, or elimination of, existing heavy, single-redundant, distributed aircraft electronics cooling systems. To insure aircraft electronics reliability and longevity, the junction temperature of conventional Si-based devices is currently maintained within the MIL-STD temperature range of -55°C to 125°C. The higher temperature capability of wide-bandgap devices (operation has been demonstrated up to 600°C) will significantly improve device reliability and enable use in remote locations where cooling is impractical, e.g., mounting of electronics on engines for control and sensing, in aircraft skins for sensing stress and temperature, and in electronic warfare or other "stores" attached to the fuselage or wings of the aircraft. Also, the efficiency of thermal transport between the electronics and heat sink increases with increasing temperature, further reducing the required capacity of the cooling system. Consequently, wide-bandgap (WBG) high-temperature electronics (HTE) are expected to play an enabling and vital role in the design of current and future military aircraft. The system-level benefits of employing WBG-HTE on military aircraft include a reduction in flight control system weight and improved reliability; a reduction in size and weight, or elimination of, the environmental control system (ECS) required to cool power management and distribution (PMAD) and flight control electronics; a reduction in engine control system weight and increased reliability using a distributed processing architecture; and the improved reliability and maintainability of stores management system (SMS) avionics. The following sections address important aircraft subsystem high-temperature electronics applications, the temperature range in which electronics will be expected to operate if they are to be un-cooled, and a description of WBG-HTE components desired for use on military aircraft.

8.2.1 More electric aircraft (MEA)

High-temperature electronics represent an important technology in the development of the More Electric Aircraft (MEA). They are critical to implementing both a distributed flight control system and reducing the PMAD heat load that the aircraft environmental control system must dissipate. Recent advancements in high-power solid-state switches, converter circuit topologies, motors and generators, and the evolution of a fault-tolerant electrical power system coupled with electrically driven actuation, have generated renewed interest in MEA. In the MEA concept, electrical power is utilized to drive aircraft subsystems that have historically been driven by hydraulic, pneumatic, and mechanical systems [Quigley 1993]. Subsystems such as hydraulic-driven flight control actuators, engine-gearbox driven fuel pumps, and air-driven environmental control system would be powered electrically via electric motors [Weimer 1995]. Studies on F-16 and F/A-18 fighter aircraft have shown that the MEA offers many subsystem level benefits in areas of reliability, maintainability, supportability, and overall cost [Eicke and Hodge 1992, Shah 1992].

In recent years, a series of Air Force/Navy technology programs have been initiated at Air Force Wright Laboratory in support of MEA. The major MEA subsystems under consideration are shown in Figure 8.9. In this concept, aircraft power is produced by an internal starter generator (ISG) that is directly driven by the main engine. The ISG design under consideration has the potential to produce up to 375 kW when driven by an F-110 after-burner fighter engine. The ISG power is fed to a fault-tolerant power management and distribution (PMAD) electronics network that drives all aircraft electrical subsystems. Major subsystems include those for electric actuation, engine starting, braking, environmental control, anti-icing, and fuel pumping. An integrated auxiliary/emergency power unit (IPU) and battery system provides power for redundancy and engine start-up.

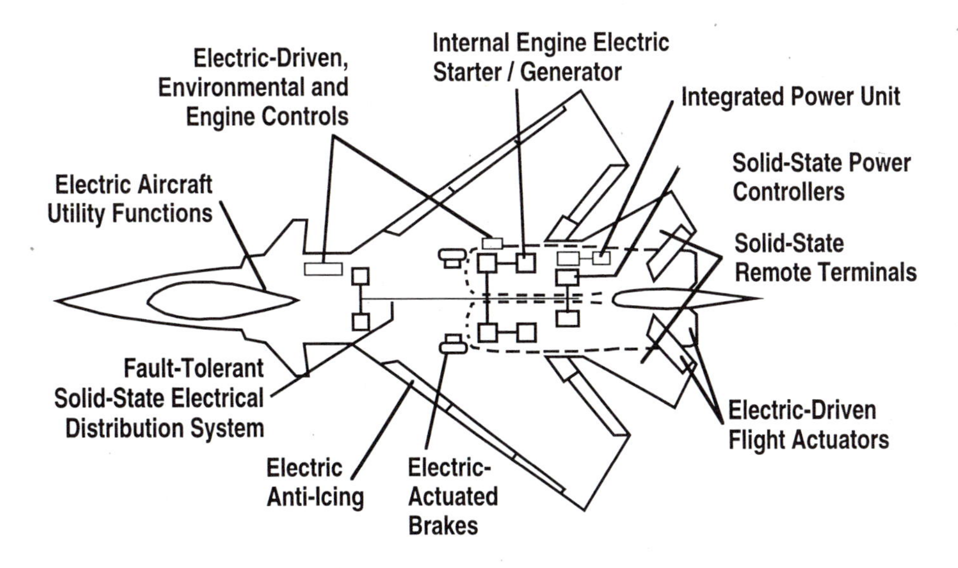

Figure 8.9 Conceptual perspective of More Electric Aircraft (MEA) subsystems.

8.2.2 MEA distributed flight control

Significant improvements in aircraft flight control system reliability, mass, volume, and reduced dependency on environmental control can be realized through the use of wide-bandgap electronics. In the MEA architecture, conventional hydraulic-driven flight control actuators will be replaced with electric motor driven actuators and a distributed flight control (DFC) electrical system. Electrically driven flight control actuation offers subsystems benefits in fault tolerance, redundancy, reliability, and power density. A distributed flight control system eliminates the major drawbacks of centralized control, in which actuators and co-located control electronics are connected via a data bus to a centrally located flight control computer. In a centralized control architecture, all power control electronics would be displaced from the actuators and centrally located in the aircraft. The DFC architecture has several major advantages: (1) It eliminates long and heavy wiring/shielding runs to actuators and sensors; (2) It increases reliability because of the reduction in the number of connector pins between control electronics and sensors and actuators; electronics system reliability is estimated, in part, as a function of connector pin count (MIL-HDBK-217F); (3) It increases survivability since centralized control electronics would be located in one location; and (4) It reduces or eliminates active cooling. Although none of the drawbacks of centralized control are prohibitive, a distributed scheme offers a significant opportunity for improving the performance and economic viability of the flight control system.

In present-day fighter aircraft, a centralized hydraulic system provides both flight control power and a thermal transport medium to remove heat from the actuators. However, in the electrically based MEA distributed flight control architecture, power control electronics will be mounted on or near the actuators without the benefit of hydraulic assisted cooling. Since it is impractical and highly undesirable to distribute an active, closed-loop network to cool remotely located electronics (e.g., wing and empennage) because this would offset the benefits of eliminating the hydraulics, "uncooled" electronics will be required. Power control, sensor, and interface and data-bus electronics are needed to provide reliable operation at ambient temperatures well in excess of the Si-device MIL-STD temperature limit of 125°C.

Unlike Si-based devices, WBG electronics have good potential for operating in "uncooled" environments because they can (1) generate significantly less heat due to increased efficiency and (2) tolerate higher temperatures because they liberate smaller leakage currents (I_L). Wide-bandgap devices rival silicon devices in efficiency through reduced conduction and switching losses. In the case of power devices, conduction losses depend on the thickness (W) and doping level (N) of the semiconductor layers used for blocking reverse voltages, where conduction losses decrease with smaller W and larger N. In the case of a p/n junction, which is the fundamental voltage blocking structure in power switching devices, we find $W \propto E_c^{-1}$ and the maximum value of $N \propto E_c^2$, where E_c is the breakdown field strength for the semiconductor. The value of E_c, in turn, increases with semiconductor bandgap (E_g) as approximately $E_c \propto E_g^{3/4}$. Consequently, the ideal resistive conduction loss for power switching devices using 4H-SiC with $E_g = 3.3$eV will be < 80X than that for Si devices with $E_g = 1.1$eV. Similarly, wide-bandgap devices will exhibit lower switching losses than do Si devices; which primarily depends on device size through the magnitude of device capacitance. Increasingly smaller-geometry power switching devices can be designed using WBG materials, resulting in smaller capacitance and switching losses due to the benefits of reduced conduction losses. Device conduction losses vary as $\propto$ (device area)$^{-1}$, whereas capacitance varies as $\propto$ (device area). Hence, WBG devices can be made smaller to reduce switching losses while maintaining acceptable conduction losses that are substantially lower than for Si-devices.

Wide-bandgap devices can also tolerate higher ambient temperatures compared to Si-devices because they generate smaller leakage currents. The maximum device operating temperature is limited to maintain a sufficiently small leakage current (I_L), such that the ratio of device "on-current (I_{on})"/$I_L > 10^5$-10^7. In the case of the Si or SiC p/n junction, the leakage current decreases exponentially with semiconductor bandgap (E_g) according to $I_L \propto \exp(-E_g/2kT)$, where k is Boltzmann's constant and T is the temperature. This relation shows that WBG devices, such as SiC with $E_g = 3.3$eV, are capable of operating at temperatures ~ 3X greater than for Si-based devices with $E_g = 1.1$eV.

In the past, the Department of Defense system development directives have specified a maximum Si-device junction temperature of 110°C to ensure reliability. Several major aircraft programs have even limited Si junction temperatures to less than 70°C to yield major reliability improvements. A conservative estimate for the highest ambient temperature to be encountered by actuator-mounted electronics (without cooling) in supersonic aircraft is ~ 200°C. At present, however, high-power electronics capable of long-term reliable operation at temperatures greater than 200°C do not exist. As a result, the MEA DFC baseline design is currently forced to employ Si-based electronics that require active cooling. Consequently, the full potential of distributed flight control will not be realized until power electronics capable of reliable operation at 250°C or greater become available. Fortunately, as shown above, WBG-based power devices have good potential of providing reliable operation at temperatures exceeding 300-350°C, that would meet MEA DFC requirements.

8.2.3 MEA fault-tolerant PMAD

Wide-bandgap electronics can also significantly reduce the total power management and distribution (PMAD) heat load that the aircraft ECS must handle. This would allow a reduction in the massive size and weight of the aircraft ECS (along with other advantages discussed later) and improve overall aircraft flight performance (e.g., range, maneuverability). The envisioned MEA concept imposes increased demands on existing PMAD technologies in the areas of power handling, fault-tolerance, and reliability; these in turn, require innovations in power generation, distribution, and source and load management. In response to these demands the Air Force has initiated the Power Management and Distribution for a More Electric Aircraft (MADMEL) ground demonstration program [Maldonado et al., 1995).

In this effort, a hybrid power system will supply 270 VDC, 28 VDC, and 115/400 V 3ϕ at 400 Hz power to MEA-type simulated loads [Weimer 1995]. Electrical load management centers (ELMCs) will distribute remotely controllable power to high-power, medium-power, and rotary vane actuators, and engine start, radar, resistive bank, and avionics load simulators. Important electric subsystem functions within the ELMCs include 270 V to 28 V DC-DC conversion and 270 V to 115/400 V DC-AC inversion. MEA-type loads require 270 VDC power for driving actuator motors and 28 VDC for driving motor control logic. Most Stores Management Systems (SMS) avionics are driven by 28 VDC, and the envisioned active phased-array radar will require 270 VDC. The 400 Hz power is only required for some externally mounted weaponry.

Silicon-based switching devices and driver control electronics will be employed in MADMEL's baseline design. They will be cooled by a poly-alfa olefin (PAO) and vapor-phase heat-exchanger ECS. A similar ECS design also provides electronics cooling for the advanced F-22 fighter aircraft. The junction temperature of silicon-electronics utilized in both MADMEL and on the F-22 is maintained below 90°C by the ECS to ensure reliability. As previously mentioned, the advantage of employing WBG power devices in future PMAD or MADMEL subsystems will be to reduce the electronic heat load that the ECS must dissipate, thereby reducing the required ECS capacity. The ECS could then be made smaller and lighter. A reduction in PMAD electronics mass and volume can also be expected using WBG power devices; reduced conduction and switching losses will result in greater device power densities. A further advantage of employing WBG high-temperature electronics results because the heat transfer rate (g) between the electronics package cold plate at temperature T_1 and the ECS PAO transport medium at temperature T_2 are related as $g = h(T_1 - T_2)$, where h is the heat transfer coefficient, and the temperature gradient between the source and sink drives the heat transfer.

Increasing the allowable electronics junction temperature and corresponding package and cold plate temperatures increases the efficiency of thermal transfer to the PAO. This would further reduce the necessary capacity of the ECS. Additional benefits of reducing or eliminating the dependency of aircraft electronics on the ECS are discussed below.

8.2.4 Environmental control system concerns

An important long-term goal is to reduce the total electronics heat load that the aircraft environmental control system (ECS) must handle in order to reduce its size, weight, and cost. As mentioned, the use of wide-bandgap electronics can reduce or even eliminate active cooling of remotely located actuator control electronics and PMAD subsystems. The current approach used to cool most avionics and power control electronics on fighter aircraft employs a closed-loop ECS. Greater than 90% of the ECS cooling capability on modern fighter aircraft is utilized by the electronics; this includes cooling of radar electronics. Hence, it is highly desirable to reduce the total electronics heat load. Eliminating the aircraft ECS altogether may not be feasible, since it is needed to handle the few hundred watts generated by the pilot.

Today's ECSs are large (in size and weight) and consume a significant amount of power, which adversely affects aircraft performance. Existing fighter aircraft electronics can generate up to 50 kW of heat. The ECS in turn requires approximately 50 kW of power from the engines to dissipate this heat. The ECS is single-redundant because of its massive size and its failure results in an aborted mission or catastrophic failure. A significant reduction in ECS size could make redundancy feasible. Today's ECSs are also very expensive, and with the anticipated growth of tomorrow's electronics systems, projected future costs are enormous. Further, the ECS aboard many fighter aircraft transfers the electronics heat to the aircraft fuel via a heat-exchanger, which in turn transfers the heat into the environment around the aircraft. This approach has several drawbacks. The aircraft must reserve a finite amount of fuel to provide an adequate heat sink for the ECS, and maintaining the fuel reserve decreases the aircraft range. An eloquent solution to many of these issues would be to significantly reduce the required capacity of the ECS through employing wide-bandgap electronics.

8.2.5 More electric propulsion control

High-temperature wide-bandgap electronics can also significantly contribute to the performance of the "more electric engine (MEE)," a design that seeks to eliminate the use of hydraulics on an engine by employing electromechanical hardware for nozzle, guide-vane, and metering-valve actuators and fuel pumps [Przybylko 1993]. High-temperature electronics will be needed to control electrical power to the electromechanical devices and communicate over a distributed data bus because it will not be feasible to cool remotely mounted devices with either air or fuel.

Aircraft engine control systems have employed electronics for many years. The full-authority digital electronic control (FADEC) processor currently controls all fighter engine functions and provides onboard, real-time monitoring and diagnostics. As the engine workload increases in the future, the required capabilities of the FADEC can also be expected to increase. Analog-to-digital (A/D) and digital-to-analog (D/A) signal processing have become an increasing part of the FADEC's computational workload, challenging its throughput capability. Consequently, as the functional complexity of the FADEC grows, so too will the size of its housing and number of outside connectors. This results in major maintenance and weight issues, due to long cabling between the FADEC, sensors, and actuators. The solution to these problems is to employ distributed processing.

In distributed processing, a "local engine network" provides communication between the FADEC, actuators, and sensors. "Smart electronics" remotely located at the actuators and sensors perform signal processing functions, loop-closure calculations, and diagnostics. Autonomous control of some subsystems may even be possible. The local network would significantly reduce the number of connectors, the total cable length, and the computational requirements of the FADEC. The result would be a reduction in the total weight and physical and functional complexity of the propulsion control system.

Since it is not feasible to cool remotely located engine-mounted electronics with fuel or ram air, they must endure the high-temperature environment of the engine. The projected engine-case temperature experienced by electronics mounted on a F-100 class of afterburner fighter engine is as high as approximately 315°C at Mach 2.5. This temperature clearly exceeds the MIL-STD high-temperature limit of 125°C for Si devices. In the absence of cooling, it is clear that Si-based electronics cannot be employed in the distributed "local engine network" control architecture. High-temperature electronics capable of reliable operation at temperatures of 350°C or greater will be required.

Existing engine-mounted Si-based FADECs are typically cooled (derated) with air or fuel to approximately 80°C to ensure reliability. It is anticipated that the FADEC will remain Si-based in the near future, due to the high level of technology required to develop the central processor. However, when the development of high-temperature WBG devices and integrated circuits progresses to the stage of VLSI technology, the FADEC will also be based on high-temperature WBG materials.

8.2.6 Stores management systems avionics

High-temperature electronics also hold a number of important applications in Stores Management System (SMS) avionics for supersonic military aircraft. SMS electronics control all weapons, electronic warfare, and other stores (e.g., fuel tanks) attached to the fuselage or wings of the aircraft. Supersonic flight causes high aircraft skin temperatures due to aero-heating (friction). The heat generated is transferred to the various compartments within the aircraft fuselage and wings via aircraft skin and frame conduction and compartment air convection. Consequently, Si-based SMS avionics housed within these compartments require cooling to maintain the upper MIL-STD temperature limit of 125°C.

The interface between the SMS and the store itself is located, ideally, at the store interconnect to allow for simple addition or modification of new stores. The result is electronics remotely located on the aircraft wings, which are nearly impossible to cool. In some cases the SMS electronics are stored in pylons attached to the wings; the pylons provide mechanical support for the store. Under certain emergency conditions, it is desirable to jettison the pylons to significantly improve aircraft flight characteristics. However, providing a pylon jettison capability practically eliminates the ability to cool the SMS electronics. If the pylon were cooled, the cooling system would be momentarily opened during jettison until a self-sealing valve closed. If the valve failed, the ECS coolant would drain. Since this could produce failure of other flight-critical systems, centralized ECS cooling of pylon stores electronics is not considered feasible.

Current approaches for cooling SMS electronics include ram air, self-contained ECS conditioning, and conductive transport to the store skin or case. Various electronic countermeasure (ECM) stores use forced air obtained from the aircraft's jet stream. While this is a logistically simple approach, it also bleeds air from the jet stream, which creates aerodynamic drag that reduces aircraft performance. The Low Altitude Navigation and Targeting Infrared and Targeting System for Night (LANTIRN) pods employed on F-15E and F-16 aircraft utilize ram air and a small self-contained PAO/Freon ECS to condition electronics. The drawbacks of this hybrid approach include ram air aerodynamic drag, added ECS weight, and single-point failure. Many deployable smart weapons and missile and bomb stores simply dump electronics heat to the outer skin via thermal conduction; electronics boxes are mounted on or near the armament case. In each of these scenarios, the stores would benefit through the use of high-temperature WBG electronics. As in the case of DFC and PMAD electronics, benefits would derive from the reduction in electronics heat generation and high-temperature

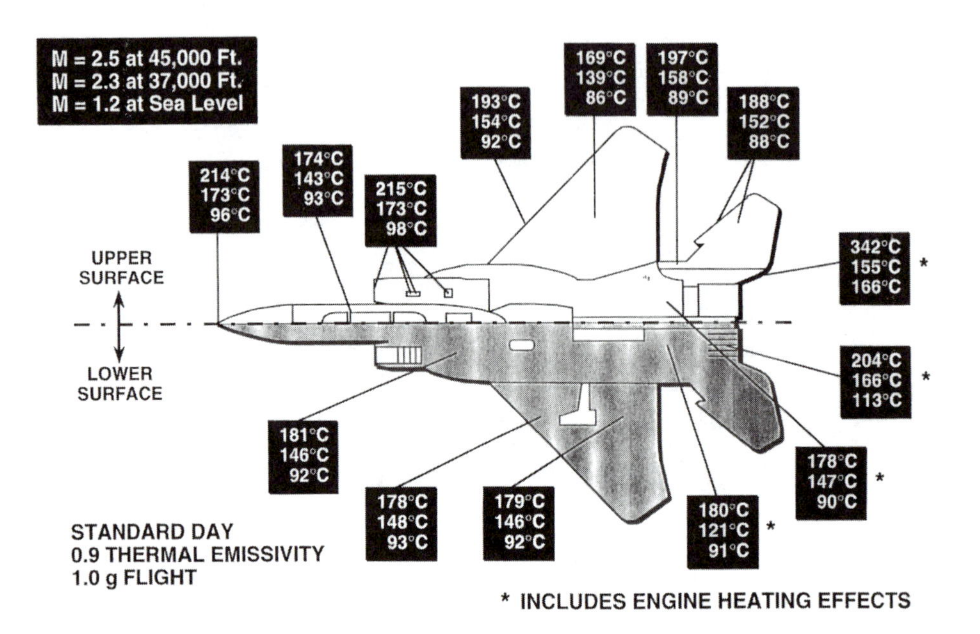

Figure 8.10 Calculated skin temperatures for a supersonic F-15 fighter aircraft.

capability. Potential advantages include reducing stores aerodynamic drag, eliminating ECS conditioning to reduce weight and improve reliability, and increasing armament electronics design flexibility; moreover, electronics mounting would not be restricted to the outer case of the weapon.

8.2.7 High-temperature environment

It is difficult to predict with precision what the exact environmental and solid-state device junction temperatures will be for the various aircraft electronics subsystems. Many factors, such as electronics location, maximum aircraft speed, duration of flight at that speed, altitude and ambient air properties, electronics packaging and mounting techniques used, and internal electrical losses, determine the final operating temperatures. However, under assumed aircraft performance envelopes, it is possible to make conservative predictions for the highest ambient temperatures encountered.

A projection for the operating temperature of actuator flight control, PMAD power control, and SMS electronics can be obtained for a modern advanced fighter aircraft by considering the calculated skin temperatures given in Figure 8.10 for an F-15 fighter aircraft. The skin temperatures shown represent the stagnation temperatures calculated for 1.0 g flight at the altitude and speed indicated. Ignoring the regions adjacent to the engines, the skin temperatures shown range between approximately 90°C - 200°C. If it is assumed that no active cooling is provided to the electronics via the ECS, the only available heat sink for the electronics boxes will be the environment around the aircraft, reached via the aircraft frame/support structure and skin.

Heat is generated in all electronics by internal resistive losses. This heat must be efficiently dissipated to prevent catastrophic device failure due to the degradation of the metal contacts and packaging structure, and device degradation due to substrate and p-n junction leakage currents. As mentioned earlier, an adequate temperature gradient must exist between the heat source (the electronics) and the sink (ultimately, the aircraft skin) to drive heat transfer to the environment around aircraft. Obviously Si-based electronics with a MIL-STD maximum

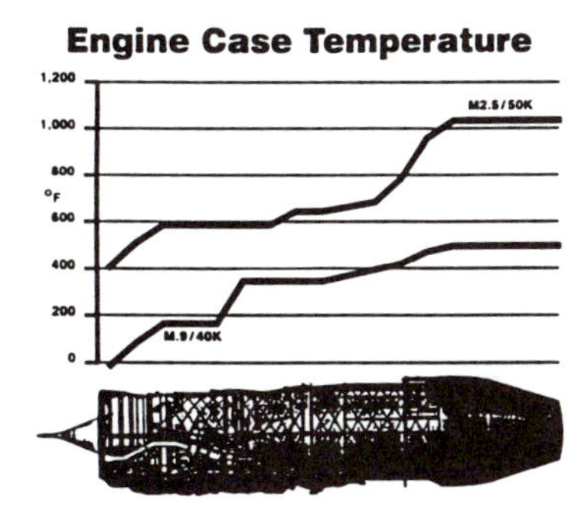

Figure 8.11 Temperature distribution for a F-100 class afterburner fighter engine, courtesy of S.J. Przybylko, USAF Wright Lab Turbine Engine Division.

temperature limit of 125°C (typically derated to < 100°C to ensure reliability) cannot be used in this uncooled supersonic flight scenario. Due to thermal resistance between the solid-state chip containing the devices, the device package, the electronic box, the aircraft frame/support structure, and the aircraft skin, solid-state devices capable of reliable operation at junction temperatures of at least 250°C will be required to support PMAD, actuator control, and SMS electronics in the absence of active cooling. A more detailed analysis may show a reduction in the maximum required operating temperature; however, at present the conservative assumption of 250°C is useful in defining the electronics challenges that lie ahead.

In the case of the electronics required for propulsion control, the projected engine-case temperatures for an F-100 class afterburner fighter engine are shown in Figure 8.11. Engine-case temperatures at locations where the electronics would be mounted range from approximately 175°F (80°C) at Mach 0.9 and 40,000 ft. to 600°F (315°C) at Mach 2.5 and 50,000 ft. This temperature range clearly exceeds the MIL-STD high-temperature limit of 125°C for existing Si-based devices. Hence, once again it is obvious that existing Si electronics cannot do the job in the absence of active cooling. It is assumed that electronics capable of reliable operation at temperatures of at least 350°C will be required to enable the local engine network distributed control architecture.

8.2.8 Required high-temperature components

In order to implement these advancements in aircraft electronics subsystem performance, a wide variety of components capable of high-temperature operation must be developed. The electronic components required to enable each of the four aircraft subsystems (DFC, PMAD, MEE, and SMS) are very similar. Electronics for power control and digital signal processing employ nearly identical components in each application. The DFC system requires electronics for actuator power control and smart actuator functions. A typical resonant link inverter circuit used for actuator power control will require capacitors, switches (MOSFETs, IGBTs, and MCTs), and rectifying devices (p/n and Schottky diodes). Expected power requirements include 50, 100, 200, 400, and 800 amps at voltages up to 1000 V. Smart actuator electronics are necessary for actuator position and pressure transducers, servovalve and solenoid drivers, and data bus interfaces. The minimum requirement for these electronics is a chip-module containing a 16-bit microprocessor with on-chip ROM, RAM, and I/O, a high speed bus serial interface, and D/A and A/D signal converters. The power control electronics required for PMAD system DC-DC and DC-AC power conversion and MEE distributed network engine control are comprised of the same fundamental components employed by the DFC system; capacitors, switches and rectifying devices. Digital signal processing requirements necessary for driver circuits, sensors, and diagnostics are essentially the same as those listed above.

8.3 Commercial Automotive

The product design drivers for non-military vehicles are cost, delivery time, quality, and reliability. The competitive nature of the vehicle business dictates that automotive electronics be inexpensive, especially for small, low-priced vehicles. In some cases, an electronics device, and thus an additional cost, is mandated by federal or state legislation. While most vehicle applications for electronic products require thousands of units, this high volume sometimes can be an advantage for some parts and materials. Custom parts, on the other hand, have high up-front development costs that must be amortized over many production parts. Up-integration of individual parts can lessen this problem, since it leads to fewer custom parts overall. Often, higher priced parts and materials are available, but their cost cannot be justified for the feature or function required. For example, active cooling systems can rarely be justified. This reliance on passive cooling solutions, however, puts substantial pressure on the product designer to use the required parts and materials. A significant number of products and product features appear now on vehicles because their decreasing costs have justified their use. In this case, the price has helped to make the market, a trend that is likely to continue.

The use of standard parts, or at least standard part packages, is almost a given. In spite of the large quantity of individual parts used, the automobile market seldom drives a technology. Rather, the vehicle business tends to adapt and use existing technology, a design disadvantage since many devices used in cars were developed for more benign environments than automotive applications. Examples of such products include personal office computers, home entertainment systems, and indoor telecommunications. Product and process engineers are responsible for ensuring the function, quality, and reliability of these products for automotive use. While it would be inaccurate to call electronic automobile parts "commercial off-the-shelf" (COTS), as opposed to Military Standard Specifications, they can be described as custom-designed and built from the same COTS technology parts and materials. Their production thus differs little from the military production of custom-designed and built parts. At the same time, the major environmental factors that compromise reliability — temperature, vibration, humidity, electrical transients, and salt exposure — are absent or minimized in the home or office environments in which designers of vehicle electronics operate. The countervailing market requirement for electronics of higher reliability nonetheless directs pulls the designer toward more expensive parts, materials, and designs, while pressure builds to use smaller and lighter parts, this is the classic engineering dilemma — trading one factor for another.

This is further complicated by great variance in the usage profiles of vehicle customers. For example, the need to accommodate both multiple, short, low-speed trips and long, infrequent, high-speed trips requires significant design variables. The customer expectations of passenger car owners vary considerably from those of truck owners.

Similarly, the external environment encountered just in North America alone requires consideration of a great range of design maximum and minimum "corners". The designer must know the range, frequency, and severity of the environmental stresses the product will encounter in order to design and test products. The distribution of these stresses tells the product designer what safety factor must be designed into the product strength. The designer must strike a delicate balance between over-designing or under-designing; the former incurs an unnecessary cost penalty, while the later allows field failures. Somehow, a balance must be struck to ensure both a good product and a successful business.

Temperature, as an environmental factor, plays a major role in the design of cost-effective electronic parts and materials, virtually dictating what technologies can be used in the various application zones of a vehicle. The four most commonly specified zones of a vehicle are in

Table 8.1 Current and Projected Required Operating Temperature for Automobile Products

Zone temp. range	Current temp. range (^{0}C)	Future (^{0}C)
Passenger compartment	-40 to +85	-40 to +85
Underhood	-40 to +125	-40 to +165
On-engine	-40 to +140	-40 to +165
Wheel-mounted	-40 to +250	-40 to +250

order of increasing temperature demands, the passenger compartment, under the hood, on the engine, and on the mounted wheel (see Table 8.1).

The -40°C lower limit requirement comes from North American FAA Weather Station data, and is usually considered the storage and an operational requirement. On occasion, storage requirements are set at different levels. For example, a radio in inventory may have one storage temperature requirement and another for "storage" when the radio is turned off. The upper limits depend on the proximity of the products to heat-generating zones, and are almost always operational requirements.

The vehicle layout and packaging may limit the effectiveness of the passive cooling. For products with significant self-heating, this can be critical.

8.3.1 Automotive underhood electronic systems

Among the underhood electronics currently available are ignition systems, voltage regulators, pressure sensors, engine-control modules, and anti-lock brake systems (Figures 8.12 to 8.17). Many of these are located on the engine, with the others located on the chassis around the engine. The circuit complexity varies from very complex microprocessor circuits to simple high-gain temperature-compensated sensor circuits. These products must be designed to operate in the underhood environment.

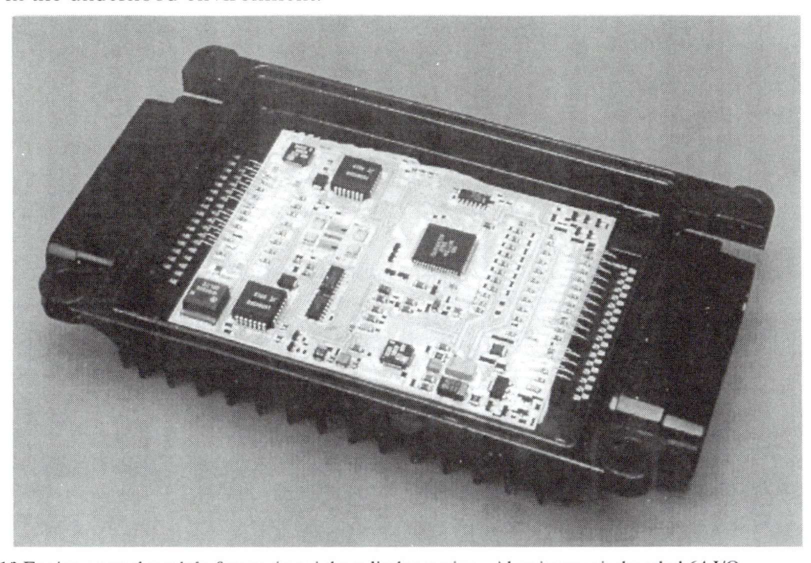

Figure 8.12 Engine control module for marine eight cylinder engine. Aluminum wirebonded 64 I/O.

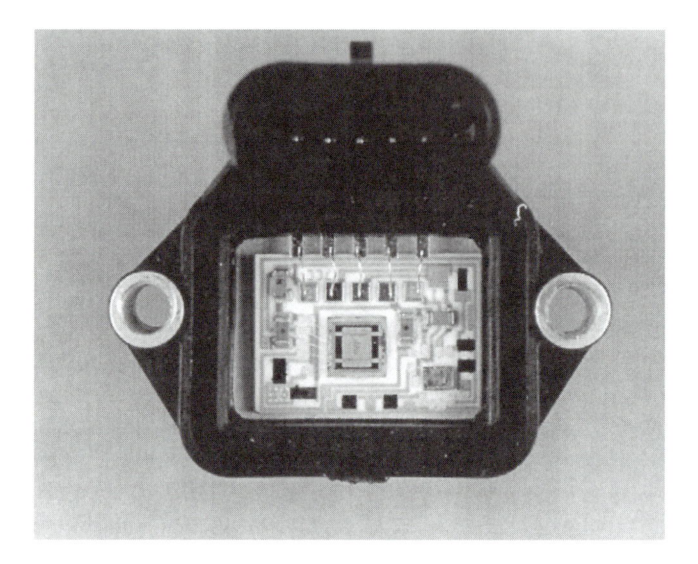

Figure 8.13 Accelerometer for air bag deployment; silicon micro-machined sensor shown in the center.

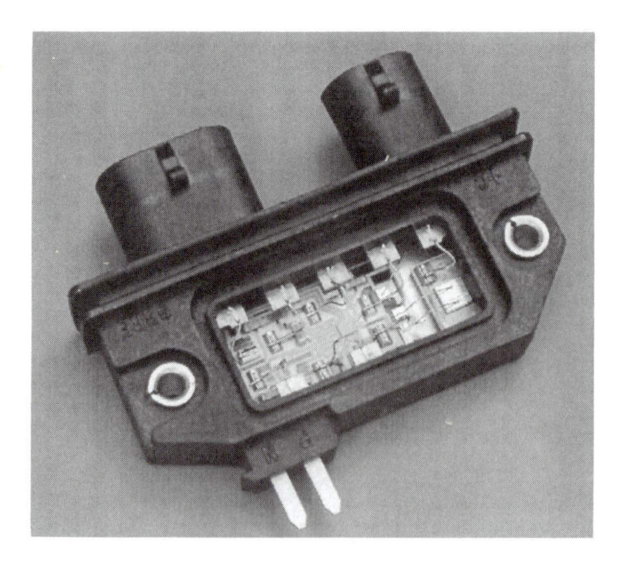

Figure 8.14 Distributor ignition module; for mounting inside the distributor; showing resistance welded interconnect and flip chipped IC.

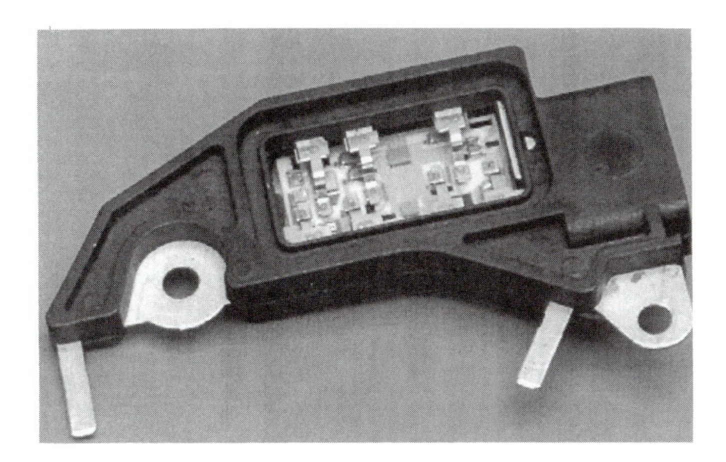

Figure 8.15 Voltage regulator; designed for mounting internal to the generator; shown with laser welded interconnect system.

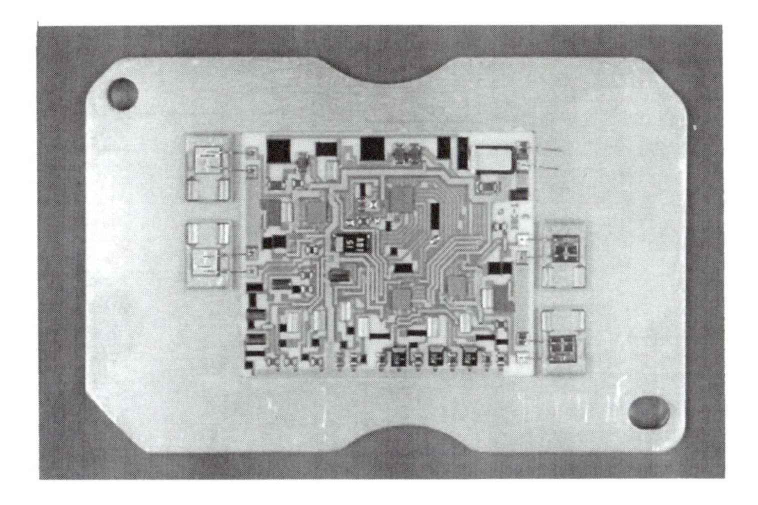

Figure 8.16 Direct ignition; for eight cylinder engine; four power drivers on separate substrates of BeO ceramic.

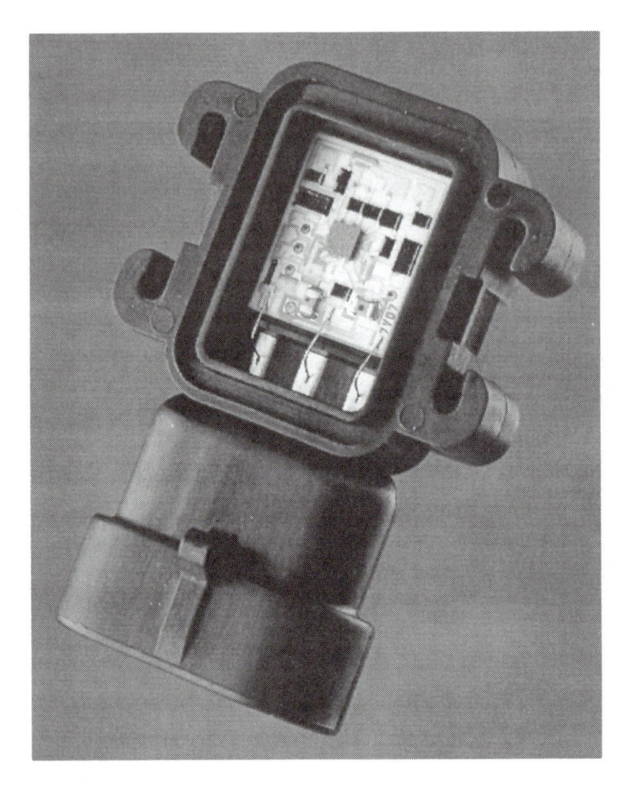

Figure 8.17 Pressure sensor; designed for mounting on engine.

The physical complexity and compactness of the engine compartment environment prevent a clear definition of the environmental conditions. At some locations under the hood, the temperature gradient is over 200°C due to the difficulty of establishing and maintaining air flow throughout the engine compartment over the wide operating speed and load ranges of the automobile. Speed may exceed 80 miles per hour, with loads varying from none (other than the weight of car and driver) to heavy trailers being pulled up steep mountain roads.

Vibration is dependent on the structure to which the product is secured. Chassis mountings are characterized by lower frequencies, while engine mountings have higher frequencies. In either case, the amplitudes and frequencies are dependent on the specific location, orientation, bracket design, and engine.

A product can be designed for a specific vehicle; this is difficult for a new vehicle because the specific mounting environment characteristics cannot be verified until after a prototype is built and tested. Early modeling can provide environmental estimates, but these may change before the vehicle reaches production due to design changes during development or option changes such as different power trains. Other changes can take place any time during the product life.

The variability in the environment definition increases when the user's operational mission profile is considered. Wide variations in the underhood environment are created by different climates and vehicle applications. Summer or warm climates raise the operating temperatures;

city driving and trailer pulling in the mountains typically produce peak underhood temperatures due to high thermal loads and low air flow.

The challenge is to produce a product with the widest possible operating specifications. This requires that products be designed using a system of materials and technologies that can function with high reliability at high-temperatures with minimum additional expense. It also requires that the designs and technologies be vibration hardened to a wide range of frequencies and amplitudes so that the effect of the mounting location is minimized.

Hybrid circuit technology, often using ceramic substrates, has long been known for its high-temperature capabilities and its immunity to shock and vibration; indeed, it has been used under the hood and on engines for twenty-five years with extremely good reliability. Hybrid technology has higher operating temperature capability than other circuit-fabrication technologies. However, to continue to meet the demand for even higher operating temperatures, all the parts and materials that make up the system need to considered and redesigned. (see Figure 8.18). All parts and materials must be capable of performing their functions during the operation of the product at all of its operational environments.

The electrical system dictates the function of the product, providing the necessary electrical output for a given electrical or other type of input. The characteristics of active and passive components must remain sufficiently stable over the specified operating temperature range so that the electric circuit can respond properly to a given set of inputs. For example, resistors and capacitors change value with temperature, and transistors change gain. For circuit operation at high-temperatures, all parameters that change with temperature must be compensated for in the design or operation of the circuit.

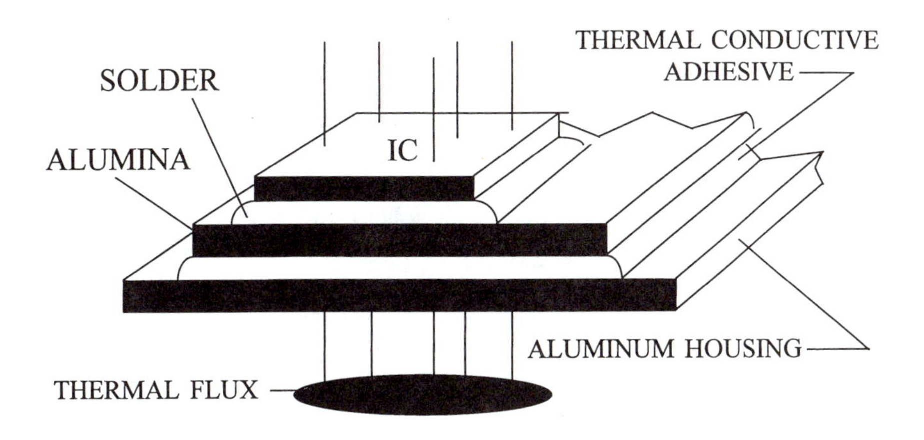

Figure 8.18 Typical hybrid construction.

Among passive components, resistors and capacitors have temperature coefficients of resistance and capacitance, respectively. The resistance change with temperature is small and only important in critical circuits. The capacitance coefficient can be non-linear with some types of capacitors and very linear with others. Hybrids use ceramic capacitors made of various material compositions; the NPO type is very linear, while the X7R type is very non-linear at higher temperatures. Ceramic capacitors are electrically stable, but electrolytic capacitors are unstable and must be voltage-derated with temperature. For example, high-value tantalum capacitors are temperature-stable. Stability is a reliability issue that must be addressed by either eliminating the need or developing a capacitor that will meet the high-temperature requirements. Other passive components, such as inductors, varistors, and crystals, will have similar temperature-sensitive characteristics that must be addressed to assure that the circuit will function properly in the anticipated environment. These components do not have problems with electrical temperature characteristics, but they maybe physically limited by the packaging of surface-mounted devices.

Nearly all of the active devices used in the underhood hybrids are made of silicon. These are mostly integrated circuits (ICs) of various types and devices for actuating the product's output functions. Some products have a processor IC and some form of memory, either on the processor or as a separate IC. Others use special-purpose ICs, such as pressure sensors. Most ICs are designed and characterized for a maximum junction temperature of 150°C. At higher temperatures the gain and the junction leakage currents will increase which may cause latch-up or other changes in device characteristics. To provide increased operating temperature, each IC must be characterized at the higher temperatures, with design changes as necessary. In power devices where the losses are greater with increasing temperature, adjustments must be made in the thermal structure or in the IC size to avoid an increase in power dissipation. The greatest challenge is in processor-based products, where the usual maximum design temperature is 125°C for the processor and memory. Memory retention time may be a problem.

Thermal systems within and outside the product are relatively independent of the operating temperature of the electronics. The internal thermal design is not affected by temperature unless there has been a structural change in the product design. There are two possible reasons for changing the thermal structure: (1) if the higher temperatures increase the power dissipation of some devices or (2) if the temperature affects the structural materials. In this case either a different or redesigned materials must be used. Since the internal thermal impedance is highly dependent on the IC size (see Figure 8.19), which determines the thermal flux density (watts/sq. in.), each active device is a special case for which the thermal path must be evaluated. Some active devices are substrate-mounted, while others with very high thermal flux density are located off the substrate on special thermal structures. The external thermal design would reflect the specific product application; for example, the product could be designed as an absorber for a conductive heat sink, or as a convective heat sink to the ambient air. In either design, the thermal impedance of the case to ambient or heat sink will not change with a change in the sink or ambient temperature. A worst-case thermal design will maintain the module case temperature at its maximum value, with maximum product loads at the maximum environmental temperature. Seldom do products operate under worst-case conditions, but they must have the capability to do so. For applications in which high heat-sink temperatures are not encountered, high-temperature technology can be used to decrease the case size.

Mechanical designs are not temperature-dependent, and therefore do not require re-design for increased operating temperature. However, other factors could necessitate some mechanical design changes for higher operating temperatures. This includes the dependence of material properties on temperature, the extended temperature excursion range produced by increased temperature. Differences in the coefficients of thermal expansion (CTEs) of attached materials create a reliability issue requiring careful design and testing to avoid a failure mode. This is problem in nearly all forms of electronic packaging due to the very low expansion coefficients of the active devices and the much higher coefficients of packaging materials; temperature is at least a linear stress factor and may be much greater, depending on the materials involved. These difficulties can be resolved with mechanical design and/or material changes. Since these changes must still be affordable, this may force a design trade-off. For some current products this may not be an issue; others will require extensive development for higher temperature operation.

Material systems used in the packaging of hybrid circuit products have a definite

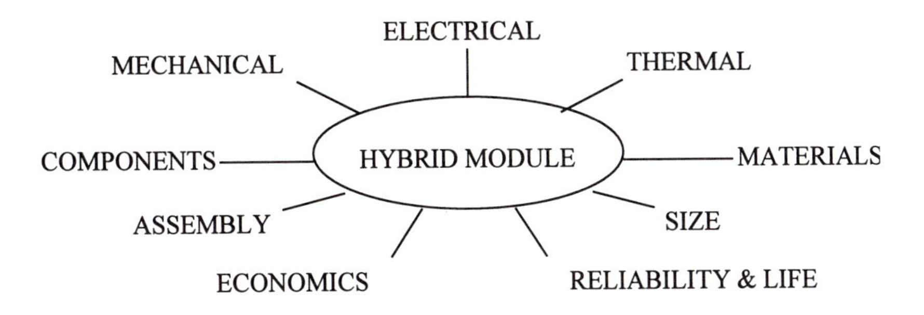

Figure 8.19 System picture.

temperature sensitivity, since all material properties are a function of temperature. The properties of some materials will change much faster than others; for example, in the temperature range of interest (-50°C to +200°C), metals such as aluminum have very little linear change, while metals used in solders have pronounced property changes. Typically, the properties change faster as the melting point of the material is approached. Many plastics also fall into the fast-change category in this temperature range. Organic materials typically used for insulators, housing bodies, and covers may lose strength and become soft; these would need to be replaced with a different material. Many of these organic materials go through a glass transition phase change in this temperature range, which drastically changes the properties; the strength, hardness, and expansion coefficient all change in undesirable directions. Housing materials of polybutylene terephthalate (PBT) may need to be replaced by polyphonylene sulfide (PPS) material for high-temperature stability. This is not a major issue for high-temperature products since many available materials would provide the required characteristics; cost drives the selection. Other materials that could have problems related to higher temperature are the adhesive and passivation materials. Many of these materials are from the silicone family and are typically stable up to about 200°C. Other common adhesives are from the epoxy group; these experience property changes or loss of adhesion at higher temperatures. Careful characterization is required to use these materials.

The most problematic material at temperatures above 150°C are the solders used in surface-mounting components on the ceramic substrate. The solders provide many functions: they conduct electrical current as part of the circuit; they secure the component to the conductors on the substrate; and they provide a compliant attachment to prevent a CTE mismatch between the substrate and component from overstressing to the point of mechanical failure. Achieving a compliant attachment is very difficult, since it must also have a melting point above the operating temperature of the components and below some higher temperature that would degrade the components being assembled. At the same time the solder must have a reasonable cost. Solders can be composed of several metals, such as lead, tin, indium, and antimony, in an infinite number of combinations. The chosen solder must also be compatible with the materials with which it interfaces. It must wet these materials and not form undesirable compounds (e.g., some intermetallics) that may adversely affect the compliance or strength of the material. The solder must have sufficient fatigue life to survive the rigors of automotive service.

Assembly methods are driven by costs, labor, and capital expense. The most effective system requires minimum labor and equipment; the ideal system assembles itself without any equipment. The trend is toward parts that are self-aligning and self-securing, eliminating the need for aligning and securing equipment (curing ovens, screwdrivers, etc.). This focuses the technology on attaching parts that stay secured when placed in position. Currently, a typical manufacturing system uses two axis robots to apply adhesive that is cured in place. The curing ovens are a capital expense that require power and occupy floor space, and the time required for curing interrupts continuous line flow. This problem is being addressed by developments in polymer adhesive technology, which must produce materials that adhere to various others used in product construction, while retaining adhesion and mechanical properties at the higher temperatures and long life demand by automotive products.

Hybrid products have historically been small. In the early 1970s, the only available ceramic substrate was small, as were early circuits using the relatively high circuit density the hybrid can offer. Over the last twenty-five years, the circuits and applications have become much more complex and the push toward smaller size has continued. Size, like operating temperature, is driven by the very limited space conditions under the hood. This is a consideration when changes increase the operating temperature. Size reduction increases the package thermal impedance, which increases junction temperatures. In high power-dissipation products, this requires a trade-off between product size and operating junction temperature—another reason for operating at higher junction temperatures. The other trade-off is the higher junction temperature in some devices, which increases their thermal losses. The drive for higher complexity and smaller size has produced circuits of much higher density and more layers. The substrate size is determined by the area of surface-mounted components and ICs, even when resistors and capacitors are buried in the layers of circuits.

Using higher temperature electronics is only one of several methods of solving the problem of operating electronics in high ambient temperatures. Active cooling methods such as fans, heat pipes, and refrigeration, can also be used to solve the temperature problem. This forces high-temperature electronics to be economical to compete with other methods of enabling high-temperature operation. Automotive applications are relatively high volume and very cost-competitive by almost any standard. Again, the key to success with high-temperature electronics is low without sacrificing required functionality, small size, and high reliability.

8.3.2 Semiconductors

During the past several years, the semiconductor industry has begun to understand the subtleties of semiconductor design, materials, and manufacturing. Many major changes have taken place:

- significant advances in semiconductor design, processing techniques, and equipment
- reduction of particulate contaminants
- incorporation of statistical process control
- advances in the packaging process, including wirebonding and molding material characteristics (improved adhesion, lower thermal expansion, lower stress, and improved moldability)
- elimination of chlorides from the plastic packaging environment (molding compound and die bond adhesive)

These changes have addressed the primary failure mechanisms in semiconductors, especially in plastic packaged devices, and have helped propel the electronics industry into unprecedented growth.

Because of the fierce competition in the industry, automotive electronics manufacturers are acutely aware that the performance of their products is, to a large degree, tied to the performance of the semiconductor devices used in particular integrated circuits. Several years ago, Delco Electronics Corporation began tracking the repair data on these products with warranties for all parts. The data from certified repair facilities were based on computerized tracking of all part replacements required to repair product warranty failures. The facilities also used another valuable strategy: they counted a part as a failure only when replacing the part fixed the product. This data, combined with part usage information, enabled them to determine and project failure rates for the five-year/50,000-mile warranty period of the engine control module (ECM). This performance data, involving hundreds of millions of units per part and analyzed on a yearly basis, provides a true indication of the field reliability of parts and the effectiveness of improvements that continue to be made.

Figures 8.20 through 8.23 show the cumulative failure rate for all electrical parts used in Delco Electronics Corporation products for the years 1990 through 1994. Note that the curves clearly indicate the infant mortality and constant failure-rate segments, both of which agree with the classic reliability "bathtub curve". Note also that infant mortality decreases between 1990 and 1994, and this change coincides with a decrease in the constant failure rate.

The four-year performance of semiconductors in particular is given in Figures 8.24 and 8.25.

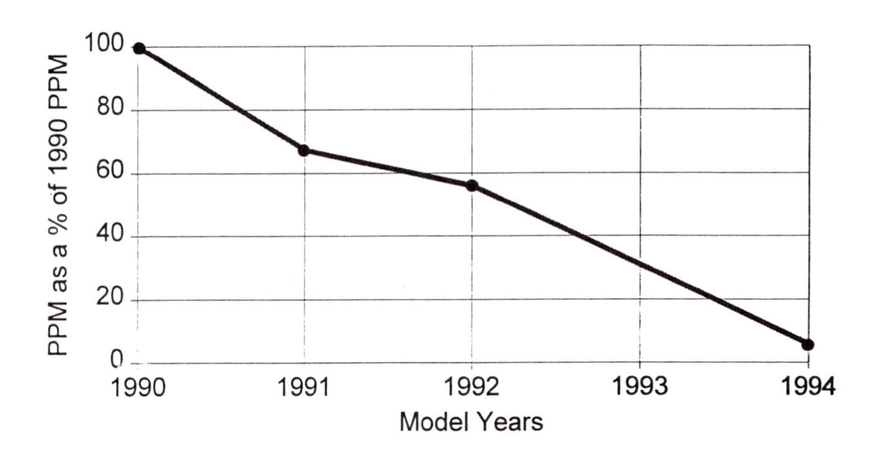

Figure 8.20 Average electrical parat failure rate for engine control modules.

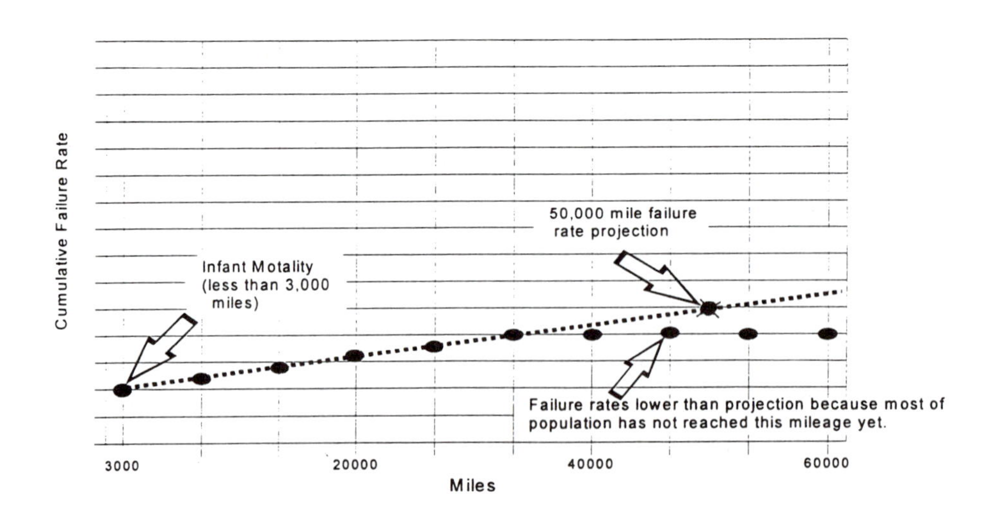

Figure 8.21 Typical failure rate curve for all electrical parts used in engine controls.

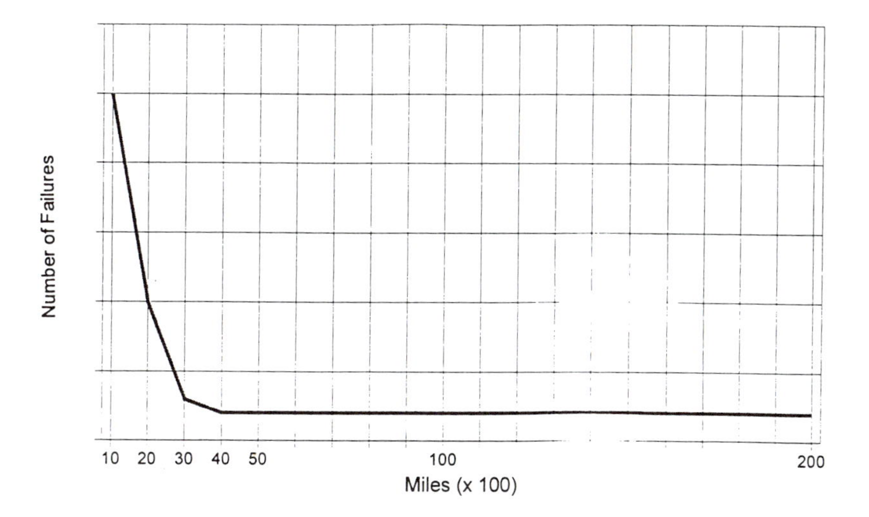

Figure 8.22 Classic reliability bathtub curve.

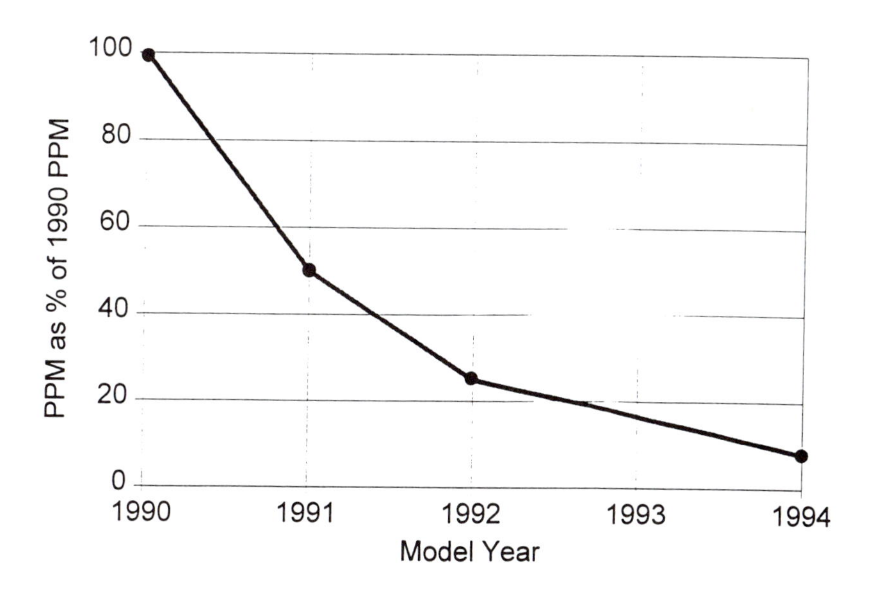

Figure 8.23 Changes in infant mortality (0 to 3,000 miles) as a function of model year.

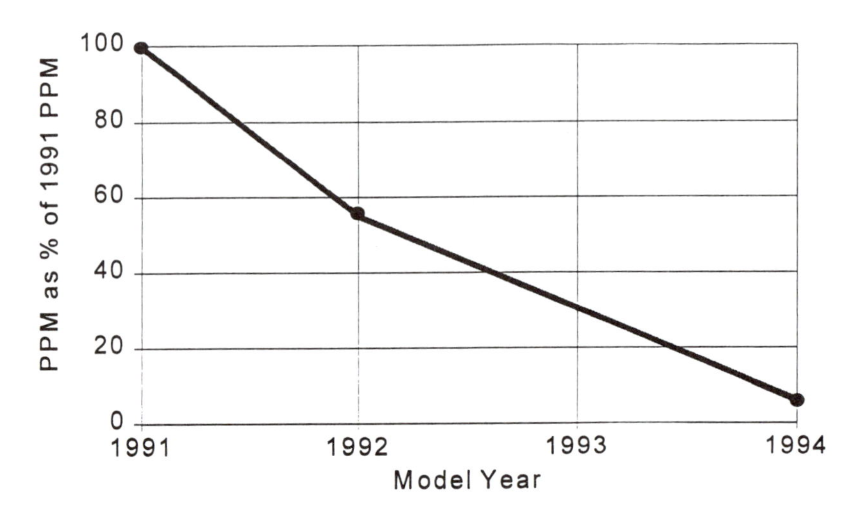

Figure 8.24 Failure rate for surface mount Zener Diodes.

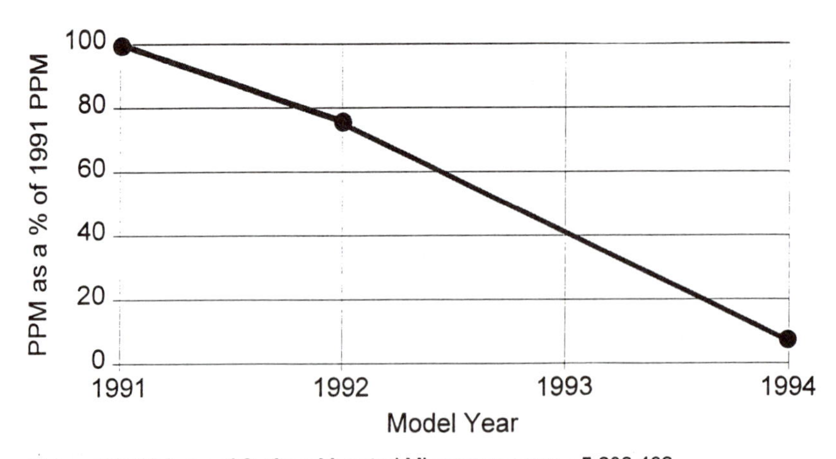

Figure 8.25 Failure rate for surface mounted microprocessors.

The complexity of these parts continued to increase during this period, as shown in Figure 8.26, but the failure rate continued to decline. Comparisons of the failure rates experienced in real applications with data from MIL Handbook 217 clearly show the data and assumptions presented in 217 are, at best, obsolete.

The performance of all individual semiconductor parts is not necessarily equal to devices of the automotive electronics. The industry has demanded qualification requirements that these devices must pass before being used in products.* Other commercial parts designed and constructed to meet less demanding requirements may not demonstrate the same degree of reliability.

The potential for vehicle applications of high-temperature electronics continues to grow. New applications are creating new opportunities with new challenges in what will continue to be a very competitive worldwide market.

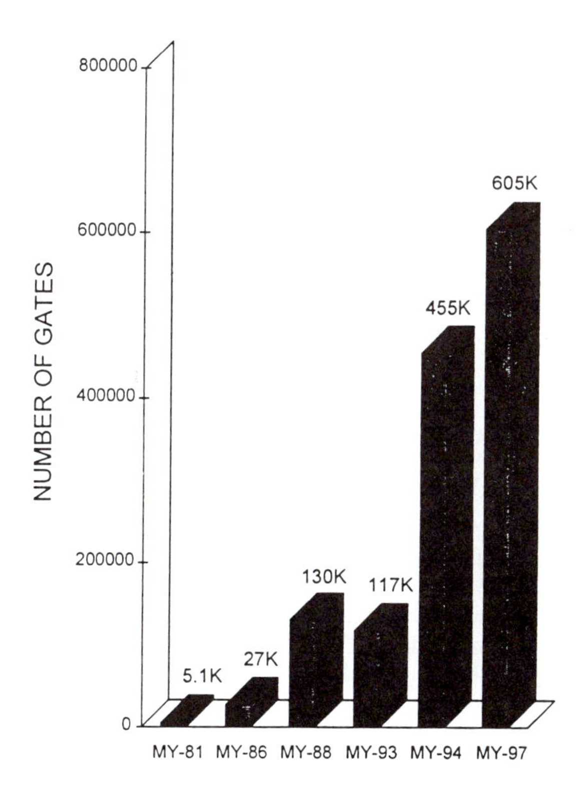

Figure 8.26 Changes in microprocessor gate count.

*These requirements are detailed in CDF-AEC-Q100, Rev. A, May 19, 1995, "Stress Test Qualification for Automotive-Grade Integrated Circuits." This document was prepared jointly by representatives from Chrysler Corporation, Delco Electronics Corporation, and Ford Motor Company.

8.4 Military Vehicles

The U.S. Army's interest in high-temperature electronics is primarily focused on their use in ground vehicles. Future Army ground vehicles will rely more and more heavily on electric power to accomplish tasks that are now accomplished by mechanical and chemical means.

The most commonly envisioned system of this kind is the All-Electric Tank. A view of one such unit is shown in Figure 8.27

Although the term All-Electric Tank is a misnomer in that an engine is used to generate power, that power is converted to electricity and powers the vehicle's propulsion system, its main weapon, the actuation of the turret and suspension, and several proposed defensive systems. Since the battle tank is the foremost Army ground vehicle, it has received the most attention in the effort to adapt a vehicle to the extensive use of electric power.

In a vehicle like a tank, interior volume is a very valuable commodity and is heavily protected. Indeed, the interior volume of a tank is more important than its weight as a figure of merit. This implies several considerations in the design process.

The first is that systems are competing with each other for space and must therefore be made as small as possible. Furthermore, any proposed new system must have minimal requirements for auxiliary systems. For example, electromagnetic guns have been proposed as tank main weapons. These devices are thought to be more effective than conventional, chemically propelled guns. However, the large energy generation and storage requirements of such systems have led to the suggestion that a second vehicle, in effect a trailer, carry the additional components. This is clearly impractical. Any proposed system must not incur an excessive space burden.

All the electric power systems on an all-electric tank require power conditioning semiconductors . Since the power levels range from hundreds of kilowatts to megawatts, such systems, even if conventional low-temperature power electronics were used, would require extensive cooling systems to operate at temperatures less than 100° C.

For these power levels, natural convection or even forced air cooling systems are not practical. A more practical liquid coolant system requires liquid-to-air radiators, pumps, piping, and controls, both mechanical and electrical, in addition to device-to-coolant heat exchangers (cold plates). Each of these components extracts a heavy space penalty. The liquid-to-air radiators are notoriously large, since there is little temperature differential between the liquid coolant and the ambient air. This is particularly true if the system must operate in a desert environment, as has been the case in recent history. This lack of efficiency translates into a radiator with large surface area and large volume.

The ideal solution, which would enable power semiconductors to be used in the systems of the All-Electric Tank, would be to have the power electronics reject their heat to an existing vehicle fluid at a temperature substantially higher than the highest anticipated ambient temperature.

A suitable fluid already exists on tanks and other military ground vehicles, the engine lubricating oil. Operating at temperatures of about 200° C, the oil is commonly pumped, piped, and cooled via a liquid-to-air radiator. Additional cooling would require only an increase in the capacity of these components, rather than an additional new cooling system.

High-temperature power components, capable of operating at mounting plate temperatures in excess of 200° C, would be ideal for such an application. Other properties of high-temperature power electronics, such as higher operating voltages, could also be effectively utilized on Army ground vehicles. These would tend to reduce conduction losses by requiring less current for a given power requirement. These lower currents could flow through devices with smaller cross-sections, due to their higher allowable current densities. This results in more compact devices.

It is useful to examine the individual application of power electronics to a vehicle such as

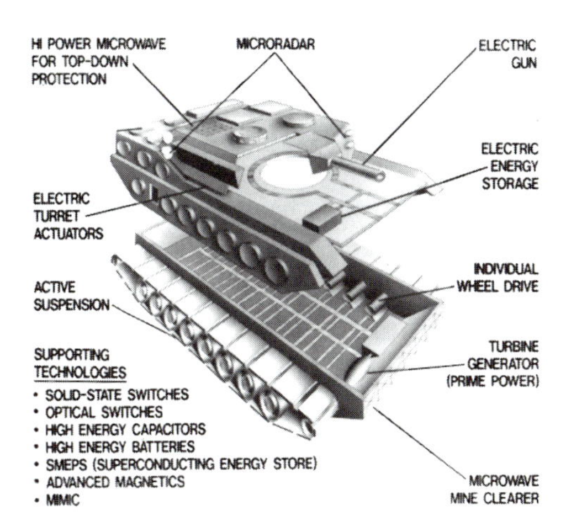

Figure 8.27 All-electric tank concept

the All-Electric Tank, first addressed by Braun and Podlesak [1993]. The most widely discussed application is for propulsion. Electric drive motors on each tread sprocket, or even each wheel, would allow the tank engine to operate at a constant speed, increasing fuel efficiency and reducing acoustic and chemical emissions. This is more critical on the battlefield than it would be in civilian applications, since such emissions may reveal the tank's presence to the enemy. These drive systems, with single-drive motors for each tread, transmitting via the sprocket, use on the order of 300 kW per motor. The motor drives and their associated semiconductors would have to be able to handle such power levels.

High-temperature power semiconductors may be mounted directly on their loads, in this case, the motors. These motors will need oil for lubrication and, probably, cooling. This oil would also be available for cooling the electronics. Motor temperatures well in excess of 100°K above ambient are very likely; semiconductor switches mounted in such a fashion could easily experience mounting plate temperatures of about 200° C.

These motor drives would also require components other than power semiconductor switches. Capacitors and inductors would be needed as filter elements of the DC links proposed for such systems. Auxiliary components, such as resistors and diodes, would protect the power semiconductor switches. And all of these systems would require low-level electronics for control and for highly desirable built-in diagnostics. If the power semiconductor switches were mounted on the motor load, then all of these components would be subjected to the same conditions and would have to be high-temperature components.

Mounting semiconductors directly on their motor loads is not restricted to just the large drive motors. Currently, actuation of various tank systems is done largely with hydraulic systems. These systems are complex, and maintenance intensive, and the hydraulic fluid is a fire hazard. Replacing such systems with electric motors would alleviate these problems. Such applications include actuating of the tank turret, aiming the main weapon, or use in an active suspension, with motors acting like springs and dampers, adjustable to the varying terrain and situation. Again, directly mounting the power electronics on the motor itself would save critical volume.

Perhaps the most critical applications of these proposed systems concern the tank's ability to defend itself. This includes the use of electric guns, rather than conventional chemical-based munitions. There are several types of electric guns, ranging from electromagnetic guns, which accelerate their projectiles magnetically much like electric motors, to electrothermal-chemical (ETC) guns, which inject electrical energy into a burning propellant, increasing the delivered energy to the projectile and improving its effectiveness. The various electric guns are more efficient than their chemical counterparts. They have several serious technical problems. In addition to the volume necessary to store the electrical energy, another key factor is the control of such large amounts of energy.

A series of workshops conducted by the Pulse Power Branch of the Army Research Laboratory at Fort Monmouth, NJ in the early 1990s came to the disturbing conclusion that to operate electric guns as required by modern warfare, such large amounts of electrical energy would have to pass through the power semiconductor switches in such a small period of time that there would be insufficient time to remove heat from the devices during normal operation. Therefore, provision must be made to operate such switches in an adiabatic mode. Consequently, the temperature of the devices will rise dramatically, easily exceeding the temperature possible with conventional low-temperature semiconductors. This is an ideal application for high-temperature devices.

The main battle tank is not the only vehicle that could benefit from these technologies. Armored infantry fighting vehicles, lightly armed vehicles capable of carrying troops, could use the drive, actuation and, if practical, electric weapons systems derived from the tank. Armored personnel carriers, used to transport troops on the battlefield, would also need drive and suspension actuation. Various transport vehicles, such as trucks and HMMWVs (Highly Mobile Multiwheeled Vehicles) may also benefit from electric drives, particularly if the engine-driven generator could be used for generating power to run electrical equipment, either mounted on the vehicle or external to it. In other words, the vehicle could be used as a power station. Here again, the advantage is reducing emissions to better conceal the vehicle from hostile forces. Similar ground vehicles may be used by the Marine Corps, with additional requirements for amphibious vehicles.

The need for high-temperature electronics is not confined to ground vehicles. The Army maintains a large fleet of helicopters, and electrical systems similar to those for ground vehicles have been proposed. In these applications, not only volume but also weight is critical. The challenges for fixed wing aircraft also apply here. The rotating wing of the helicopter may be electrically driven, eliminating the need for complex, heavy mechanical transmissions to transmit power from the engines to the rotor. Electric actuation systems could eliminate hydraulics, and helicopters carry weapon systems. One interesting aspect of having both Army ground and air vehicles use electric technology for major onboard systems is that it could lead to commonality of systems, perhaps by using the same engine for both an Army ground vehicle and an Army air vehicle. This would simplify the logistics of maintaining a force in the field, reducing the number of unique components and service personnel for individual systems. Such commonality is being pursued in other areas. For example, the Army is trying to run all of its vehicles in a particular area on one type of fuel. Common systems designed to minimize both size and weight would be extremely advantageous to the twenty-first century Army. Systems based on high-temperature electronics could contribute significantly to this goal.

8.5 Geothermal Measurement

One important application of high-temperature electronics is their use in the borehole logging tools typically required by the geothermal and deep oil and gas industries. The authors have worked on the development of high-temperature borehole instruments for many years; in this section they explain some of the practical issues associated with the implementation of these systems.

The principle issues to be considered when designing a new borehole tool can be divided into three interrelated sections :

1. Electronic systems
2. Thermal systems
3. Mechanical systems

8.5.1 Electronic Systems

Electronic systems can be designed and built for continuous operation at 200°C. This method requires no cooling and relies upon semiconductor technology to withstand the ambient temperatures within the borehole tool.

This technique enables indefinite operation at high temperatures with mechanical simplicity. There is no benefit in combining such a system with passive cooling since this would limit the time the tool can operate and duration. However, the use of active cooling systems can extend the operating temperature, but with increased complexity.

High-temperature electronics are still very much in their infancy. The technology was advanced somewhat in the late 1970s when the U.S. Department of Energy commissioned a number of studies, principally for use in the geothermal industry. The work demonstrated that certain semiconductor technologies can function at temperatures up to 300°C. Most of the work addressed component performance, not the operation of systems in which component specification is critical. As a result, it was possible to generalize about the temperature performance of a specific technology, such as metal oxide semiconductors (MOS), but difficult to obtain a suitably fabricated part with the required specification performance.

Since then, semiconductor manufacturers have produced high-temperature (200°C) integrated circuits of varying complexity but the small market for the products has limited the actual product range. High-temperature electronics above 220°C are still very much in the laboratory.

The current upper limit for continuous operation is probably 200°C. This can be extended by using an active cooling system to approximately 250°C, using thermoelectric heat pumps. Complex mechanical designs using compressor systems may extend this beyond 250°C, though with a loss of system reliability and increased size and cost.

Components. For the past ten years, we have been researching and developing electronic and allied systems that can operate reliably at 200°C. The original applications were for geothermal seismic monitoring instruments that had to operate for long periods at these temperatures. Consequently, a database of electronic parts that can operate continuously at 200°C has been developed. The electronic component evaluation program has also been developed and was initially based on the published work of Sandia laboratories, but was finally extended to cover a broader range of component types and technologies.

These components have been used to build a range of electronic systems that operate at high temperatures. For example, computer-based data acquisition, control, and telemetry systems that can operate continuously at 200°C.

Components considered suitable for a particular application were tested for function and parameter variation at temperatures up to 220°C. The testing concerned itself specifically with the selection of parts for the application being investigated. Therefore, many inferior aspects of a particular device's function were accepted as adequate for or irrelevant to the application. Life tests at 220°C were carried out for periods in excess of 1000 hours.

The basis for the electronics construction was component evaluation at 220°C; electronic systems construction, testing, and evaluation at 200°C; and long-term tool operation at 180°C.

Tool management systems. A data acquisition, control, and telemetry unit has been constructed for continuous operation at 180°C. The system uses CSMA qualified standard electronic components and a limited range of high-temperature parts available from special manufacturers.

A block diagram of the system is shown in Figure 8.28. It combines six channels of high-bandwidth (4 kHz) seismic data with a tool management data acquisition and control system. Seismic data is telemetered in a frequency modulated (FM) format with a modem operating at the low end of the multiplex sending and receiving tool management data.

Tool management is critical to the operation of borehole tools in high-temperature environments. Typical information required of the tool operational conditions are: power supplier states; internal temperatures (various); environmental temperatures and circuit reference points. Control outputs can be achieved locally under software control or from the surface.

Future electronic systems. Considered as a whole system the temperature at which present electronics can achieve function and reliability is probably 200°C. New processing techniques with silicon (SOI) may increase this to 250°C. However, moves to 300°C plus will probably only be achieved by new technologies - silicon carbide (SIC), diamond, and so on. This is a fairly distant prospect, although some limited SIC parts are available now.

Less glamorous passive components - circuit boards, connectors, wiring, and so forth - will need more attention in the future for any of the new technologies to find commercial applications.

8.5.2 Thermal systems

The thermal system has to control the temperature distribution within the tool's internal environment. This statement may seem obvious, but it is the fundamental task and goal of the design. Subject to the temperature limits of the electronic components, the "thermal system" could be as simple as a heat sink to remove a hot spot on a discrete component or as complex as a full heat shield and thermal transport / heat management system.

Sections describe the various "components" that can be combined to form a comprehensive thermal system.

The heat shield. The heat shield, or dewar, is the fundamental component in a thermal protection system. The dewar usually consists of two concentric, thin-walled, stainless steel tubes joined together and sealed at the ends, with the annular volume between them evacuated.

All dewars for borehole tools are custom-made, and though the cross-section may be generally circular, the end details vary considerably.

A typical dewar may be 75 mm (3 inches) in outside diameter with a 4 mm wall, giving an internal diameter of 67 mm. One end of the dewar will be fully open, allowing access to the full internal diameter. The other end may be completely closed or might have a small diameter hole to allow access for wiring and other services. It is usual for the closed end of the dewar to include the bellows assembly (convoluted tube) that forms part of the evacuated shell and allows axial deformation to take place. The bellows are essential because the dewar can have very large internal and external temperature differentials with the external surface hotter during the run into the borehole and the internal part hotter upon recovery to surface. A total temperature range on the order of 500°C is not uncommon. Figure 8.29 shows a typical dewar in cross-section.

The thermal performance of dewars varies from one manufacturer to another and from one dewar to another. In general, dewar suppliers quote minimum performance figures, average values that apply to the dewar as a whole. While these values are useful for summary

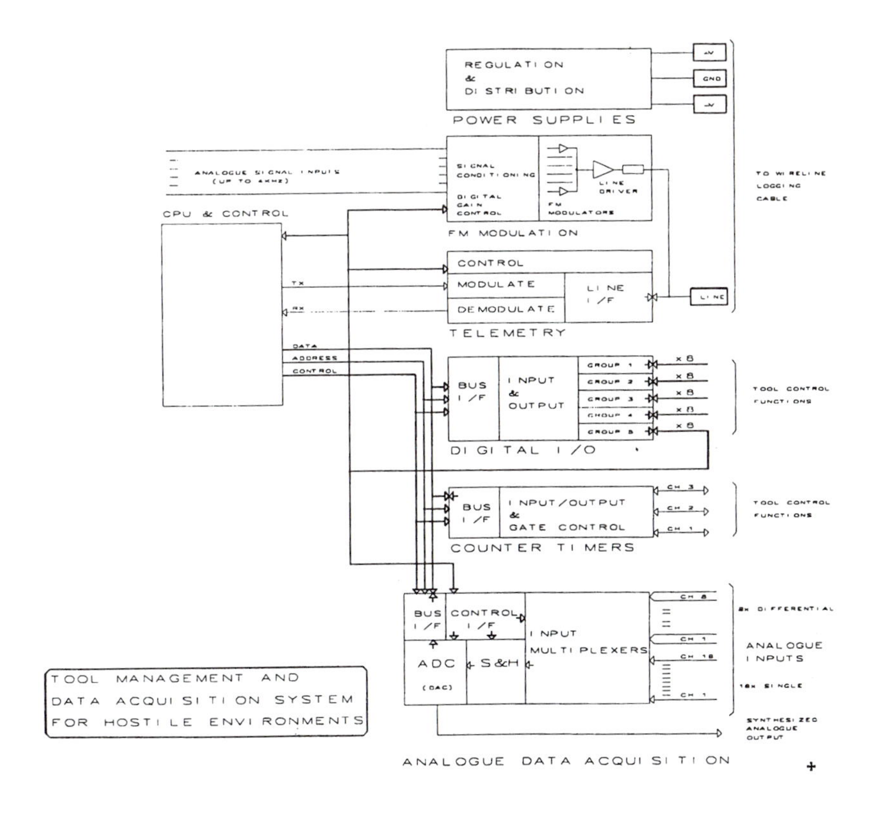

Figure 8.28 Tool management system.

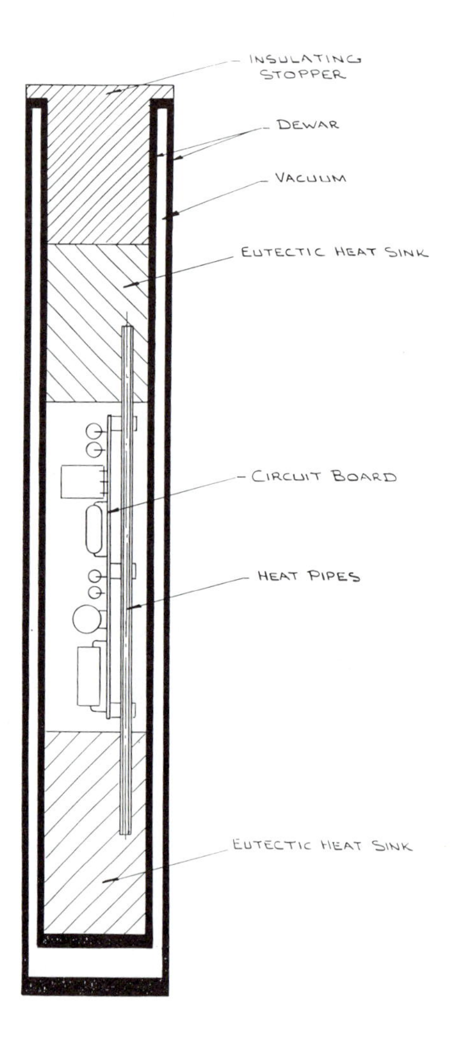

Figure 8.29 Dewar assembly - typical cross-section.

estimates, they are of limited use for critical or thermally sensitive applications because the heat leakage into a dewar is usually greater at the ends than in the middle. Although the dewar provides a very effective barrier against heat flux, great care must be exercised when designing the internal system layout to avoid hot spots.

While dewars are very good at keeping heat out of an enclosed system, the reverse is also true. Once the system has increased in temperature, it can remain hot for a long period unless the internal heat is allowed to escape by removing the system from the dewar or by extracting the heat (see section thermosyphons).

Heat stores: Using a dewar to reduce the heat flux entering a system only reduces the flux to manageable proportions. To control the increase in temperature, it is necessary to ensure that the system has an adequate thermal mass to absorb the heat flux. It is common practice in downhole tools to incorporate discrete blocks of metal as heat sinks.

8.5.3 Solid-solid heat sinks

As the name implies, these materials remain as solids throughout their usable temperature range, unlike solid-liquid materials (see Section 2.2.2).

The properties required for a suitable heat sink material are high specific heat capacity, high thermal conductivity and high density. In addition, cost, availability, and machinability must be reasonable. Table 8.2 lists some candidate materials.

Unlike aerospace applications, borehole tools require weight to aid their downhole deployment and to ensure that the use of heavy materials does not create any problems. However, volumetric space is limited, usually by the need for a small outside tool diameter and the practicalities of handling long tool bodies. Both these factors point to the use of dense materials.

It would appear, from observing many commercial designs and by recommendation from dewar manufacturers, that a common mistake when choosing heat sink materials occurs with the selection of aluminum. While it appears to have an excellent specific heat capacity, the better parameter to evaluate is its volumetric heat capacity. When this is considered, the low density of aluminum reduces its attractiveness since larger volumes are required than for, say, copper.

Table 8.2. Potential Heat Sink Materials

Material	Specific heat capacity J/Kg/K	Thermal conductivity (W/m/K)	Density (Kg/m^3)	Volumetric heat capacity ($J/m^3/K$)	Linear expansion (K^{+1})
Aluminum alloy	910	240	2710	2466	23
Copper	397	395	8930	3545	17
Monel	448	24	8800	3492	14
Stainless steel 18/8	519	16.5	7930	4116	16

Depending on the specific application, it is preferable to use copper or nickel alloys in conjunction with heat-flux boosting devices (see section on Heat Pipes).

8.5.4 Solid-liquid heat sinks

This group of materials enables the thermal flux-absorbing quality of the latent heat of fusion to be used as part of the overall heat sink capacity. The latent heat of fusion arising from the phase transition as the material passes through its eutectic point and changes from solid to liquid offers two potential benefits for borehole tool design: an increase in the overall heat capacity over a given temperature range, and the ability to create a plateau in the temperature contour.

Examples of phase-change or eutectic materials are given in Table 8.3. An important feature is the increasing value of the latent heat of fusion with increasing melting point temperature. The ability to operate electronic systems at higher temperatures allows the improved thermal properties to be exploited. This single feature provides a powerful argument for the use of high-temperature electronics in very high performance, demanding applications.

Figure 8.30 shows typical thermal histories for systems using solid-solid and eutectic heat sinks. The temperature of the plateau produced by the use of eutectic materials can be selected by the use of the appropriate melting point alloy, the gradient can be controlled by the mass of material and the use of heat pipes. The ability to create a plateau can greatly enhance the stability of electronic systems and their resulting accuracy.

The fundamental disadvantage of using eutectic materials is the added complexity of the engineering. The fact that the alloys melt means that fluid-tight containers are required and that the sealed containers must cope with the volumetric expansion of both the alloy and the trapped air. This may seem trivial, but many precision welds and potential leakage points result when tubes for wiring looms must pass through and closed-end inlets for heat pipes are included. In addition, the containers have to be sealed after they are filled and the final welds are very difficult to check.

Table 8.3 Typical Eutectic Material Properties

Melting point (^{0} C)	Alloy elements	Latent heat of fusion	Specific heat capacity solid/liquid (J/g)	Density (g/cm^3)
47	Sn, Pb, Bi, In, Cd	36.8	0.163/0.197	9.36
58	Sn, Pb, Bi, In	28.9	0.167/0.201	9.23
70	Sn, Pb, Bi, Cd	39.8	0.146/0.184	9.67
96	Sn, Pb, Bi	34.7	0.151/0.167	9.85
138	Sn, Bi	44.8	0.167/0.201	8.58
187	Sn, Pb, Sb	52.0	-/0.209	8.50
199	-	71.2	0.239/0.272	7.27

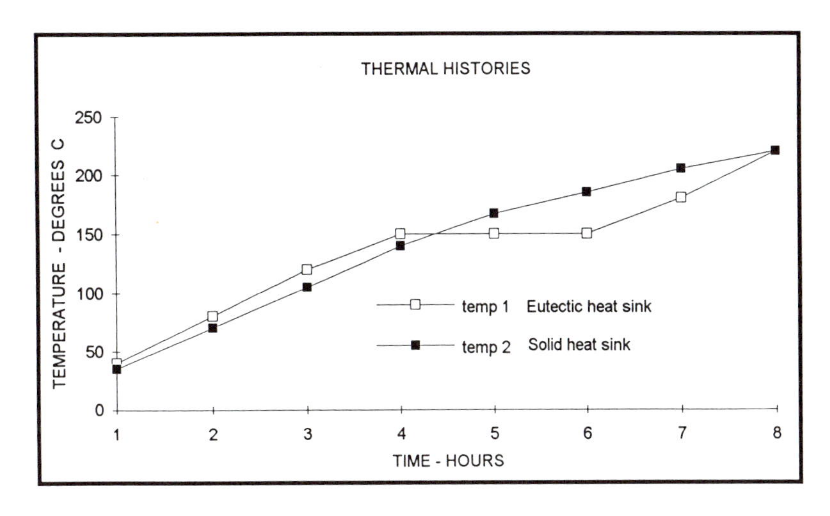

Figure 8.30 Typical thermal histories for eutectic and solid heat sinks.

Heat pipes. The heat pipe exploits the latent heat of fusion and the ability of vapor molecules to move rapidly.

8.5.5 Standard heat pipes

A heat pipe consists of a sealed cylinder or tube containing a small amount of fluid in a vacuum. When one end of the tube is warmed, the fluid boils, changing to a vapor and absorbing large amounts of heat. The vapor travels rapidly to the cooler (condenser) end, where it condenses and gives up the heat energy it contained. The cool liquid is absorbed by a wick material that lines the inner wall of the tube, and capillary action returns the fluid to the hot (evaporator) end.

The most commonly used fluids in heat pipes are water and methanol. Water allows for an operational temperature range of +5 to +300°C and methanol -40 to +150°C, with water-based pipes handling more than twice the heat flux of methanol ones. Heat pipes have thermal conductivities around a thousand times those of equivalently sized copper rods. Table 8.4 shows some typical heat-pipe power-transport capabilities. Note how the thermal power increases with the diameter of the pipe.

Table 8.4 Typical Heat-Pipe Powers

	Power - Watts			
Diameter of pipe (mm)	Differential temperature between ends (^{0} C)			
	20	40	80	120
2	8	11	13	15
4	22	27	31	37
6	72	86	94	108
8	90	108	122	134
10	112	134	152	169

The performance of a heat pipe is also dependent upon the orientation of the pipe with respect to gravity; performance generally doubles when the pipe is horizontal rather than vertical.

8.5.6 Thermosyphons

The thermosyphon is a heat pipe without a wick inside. The pipe is attitude-sensitive and only functions when not horizontal. The lower end will allow heat to boil the inner liquid which transports the heat and condenses at the upper, cooler end. The liquid then runs down the inner walls of the pipe due to gravity. The process will not work in reverse.

For borehole tools, the thermosyphon is used to allow heat to be extracted from the inside of dewars with the advantage of no return heat-flow path. The thermosyphon is connected to the upper heat sink inside the dewar, and passes out through the end to another heat sink assembly in intimate contact with the tool's pressure housing. When the tool is recovered to the surface, heat trapped inside the dewar can flow out via the tool housing to the atmosphere.

Dewar closures. The open end of the dewar permits loading and unloading the payload, but requires sealing to exclude heat entry. The seal or closure usually must accommodate the passage of wiring and other services. Great care is needed to minimize any air gaps around wires and heat-conducting materials.

Simple closures may consist of plastic plugs with suitable holes for the wiring, more complex closures can be thin-walled metal cylinders filled with an insulating material. A short, evacuated cylinder can be used as a dewar to plug the open end, but this can be an expensive solution. In general, the less metal involved, the better the closure.

Thermoelectric coolers. Electronics engineers engaged in borehole system development have long wanted to have an electric "fridge" to keep the electronics cool. Thermoelectric or "Peltier" devices come close to providing such a system.

A system has been constructed that produces a 40°C reduction in temperature from a 220°C ambient. The system was able to support a continuous 6-watt cooling load running at 180°C.

Power supplies for the cooler were designed to operate at a continuous 240°C. The heat dissipation from the power supply required it to be outside the cavity being cooled, so it had to be designed to function continuously at the downhole temperature specified as the tool's limit. Thermoelectric cooler power requirements are quite simple, requiring only a raw unregulated DC feed. This allows a fairly simple unit transforming an AC feed down to 5v DC at 10 amps. The system operated at 180°C with a 40°C differential load for over 4000 hours with no serious problems.

A cooler like this is mounted away from the electronic systems in good thermal contact with the tool body and with the well fluid. We are currently researching a 50-watt switch-mode power supply for use at 200°C with high thermal cycling for aerospace applications.

Construction materials. The thermal properties of the materials used to construct the tool systems, both inside the dewar and out, need careful consideration. The internal materials need to have adequate strength and stiffness, while offering high heat capacity. Stainless steels and bronze alloys are a popular choice. When electrical insulation is required, several plastic materials offer useful properties up to temperatures of 300°C and beyond. Table 8.5 lists some examples. The external material choice is much more complex (see Section 3.0), and the thermal expansion properties must be considered, particularly when different materials are used.

Table 8.5 Examples of High Temperature Plastics

Raw material group	Trade name	Tensile strength (MPa)	Young Modulus (GPa)	Max temp short time (°C)	Max temp long time (°C)
Polymide 66	Akulon	220-150	16-11	200	130
Fluorpolymer	Tefzel	44	0.825	180	155
Fluorpolymer	Teflon	25	0.7	300	200
Polyamide-imide	Torlon	140	6.5	350	250
Polyether-etherkeytone	Peek	92	3.6	280	250
Polybenzimidazole	Celazole	160	5.9	700	3-400

8.5.7 Mechanical systems

The mechanical engineer has to exercise careful judgment when designing a tool for high-temperature use. On the one hand, it must survive repeated trips into the very hostile environment of the borehole; on the other, it must be accessible on the bench as an assembly that allows servicing, repair, and calibration.

The basic components of a high-temperature tool are the pressure housing, the dewar, and the electronics chassis assembly.

The pressure housing has to withstand the stress due to the borehole pressure at high temperatures and resist highly corrosive fluids (a geothermal well at 400°C can produce and extremely corrosive environment). High nickel-content super alloys and some cobalt-based alloys are among the best, if most expensive, metals for the application.

Much research and experimentation has been conducted to evaluate and develop materials suitable for use in various downhole conditions, particularly for the oil and gas industries. Many materials are well-characterized for use at temperatures up to 250°C. The geothermal industry encounters higher temperatures, and current research is underway to develop tools for use at 400°C. For these applications, little if any real data is available on materials corrosion. Long-term autoclave tests are being conducted to expose the selected materials to simulated downhole conditions.

Tool design example. Tool design begins with ascertaining basic client requirements and application limitations such as the maximum allowable tool diameter; tool section length; measurement type, accuracy and resolution, downhole life, temperature, and pressure; and so forth.

Once the external diameter has been established, the pressure and temperature rating of the basic cross-section of the tool can be calculated. In the example shown in Figure 8.31, the tool was 48 mm in diameter, the pressure housing wall was 5 mm thick (it could have been thinner, but for reasons other than stress it was set at this figure), and the dewar wall was 5 mm thick with 0.75 mm clearance inside the pressure housing. This gave a dewar bore of 27.25 mm. To allow sufficient headroom for the components on the circuit boards, particularly near the edges, the boards were located 4 mm below the center line, reducing the board width to a maximum of 23 mm.

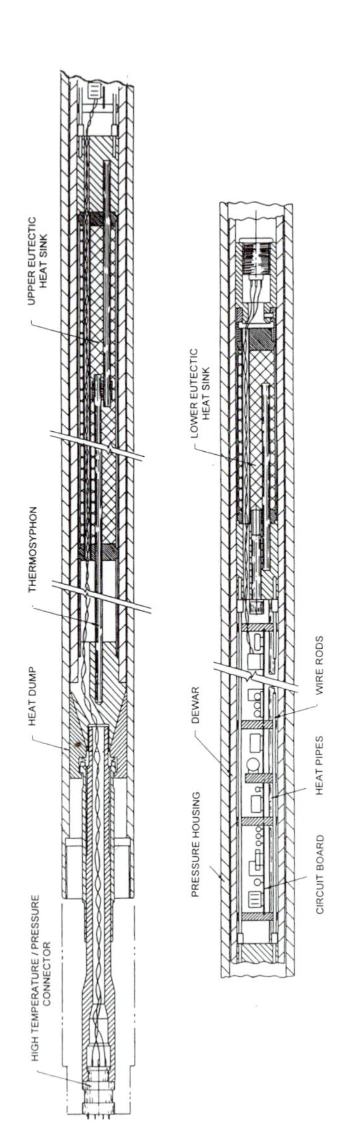

Figure 8.31 48-mm diameter tool assembly.

The maximum length of any one section of the tool system was restricted to 3 meters. Taking the tool housing connection systems into account, this allowed a dewar some 2560 mm in length with a usable internal length of 2460 mm. The thermal design requirements to achieve 10 hours of downhole life at 350°C dictated eutectic heat sinks 1350 mm in total length and a length of 870 mm for the remaining support structures. Approximately 60% of dewar length is devoted to the support systems for the electronics.

The electronic systems, analogue and digital, performed a complex series of electric field control and measurement functions, sensor-based measurements, and circuit-state monitoring. A microprocessor-based control system and a duplex modem communication system were included. The operational temperature limit for the electronics was 185°C, continuous rating, 220°C shut-down survival temperature.

The temperature in the dewar was controlled by the use of upper and lower eutectic heat sinks on either side of the electronics, with a 170°C eutectic temperature. The heat sinks were linked by two heat pipes located beneath the circuit boards, running the full length of the electronics section and halfway into each heat sink. The heat pipes ensured efficient use of the heat sinks by conducting heat from inside the circuit board area and rapidly transferring it to the heat sinks. The location of the heat pipes was not ideal, since the circuit board was a poor conductor of heat and tended to shield the components, but the lack of space did not allow any alternative arrangement. A thermosyphon system was fitted to the upper heat sink to allow the internal systems to cool without having to be disassembled.

The main application of the tool was in very hot sub-sea boreholes; the ability to pull the tool out of a borehole into the ocean to cool and then re-insert it into the borehole - without a time-consuming and costly trip to the surface - was a major design consideration.

Differential thermal expansion was also a major consideration in the design. The dewar was a through-wired type and electrical connections were made at each end. An electrical connector was installed in the internal end of the dewar to allow the electronics/heat sink assembly to be easily withdrawn without any trailing wiring that could easily get trapped and damaged. The connector assembly passed through the end of the dewar to a custom-made high-temperature connector (450°C rated). The differential temperature range of the dewar was 0°C inside and 350°C outside (a 350°C positive difference) and 200°C inside and 0°C outside (a 200°C negative difference), giving a total difference of 550°C in a theoretical worst case. This gave a movement range of 16 mm between the internal and the external end of the dewar, which had to be accommodated in the mechanics and wiring loom.

The circuit-board mounting structure had to allow for the differential expansion between the board material and the metals. This was achieved by fixing the boards end-to-end and applying a small compressive load using spring loaded wire rods. The assembly was flexible and compliant (if a little fragile for handling on the bench), with the boards as part of the structure; little space was being consumed by purely structural components.

The external wiring, used where exposure to full well temperature was required, was glass-insulated, multistrand, silver-coated nickel and glass-coated, single-strand silver wire. Internal wiring looms use PTFE-coated copper.

The absence of multipin connectors for use at high temperatures led to the development of a unique set of connectors. One half was an adaptation of a commercial design updated to 450°C by changing all the materials, the other half was a special development rated to at least 500°C at 100MPa pressure, using a prestressed shell and ceramic/glass-sealed contacts. It is unusual, but not exceptional, to have to develop so many basic items for a downhole tool.

Many manufacturers offer polymer-based high-temperature seals for use up to 320°C, and a few offer metal seals for temperatures beyond this. The problems associated with working with polymer seals near their limits are small compared with the difficulties in achieving a reliable total seal with metal seals. Very high quality surface finishes and high confining loads do not necessarily produce reliable results.

Research is underway to develop a limited application polymer seal for temperatures up to 500°C.

8.6 Summary

- High-temperature cycle electronics will be needed to accommodate databus terminals on the actuator PCUs in the nacelles, empennage, and wings of the HSCT.
- A bidirectional multiplex data bus protocol that is acceptable to the airplane manufacturer, customers, and certification officials (particularly from a failure mode and complexity standpoint), and which is significantly faster than ARINC 629, will be required in a high-temperature qualified and reliable system.
- High-temperature electronics can significantly benefit the overall economic performance of the HSCT flight and propulsion control systems by reducing weight and system complexity in terms of individual wires and connectors

An opportunity exists to significantly improve the overall performance and economic viability of military aircraft flight control, propulsion control, power management and distribution, and stores management subsystems through the use of wide-bandgap (WBG) semiconductor devices. WBG-based electronics are potentially more efficient, can dissipate less heat, and are capable of operating at much higher temperatures compared to conventional Si-based electronics. These attributes will help enable the development of future distributed engine and flight control systems, as well as allow the reduction (in size) or elimination of heavy, single redundant aircraft environmental control systems (ECS). Reduction in the ECS size and utilization of distributed control offers substantial payoffs in aircraft capability, reliability, maintainability, and supportability; this includes lowering fly-away and life-cycle costs, and increasing aircraft payload and range capabilities.

Chapter 9

ACCELERATED TESTING OF ELEVATED TEMPERATURE ELECTRONICS

9.1 Accelerated Testing at Extremely High-Temperatures

Although many users have had harsh environment electronics requirements for years, high-temperature electronics has yet to become part of any large manufacturer's main product line. Proffered rationales have included a lack of technology development funding, the lack of a market significant enough to warrant the development of such a technology, and the availability of alternative solutions that include cooling systems and the relocation of any required circuitry within the application to more benign environments. Perhaps it can be argued, on a case-by-case basis, that these are fair criticisms. One point appears certain: for high-temperature technology to evolve into a viable system solution for applications and market niches with harsh environment requirements, it must address a complete high-temperature application solution. That is, a single transistor, a discrete component, or even a fledgling integrated circuit process, by itself, is not a true high-temperature circuit solution. Most medium-to-large volume applications will require a combination of all of these components and more. In addition, the user will need a circuit board, a die attachment process, well-characterized wirebonding techniques, a package, and perhaps even sockets and burn-in cards. This is true not only for emerging applications with extremely high-temperature exposure requirements, but also for accelerated testing of current high-temperature products.

Accelerated testing is a process by which a piece of hardware, an individual electronic device, a subsystem, or an entire system is subjected to a harsh environmental screening whether designed to shorten its life or hasten the degradation of its performance. This is routinely done in industry to obtain product life characterization data much faster than if the hardware being tested were simply observed while operating in its normal use environment. When this data is modeled and analyzed properly, key information regarding the hardware life expectancy and long-term performance degradation may be obtained. For example, for failure mechanisms dependent on absolute temperature, it may be possible to simulate 100,000 of a product's normal operating lifetime by testing the product for only 100 at a temperature 75 degrees higher than the maximum specified normal operating temperature.

The science of accelerated testing is often complex. The researcher must have a good understanding of the physics at work in any given system to be tested at an accelerated rate in order to construct the required models, as well as to be sure that any additional failure mechanisms (beyond those explicitly being tested) are not excited. In cases where the product is designed to operate in a harsh environment and at high-temperatures in particular, accelerated testing is even more problematic. This is so because when the acceleration temperatures are greater than 200 to 300°C, the physical limit of all commercially available test fixturing is being reached, and it becomes a race to see whether the product being tested or the burn-in cards will fail first.

Test fixturing and packaging capable of withstanding long hours at extremely elevated temperatures will, of course, be required for the implementation of any deployable harsh environment hardware. However, since the field of high-temperature electronics is still in its infancy, it is just as likely that such fixtures are needed even more for the purposes of discrete and integrated device characterization. Also, because of the early state of development of the high-temperature market, a researcher in this area has, up to now, had to resort to a considerable amount of process development in order to fabricate homemade hardware suitable to the task of high-temperature testing.

Those involved in the development and characterization of high-temperature devices or systems know that there are very few commercially available components with which to construct sockets, burn-in cards, packages, or associated test fixturing capable of withstanding extended periods of high-temperature exposure. This is especially so when the test temperatures exceed 300°C. Further problems result from the conspicuous absence of almost any commercially available test hardware suitable for long-term testing when the expected operating temperatures increase to 400, 500, 600°C or more. These high temperatures might easily be reached in a number of gas turbine engine distributed control applications, or in an automotive in-cylinder pressure sensor monitoring applications. In such cases, how is the accelerated testing performed for these extremely high temperature applications? Where does the required test hardware come from? It is relatively simple to build or purchase a furnace for this purpose. But from where does one obtain the required cabling, burn-in cards, sockets, fixturing, and so on? It is clear that high-temperature accelerated testing is hampered from the outset by the lack of available test hardware.

A number of excellent references deal with the basic theory of accelerated testing, statistical modeling, construction of test plans and methods of analyzing data [see, for example [Nelson 1990]. This chapter does not review these topics, but rather tries to shed some light on how to configure experimental set-ups for the characterization, modelling, and accelerated testing of electronic components at extremely high temperatures. It considers a number of options for obtaining data from both passive and active discrete and integrated circuits. Most important to the high-temperature researcher, a number of practical solutions are outlined for the development and fabrication of useful testing hardware for high-temperature electronic device characterization and accelerated testing of existing harsh-environment hardware.

9.2 Methods for Fabricating High-Temperature Packaging and Test Fixturing

In an overall system approach it is critical that a design procedure be established so that engineers can routinely construct circuits for their high-temperature applications. This requires a database that includes circuit model parameters for incorporation in SPICE simulations. Accurate device models for even simple passive components would be much more important for the high-temperature designer than for standard military temperature-range applications. For example, in the case of a simple capacitor, the bulk capacitance and dissipation factor of the component can be expected to be much different at 400°C than at 125°C [Grzybowski and Beckman 1993]. While these parameters do vary predictably, in order to design a circuit that compensates for these variations with temperature, the particular dielectric type selected, the terminations employed, and so on, must be included as a conscious part of the component selection process. This requires that an accurate modeling capability be available for device parameter extraction at elevated operating temperatures.

In addition to the importance of accurate high-temperature device models for harsh-environment design, the reliability ramifications of a circuit designed to operate at 175, 200, or 300°C and that utilizes components designed to operate at 400°C has yet to be explored. It is expected that the increased reliability in medium-temperature applications that use such components may, by itself, make the technology worthwhile.

9.2.1 Fixture fabrication

This section explores various methods for fabricating test fixturing and packaging for high-temperature accelerated testing of electronic components. The fabrication processes for several fixtures designed to operate for long periods at temperatures as high as 500°C are outlined. Included are the basic clamping fixture for passive components, a heavy-duty 20-pin DIP socket, a four-position 28-pin DIP socket burn-in card, and a 28-pin hermetically sealable DIP package. These fixtures were all designed and fabricated with an eye toward making them manufacturable for large-volume applications. Great care was taken to select materials that were available in the commercial marketplace without having to resort to materials development in a laboratory. The hardware discussed in this section represents a very new area in electronics packaging. As such, the researcher should feel free to experiment with the approaches discussed here.

9.2.2 Clamping fixture

The most simply and inexpensively constructed compression test fixture is useful for testing any surface-mount passive components, but particularly components that are physically large, such as multilayer capacitors with values of several microfarads. Figure 9.1 shows such a clamping test fixture loaded with five large multilayer ceramic capacitors. In Figure 9.2, the basic fixture is shown in a diagrammatic cross-section. The fixtures shown have endured several thousand hours of testing at temperatures as high as 500°C. They are basically constructed from two L-shaped metal jaws and a spring-loaded cross beam. The lower L-shaped section is the support bed and the backstop for the spring loaded common wall. This common wall supports one set of device contacts. The upper L-shaped piece supports the other set of device contacts. Such a fixture might be fabricated from many suitable base metals. Among these are the more exotic high-temperature aerospace alloys, which include Hastaloy and the various Inconel alloy xxx and Inconel xxx materials, such as Inco 718 and Inco 903. However, stainless steel has worked well for many thousands of hours for accelerated tests at temperatures greater than 400°C.

Nickel-coated spring steel shims are used as the actual contacts to the surface-mounted components. For the case shown in Figure 9.1, all of the five devices shown possess separate contacts, but they could just as easily be common grounded on one entire side of any gang of devices. Each of these contacts is outfitted with a screw terminal to which the high-temperature furnace wire is attached. This arrangement is very convenient because, although the clamping fixture can be used over and over for many thousands of hours, the individual contacts must be discarded after each accelerated test. This permits a fresh, non-oxidized contact to be used at the beginning of each accelerated test. All contacts, as well as the two L-shaped jaws, are kept dielectrically insulated from one another by the incorporation of ceramic substrates as separating materials. Standard 25 mil 96% alumina substrates, like those used in the manufacture of hybrid circuits, have performed very well, are easily procured, and are easily scribed and snapped to size.

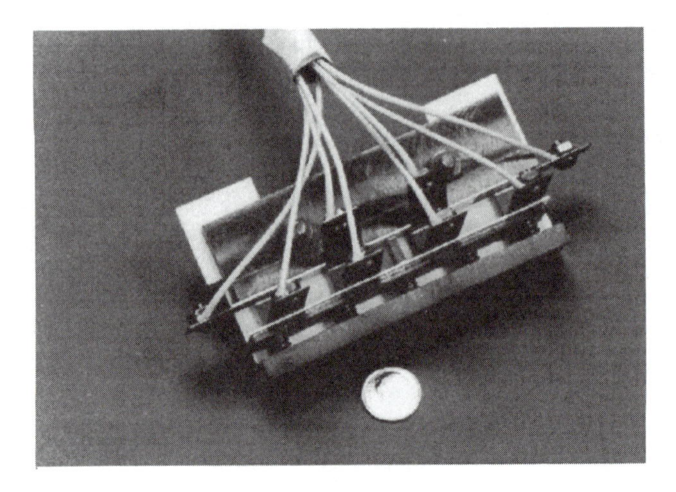

Figure 9.1 A high-temperature passive component accelerated spring-loaded clamping test fixture.

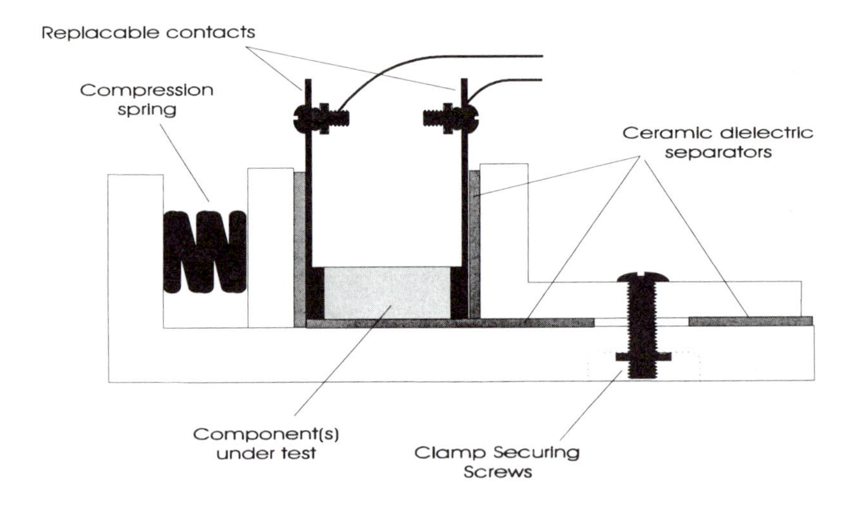

Figure 9.2 A high-temperature passive component accelerated spring-loaded clamping test fixture (cross-section).

A controlled contacting force is applied to the components by the five-high temperature steel compression springs (seen in Figure 9.1and Figure 9.2). This distributed spring arrangement allows a more even distribution of force applied to the components. In addition, because the spring constant of these coils will vary with the acceleration temperature, the use of multiple springs reduces the variation of compressive force applied to the devices as temperature is applied with time.

9.2.3 Heavy-duty dual in-line sockets

One type of test fixture that is very useful for the accelerated testing of leaded components or dual in-line (DIP) packaged integrated circuits is a DIP socket. For testing axial or radial leaded devices, individual components may simply be inserted in dual in-line arrays within these fixtures, subject them to the desired accelerated testing, and remove them in preparation for the next sample lot. If properly constructed, these fixtures can be used repeatedly for many thousands of accelerated test hours to temperatures at least as high as 500°C. A sample of one method for constructing a heavy-duty 20-DIP socket for a 0.300" wide package, is shown in Figure 9.3. The DIP style selected and the number of pins included is arbitrary and could just as easily be a 28-pin or 40-pin 0.600" wide package configuration. The socket shown measures 0.7" wide x 1.5" long.

This particular example fixture was designed to serve a number of high-temperature accelerated device testing and characterization purposes, ranging from evaluating ceramic-packaged silicon and silicon carbide integrated circuits to the characterization of various combinations of passive components. For this reason, this single position socket was designed to be quite robust, with a selected starting substrate material thickness of 0.100". Alumina was selected as the substrate material because of its high strength, good thermal dimensional stability, ease of machining, compatibility with thick-film processing, and its lack of electrical conductivity. A total of 40 holes that will envelop the custom socket pins are laser-drilled into the alumina. In Figure 9.4, the inner two rows of ten pins are positioned to accept the 0.300" wide 20-pin package. The outer two rows of ten pins are positioned to accept high-temperature signal and power supply lines routed to the devices being tested. Each package pin is then

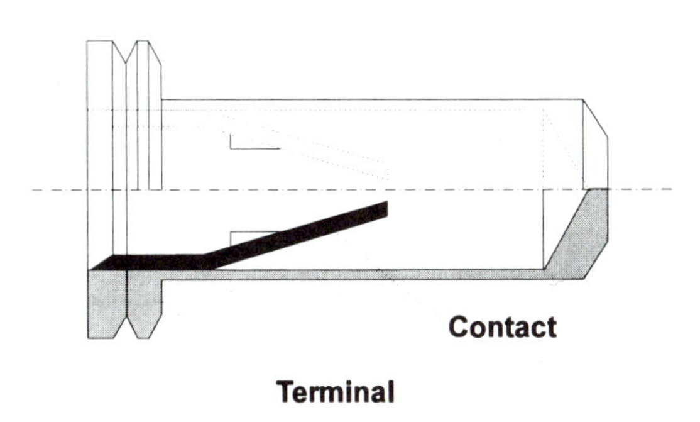

Figure 9.3 High-temperature socket-pin receptacle.

interconnected to a corresponding I/O wire receptacle, as shown. The patterned interconnects consist of a thick-film tungsten that is printed to a pre-firing thickness of between 0.0005" and 0.0007" and then fired. To add test flexibility to the socket, a field of thick-film tungsten is applied between the socket pins case a back-side connection is required for bare die evaluation. All of the tungsten metallization is subsequently nickel-plated with 60 μ-inches of electroless nickel.

The next step in the fabrication sequence requires positioning the custom socket pin receptacles in the pre-drilled holes and CuSil brazing them in place. The custom socket pin receptacles were specifically designed for the 500°C test environment. They were fabricated on a screw machine from stainless steel type 303, and subsequently nickel-plated. A heat-treated beryllium-nickel 1/4 hard alloy 44D spring is then inserted into the stainless steel barrel to complete the pin assembly. Once the brazing of the receptacles is completed, all of the metal interconnects are plated with a 100 μ-inch protective layer of electroless gold to complete the test fixture.

9.2.4 The 28-pin DIP package

One suitable 500°C environment dual in-line package, a 28-pin, 0.600" wide embodiment, is shown in Figure 9.5. Details of the top and bottom sides of the package are provided in Figure 9.6. A variation on the fabrication strategy used to build the heavy-duty socket test fixtures is used for producing this package. The base substrate is made from 0.025", 96% alumina. This substrate is prepared by laser drilling the 28 holes to accommodate the kovar pins. A thick-film tungsten interconnect pattern is printed and fired on the substrate. The pattern defined in Figures 9.5 and 9.6 was designed, once again, for maximum flexibility in the accelerated testing laboratory. It includes a large central pad for die attach and back-side biasing. In addition, the layout of the interconnect is arranged so that I/O pins can be programmed via wirebonding to their ultimate destination within the package.

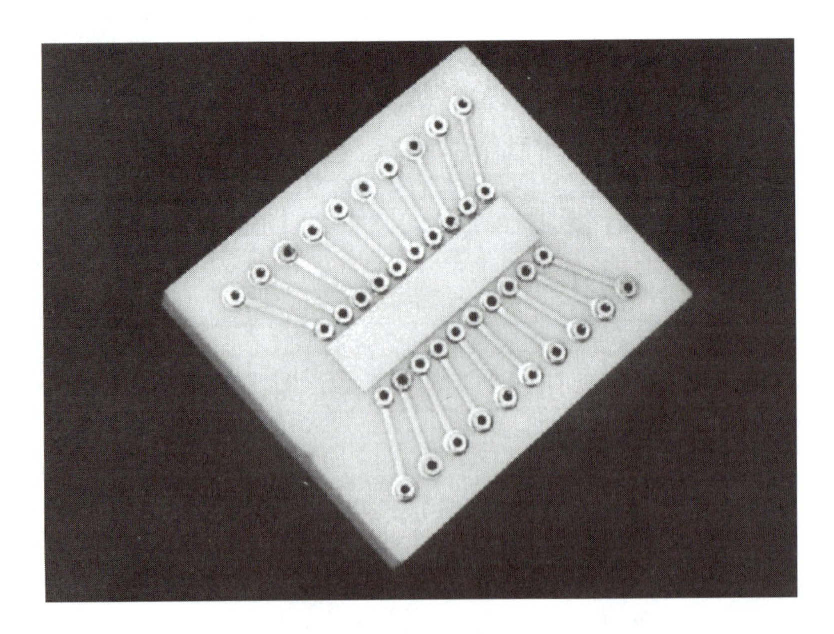

Figure 9.4 Heavy-duty 20-pin DIP socket designed for 500 °C long-term testing.

Because of the extreme temperatures at which these packages will operate, great care must be taken to match the thermal coefficients of the ring frames to the substrate material. For this reason, the ring frames and the substrate are fabricated from the same material. The package side wall is actually composed of two individual laser-machined alumina ring frames, each measuring 0.060" high. Two separate rings were used because a package depth of at least 0.100" was desired, and the limit on fine machining of alumina is approximately 0.060". When the two walls are stacked, the package depth is approximately 0.120". The approximate outer dimensions of this package's side walls are 1.45" long x 0.48" wide. The top surface of the lower ring frame is metallized with a layer of thick-film tungsten. The bottom surface of the lower ring frame, which is attached to the substrate, must form a dielectric seal to prevent the I/O traces from short-circuiting, so it has no metallization. Both the top and the bottom surfaces of the upper ring frame are prepared with thick-film tungsten. The upper surface will help form the lid-sealing surface.

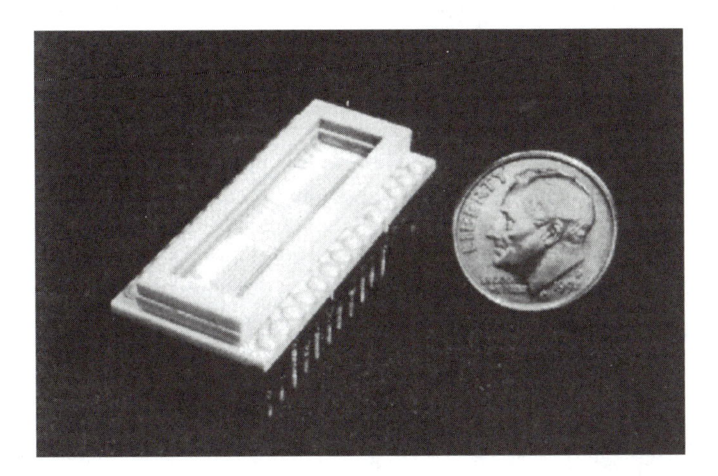

Figure 9.5 Top view of 28-pin DIP package designed for long-term operation at 500°C.

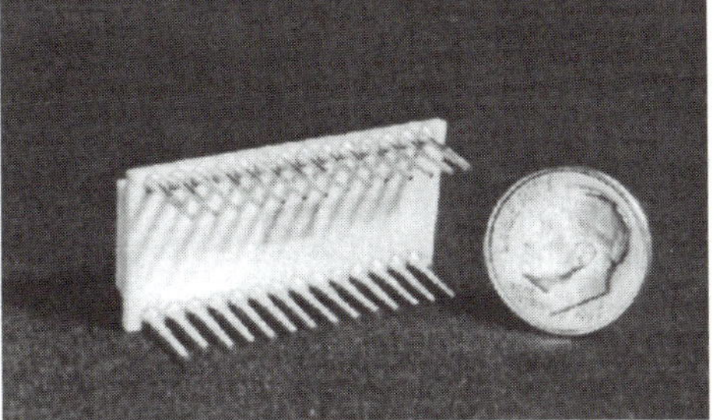

Figure 9.6 Top and bottom details of the 28-pin DIP package designed for long-term operation at 500°C.

Next, a series of subassemblies must be completed by reflowing the glass that seals the substrate to the lower ring frame, brazes the lower ring frame to the upper ring frame, and brazes the kovar package pins to the substrate in a single firing operation. The first of these subassemblies anticipates the attachment of the lower ring frame to the package substrate using a glass dielectric preform or a glass paste that is either screen printed or painted onto the substrate. This material must also be closely matched in coefficient of thermal expansion to the alumina substrate and ring frame. The second subassembly prepares for brazing the kovar package pins to the substrate. Pure copper is used as the brazing material, in the form of copper wire rings. The type of braze material selected depends on the long-term operating temperature at which the package is expected to operate and on whether the silver in the braze poses a reliability hazard. This is specifically a concern with brazes containing silver (e.g., CuSil) because, at elevated temperatures, silver may readily form a dendritic short circuit between two adjacent contacts with differing applied electric potentials. A graphite fixture is used to hold all of the pins, the braze rings, and the substrate together until the firing operation is complete. To make the package as structurally sound as possible, the copper braze rings are applied to both the top and bottom of the substrate.

The third subassembly prepares for brazing the upper ring frame to the lower ring frame. The braze material selected for this step is a foil preform oxygen-free hard copper roughly 5 mils thick. All three subassemblies are then assembled on the graphite carrier for the firing operation, which occurs at 1100°C in a non-oxidizing furnace environment.

Following the braze and reflow firing step, the assembled package is ready for nickel plating. A 60-min. coating of electroless nickel is plated over the tungsten traces, exposed copper braze, side wall frames, and package pins. The nickel plating is then fired, sintering the nickel into the tungsten to strengthen the nickel - tungsten bond. Sintering also prevents the hard nickel coating from chipping or blistering during high-temperature exposure. Proper nickel plating will prevent the tungsten from oxidizing. However, the nickel itself will experience some oxidization at very high temperatures, so the nickel plating is followed by an electroless gold plating of at least 100 min. The final step is to sinter the gold to create a niaural alloy at the nickel/gold interface. While the gold forms a non-oxidizing protective layer, it is not completely impervious to oxygen penetration. Hence, the underlying nickel plating helps to keep any oxygen that may penetrate the gold layer from reacting with the tungsten thick-film below.

Finally, once the application circuitry has been assembled into the package, a standard nickel-plated kovar step lid may be brazed to the package seal ring to complete the hermetic assembly.

9.2.5 Four-position 28-pin DIP package burn-in card

The four-position socketed burn-in card is shown in Figure 9.7. The fabrication sequence for this fixture is similar to that outlined for the single-position socket and the high-temperature package just described. The starting substrate thickness, however, is only 0.060". Once again, holes for the socket pin receptacles are laser-drilled. In addition, Figure 9.7 reveals four large holes in each card that serve as component ejector ports. These are especially important for high-temperature (400°C+) burn-in because there is often substantial intermetallic growth at the pin/receptacle interface that can make subsequent package removal difficult.

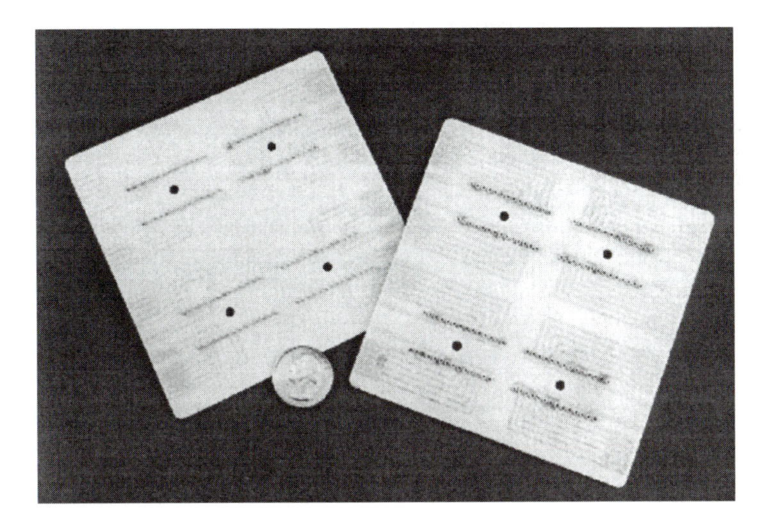

Figure 9.7 Four-position 28-pin burn-in card designed for 500°C long-term operation.

9.2.6 Special caveats for high-temperature fixture and package fabrication

This chapter cannot detail all of the mechanical and materials issues associated with fabricating fixtures and packaging for electronics intended to operate at high temperatures. Objects soaking at 500°C are incandescent, glowing red hot, and in such an environment, coefficient of thermal expansion mismatches are a considerable problem. Oxidation of all metal surfaces is also of great concern. All of the fixture interconnects described in this section were formed from thick-film tungsten, a refractory metal. Tungsten, however, will readily oxidize at high temperatures in a matter of hours if not properly protected. This is readily apparent in Figure 9.8, which shows two thick-film tungsten fingers. The one on the left was not properly plated and was completely reduced to tungsten oxide when exposed to 500°C for 24 in an air ambient. Note the substrate visible through the oxidized trace. The one on the right was properly plated and has survived without any signs of a problem.

Figure 9.8 Oxidized thick-film tungsten traces improperly protected for the 500°C environment.

Figure 9.9 tells a similar tale. The pictures on the left show a side view of the package pins and the lower braze joints before (top) and after (bottom) 24 at 500°C in an air ambient. The plating was not uniform or thick enough and the nickel plating was oxidized along with the kovar below it. Examining the copper braze joints in this figure, it can be seen that inadequate nickel plating has permitted the gold plating to diffuse into the copper. The exposed copper was then free to oxidize. This is even more pronounced in the photographs at the right, which show the upper package pin braze joints. Once again, before (top) and after (bottom) 24 at 500°C in an air ambient. The key to success with high-temperature fixture and process development lies in very careful process development, with special attention to plating quality, thickness, and sintering steps.

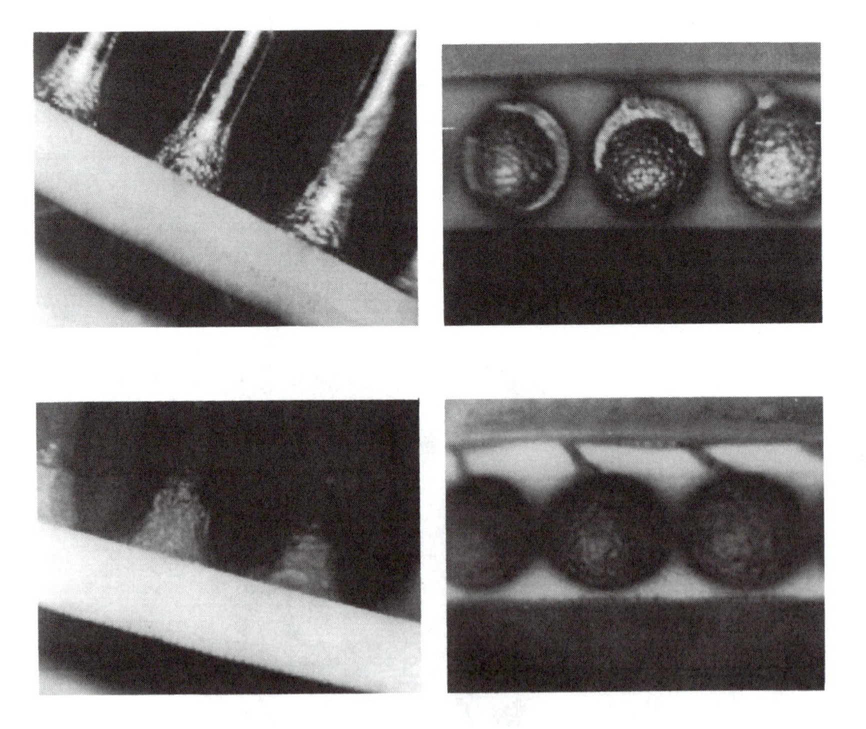

Figure 9.9 Package pins and brazing joints before and after 24 at 500°C, with flawed plating of barrier layers.

9.2.7 Results to date

At this time, all of the fixtures described have been used for device characterization to temperatures as high as 500°C. A sample of the diversity of such testing is cited in the literature [Grzybowski 1991a, b, Grzybowski and Kerwin 1994, Grzybowski and Tyson 1993]. The 20-pin single-position sockets have been exposed to 500°C for several thousand hours of testing. The four-position burn-in cards have been used in an on-going life test of package wirebonds for over four thousand hours at 400°C. Life testing and reliability assessment of the 28-pin DIP package is still in its infancy, primarily because we are continuously iterating the basic fabrication cycle to enhance manufacturability. However, to date, sealed units pass gross and fine leak tests following several hundred hours at 500°C. Life testing and reliability assessment of the fixturing described is ongoing at UTRC.

9.3 Concerns for High-Temperature Characterization of Electronic Devices

In the many research activities associated with developing an electronics technology for extremely high temperature, passive components have often been slighted in deference to integrated circuit development. Technical articles on high-temperature electronics have generally been devoted to evolutionary developments in active components such as MOSFETs, bipolar transistors, and small integrated circuits. However, once the active components are provided for a given application, they invariably must be surrounded with passive support components. All of those resistors, capacitors, and inductive components are destined to share circuit board space with the high-temperature active devices, and will have to be characterized by someone for reliable operation at elevated temperatures. High-temperature electronics is in an early state of development and there is not a lot of literature available. This section discusses some of the hazards associated with the characterization and accelerated testing of electronic components and provides a guide through some of the steps necessary to successfully conduct an examination.

Although there is enough technology development homework left to go around, a fair number of commercially available passive components have already demonstrated long-term, reliable operation at elevated temperatures. The problem encountered by researchers and designers trying to identify off-the-shelf components for a 350°C distributed control system is that all of the manufacturer's published data sheets specify guaranteed (usually derated) operating temperatures of 125, 150, or perhaps 200°C. Why aren't manufacturers providing design curves that extend temperature further? One of the principal reasons is the lack of a significant market for these components. Several component manufacturers who have already done some development in conjunction with the expected market testing and analysis have come up with the same result - the market for high-temperature components is not yet big enough to warrant the required investment in testing and qualification. Even scarcer are published results of long term aging or reliability testing data.

In many cases, however, the basic component technology used in fabricating any given passive device will survive for periods well beyond the data-sheet specified temperature limits. What generally fails far in advance of the basic component construct are the soldered lead joints used for leaded components, the epoxy dip, or one of the other packaging or encapsulating materials in the completed product. It is only fair to note that most of these devices were never intended to be used in very harsh environment applications. Traditional solders, epoxies, and plastic encapsulants would have to be replaced by reliable substitutes, another cost associated with developing high-temperature passive components. This problem has been mitigated somewhat with the surge in surface-mount technology because many of the additional processing steps associated with the packaging of traditional through-hole components has been minimized or eliminated in surface-mount components.

Assume, for a moment, that the basic materials problems associated with manufacturing and packaging high-temperature have been resolved. Even with very good components, device characteristics will not remain constant throughout the entire temperature excursion. Typical harsh-environment aerospace needs require total operating temperature ranges of -55 to 200°C and -55 to 350°C. For typical under hood automotive applications, the requirement may be reduced to -40 to 200°C. High-temperature system design is an advanced exercise in worst-case circuit and packaging analysis. To make this task feasible, components must be modeled over the entire temperature range of interest and must be incorporated with the libraries of commercially available circuit simulators, such as SPICE. Once this has been accomplished high-temperature circuit design, while always specialized, might become more routine.

To introduce the basic methodology used herto characterize and accelerate the testing of electronic components, consider a sample case. This example describes the high temperature characterization of multilayer metal oxide varistors, an especially illuminating illustration because these passive components behave like capacitors at some points in their useful lives and like semiconductors at others. In general, when semiconductors must operate at high temperatures (T > 125°C), designers must take into account the commensurate decrease in noise margin and reduced drive current. Hence, for digital electronics designed to operate at high temperatures, low-voltage transient suppression often becomes even more important than for conventional military temperature-range designs. Depending on the circuit application and the breakdown voltage protection required, transient voltage suppressors take a variety of forms. Transient suppression methods include simple diode or capacitor protection, as well as metal oxide varistors (MOVs) and metallic shielding. Of these, the multilayer MOV (MLV) is a comparative newcomer, introduced less than a decade ago. Single-layer MOVs come with two significant drawbacks: they can short-circuit in a single surge, and they are physically large. MLVs have been developed to address the transient suppression needs of 3.3 to 100 WVDC applications. MLVs can be manufactured in small surface-mount packages, eliminate the wear-out characteristics exhibited by single-layer MOVs, and generally boast more stable breakdown and clamping voltages. In addition, since the multilayer design places electrode pairs in parallel, inrush current is divided and dissipated throughout a larger effective area.

Obtaining a definitive standard methodology for testing low-voltage transient suppression devices is difficult. A long list of standards has evolved over the last 20 years that often includes different and apparently conflicting tests. In this example, however, the goal is not to test these components to any particular standard, but to determine the effects of high temperature on the varistors and to gain fundamental insight into how their behavior can be modeled in elevated-temperature circuit simulations. The original effort aimed to characterize several zinc oxide varistors at elevated temperatures and to determine their ability to protect sensitive analog or digital circuitry from low-voltage transients.

The data for this example was obtained from both axial leaded devices and surface-mounted components. If the objective is to accelerate testing of the components or to obtain basic characterization information, the device attachment means must be understood well enough to neglect it or to separate its parasitic effects from the data obtained. Test leads bringing signals from instrumentation outside the high-temperature oven to the devices inside and back again will contribute unwanted inductance. All of the connectors, joints, or sockets will contribute parasitic capacitance. In measuring reactive parameters, all of these parasitics must be minimized and calibrated out to obtain accurate measurements. In addition, because at least some of the hardware will be subjected to large temperature excursions, bimetallic effects may be a concern in sensitive current-monitoring applications. After selecting the components to be tested, the next decision is how to attach the devices to the burn-in fixture. The tested devices can be connected directly to test leads using high-temperature soldering or brazing. However, to accelerate the testing of an entire lot or a statistically significant quantity of devices, some sort of burn-in card is generally employed. These concerns are exacerbated as the peak test temperature is increased. Since accelerated testing at peak temperatures in the range of 200 to 300 can be easily performed with commercially available test fixturing, this example focuses on tests at peak temperatures of 300 to 600°C.

If the components to be tested are leaded, they have to be spring-clipped, compression-fixtured, soldered at high temperature, and brazed or socketed into the burn-in fixture. If soldering is used, the peak test temperature will have to be matched with a high liquidus solder. Most high temperature solders have a liquidus below 400°C. If the peak device test temperature exceeds 400°C, a braze must be employed. Brazes with liquidus temperatures around 750°C are easily found. One caveat: these braze materials are generally silver-based; EZFLOW 45 is an example. Silver forms dendrites very easily under the effects of temperature, voltage bias, humidity, or surface contaminants. Over time, which can be only a matter of minutes if the conditions are right, a dendritic formation can develop into a short-circuit between two interconnections. A potential difference of only 25 volts between two adjacent contacts is all that is required. Silver dendrite formation is generally *not* a problem if components are being tested without an applied bias, or if there will be no thermal cycling. The latter condition applies if, say, the devices being tested are brought from room temperature to 500°C and held at that temperature; in such a case, there will be very little water around to aid dendritic growth, because it will have quickly vaporized. The same is true of most organic materials left behind, like finger oils. However, if the parts are being thermally cycled so that condensation can form, dendrites could be a concern. Because of the problems with leakage and conduction via surface contamination or dendrite growth, all of the residues left behind by soldering fluxes or brazing must be removed.

Leaded components may also be attached to the burn-in fixtures with spring clips or sockets. The difficulty here concerns availability. There is no commercially available source of such products for the 500°C test environment. At these very high temperatures, the spring must be carefully designed so that it does not completely oxidize or lose all of its mechanical properties. It is possible to fabricate socket receptacles that work well to 600°C.

Surface-mount components may also be soldered or brazed to the burn-in cards, with the same caveats. Of course, thermal expansion mismatches between surface-mounted devices and the test fixture must be much more carefully controlled than for leaded component, in which the lead provides some leeway. I have had great success using custom-designed high-temperature compression test fixtures for evaluating components designed for surface-mounting. These are especially useful for testing physically large components.

The leaded components evaluated in this example were inserted into socketed test fixtures capable of long-term operation to temperatures as high as 500°C. The surface-mount devices were tested in a clamp fixture. How do we connect the fixtured devices through the furnace to the test instrumentation? If the acceleration temperature extends to only 200°C, then commercially available wires insulated with various polyimide insulation materials, from which a suitable test harness might be fabricated can be used. If the testing requires temperature excursions between 200 and 300°C, a teflon-coated wire would be a better selection. If the accelerated testing requirements extend into the 400 to 600°C range, the selection of commercially available wire becomes more finite, but there are still choices. Wire insulation choices now comprise glasses, ceramics, and their combinations. A number of researchers utilize thermocouple wire (platinum, 20% rhodium) that has been insulated with ceramic beads. This serves well if the set-up is very stable; if there will be any significant vibrations, fixture adjusting, or other mechanical motion, care must be taken not to allow spaces in the ceramic insulator beads to permit short-circuit failures. For a long-term test set-up, a number of commercially available mineral-insulated wires with metal jackets will operate up to 900°C. This type of interconnect is commonly used in high-temperature turbine engine applications. It is fairly rigid, but can be bent or formed within some minimum radius, much like copper tubing. It is difficult to work with, however, because the solid metal jacket must be cut to remove the powdered mineral insulation material in order to form and attach the inner interconnect wire(s). The most flexible, and least expensive, option for constructing high-temperature test harnessing is furnace wire, like 2102/20, supplied by Brim Electronics. This stranded wire is insulated with a glass-and-ceramic braid. It performs very well at these temperatures and can be formed, routed, and handled like conventional wire.

Now, the example varistors are fixtured, the interconnect wiring has been attached, and the assemblies have been enclosed within the test furnace. A number of companies sell high-temperature ovens and furnaces. For the devices in this example, a Blue-M furnace was used, capable of sustaining a controlled environment to temperatures as high as 600°C. Even though the fixturing used in this testing is unique, all traditional, carefully executed testing and measurement practices should be applied. For example, four-wire Kelvin connections should be used in all tests, from the instrument signal heads right up to the point at which the interconnects enter the high-temperature furnace port. The final length of high-temperature cable from the furnace port to the tested devices should be kept as short as possible and be made part of the initial calibration to account for the presence of parasitic contributions in the measurements.

In order to obtain the maximum adaptability from the test set-up, a flexible pulse driver circuit was constructed around the IR2110 MOS gate driver, which can easily be configured to control up to 500 V switched from a second-stage power MOSFET. The experimental set-up and instrument suite used to test the MLVs is shown in Figure 9.10. In addition to this instrumentation, an HP4140B pA meter and an HP4275A LCR meter were used to acquire leakage current and device capacitance-vs.-temperature data.

A Tektronics AGF5101 function generator was used to send single 5 µS test pulses to the IR2110 MOS gate driver. The MOS gate driver then triggered an IRF830 N-channel power MOSFET, which passed the actual test signal through the tested varistor. The rise time (t_r) and the fall time (t_f) of the test pulses used were approximately equal to 2 µS. The drain voltage applied to the IRF830 and the value of the series current limiting resistor established the current pulse delivered to the varistor. A simplified drawing of this test set-up is shown in Figure 9.11.

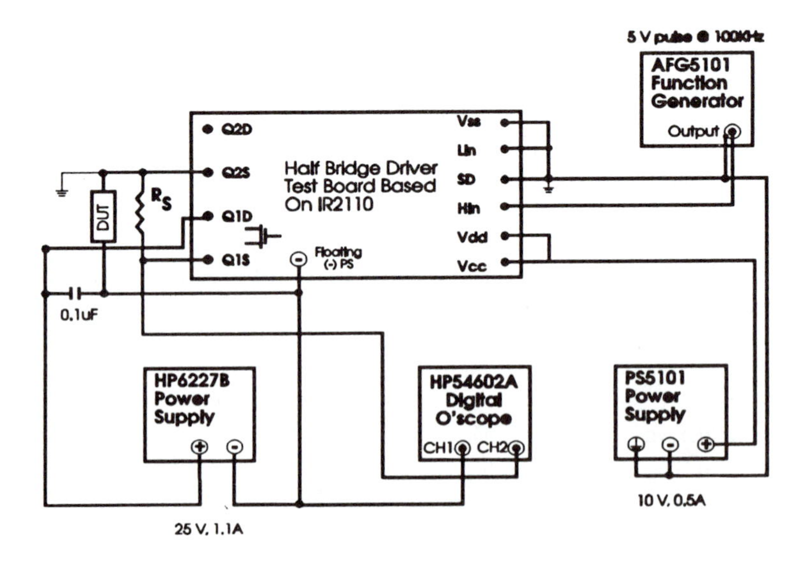

Figure 9.10 Set-up for accelerated high-temperature testing of varistors.

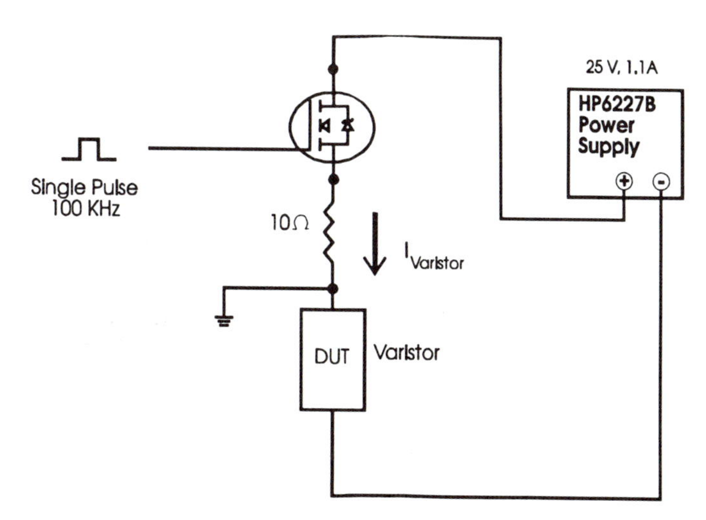

Figure 9.11 Simplified test set-up for accelerated high-temperature testing of varistors.

The data presented here was obtained from samples of two commercially available MLV devices manufactured by AVX, part numbers VA100005A150 and VA100014A300. The VA100005A150 device has a specified working voltage of 5.6 VDC (max.); a breakdown voltage of 7.1 to 8.7 VDC (max. at 1mA); and a clamping voltage of 15.5 V (max.). The VA100014A300 device has a specified working voltage of 14 VDC (max.); a breakdown voltage of 16.5 to 20.3 VDC (max. at 1mA); and a clamping voltage of 30 V (max.). The test set-up illustrated in Figures 9.10 and 9.11 is configured to test the 5.6 VDC devices. With the series resistor value selected as 10 Ω, the peak current expected through the 5.6 VDC varistors was 9.25A and the maximum expected voltage across the varistor was 16.2 V. The configuration of the instrument suite used to test the 14 VDC devices was identical, except that the HP6227B was set to deliver 60V pulses.

A first-order circuit model for an MLV is shown in Figure 9.12. The capacitance, C, can range from 110 to 4000 pF and the inductance, L, can range from 1.2 to 3.5 nH tested at low frequencies (1KHz) in typical operating environments [Demcko 1993]. As indicated earlier, the main purpose for accelerated testing and device characterization is to obtain some insight into how these components might be modeled while operating under simulated harsh-environment stressing. Contributors to the inductance value, L, will largely be affected by the device packaging scheme (devices with leads may have significant inductance, while surface-mount devices have very little) and by how the component is interconnected within the circuit. This parameter will, therefore, be regarded as primarily layout-dependent and not specifically modeled in this investigation. Since the MLV is basically capacitive in nature, of greater importance are the variations with temperature of the capacitance value, C, and the leakage current, which is a function of R_P.

Figure 9.13 shows a summary of typical MLV capacitance-vs.-temperature data obtained from both the AVX VA100005A150 and the VA100014A300 axial leaded devices at a test frequency of 4 MHz. Data for this plot was taken from room temperature to 375°C. This upper temperature boundary was established by the device packaging materials. Beyond 300°C, the soldered terminations and the dip coating of the MLV soften and soon fail.

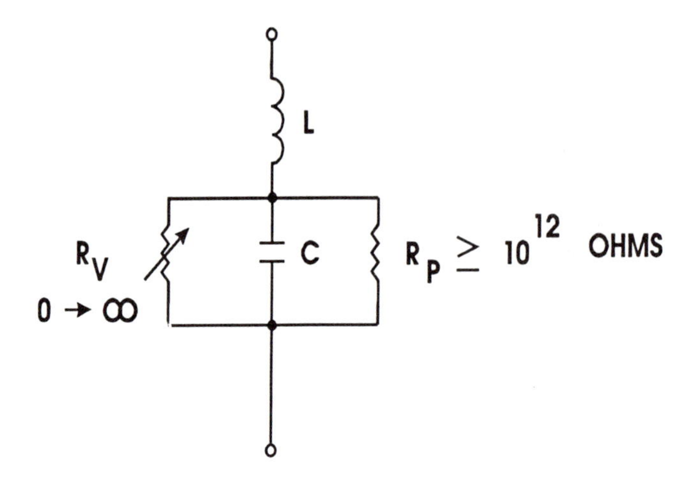

Figure 9.12 First-order circuit model for MLV.

Evidence of this can be clearly seen in Figure 9.13, in which a sudden change of slope in the capacitance-vs.-temperature curves begins at 275°C. At this point in device characterization, the researcher must decide whether the models will include this onset of packaging failure or not, because the curve fit will be substantially changed by this decision. For this case, the models do include the onset of packaging failure.

In all cases, the capacitance-vs.-temperature data could fit a simple polynomial equation of the form:

$$\frac{1}{y} = a + bx \tag{9.1}$$

A sample curve fitting for one of the 5.6 VDC device curves in this family is shown in Figure 9.14.

Surface-mount devices may undergo similar accelerated testing. Figure 9.15 shows capacitance-vs.-temperature data obtained from unleaded, surface-mounted devices that were otherwise identical to their lead-packaged counterparts. These components were attached to thick-film ceramic substrates with a higher temperature solder than that employed in fabricating the axial leaded versions. Since the solder used did not start to soften until 325°C, the slope change at 275°C that was seen in Figure 9.14 is conspicuously absent. There is also less parasitic capacitance associated with the surface-mounted devices. Of course, modeling equations similar to these may be derived for those devices as well.

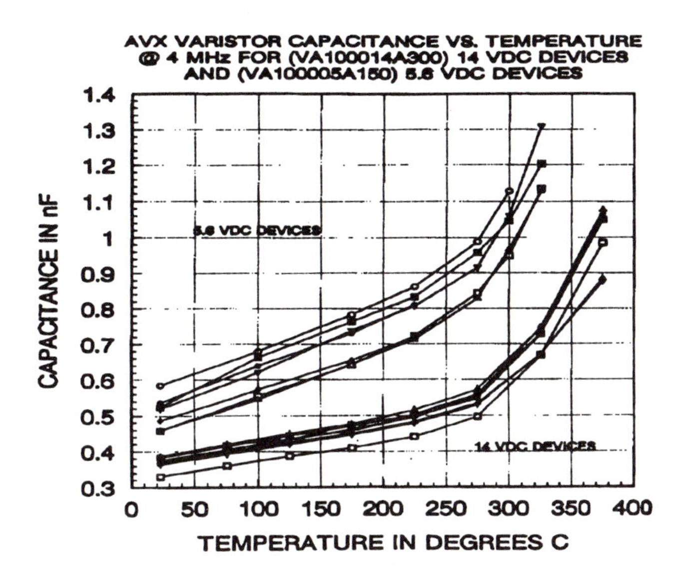

Figure 9.13 Capacitance-vs.-temperature data.

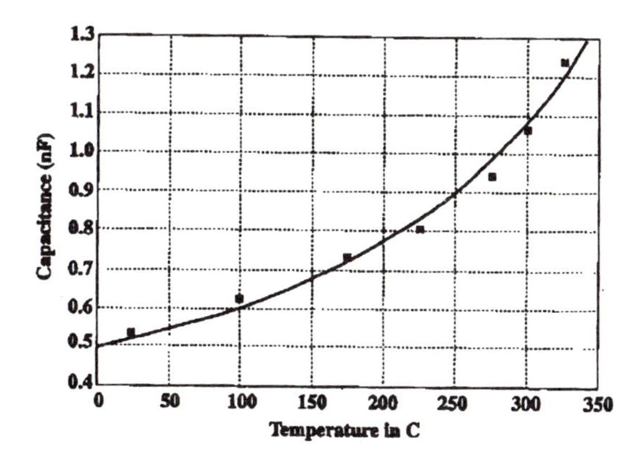

Figure 9.14 Typical curve fit for capacitance-vs.-temperature data.

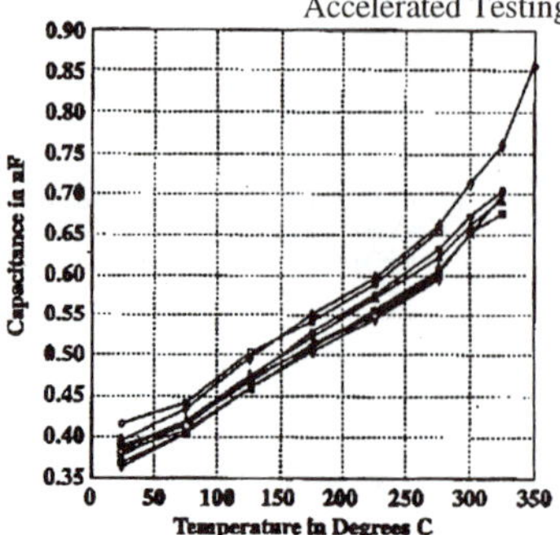

Figure 9.15 Capacitance - vs.- temperature data

Figure 16 summarizes a typical family of MLV leakage current-vs.-temperature data for the same axial leaded device. In this figure, the unit for leakage current for the 5.6 VDC devices is milliamps and for the 14 VDC devices, it is microamps. This data was obtained with a constant 10 V_{DC} bias applied and a current compliance set to 0.5A; above 275°C, these bias conditions raised the device temperature above the ambient sufficiently to melt the soldered terminations of the component. During the first several seconds of the test, leakage current increased exponentially and, ultimately, the device failed due to an open circuit caused by melting solder at the device terminations.

Once again, the propensity of the packaging materials to fail has been included in the models developed. The leakage current-vs.-temperature data for all of the 5.6 VDC devices could be fit to a simple log-normal equation of the form

$$y = a + b \exp(-0.5(\ln \frac{x/c}{d^2})) \qquad (9.2)$$

Leakage current-vs.-temperature for the 14 VDC devices could be similarly modeled. A sample fit for one of the 5.6 VDC device curves in this family is shown in Figure 17.

Once again, the results obtained from surface mount devices will be compared to their otherwise identical leaded counterparts. Figure 9.18 shows leakage current-vs.-temperature data obtained from unleaded, surface-mounted devices attached to thick-film ceramic substrates with a higher temperature solder than that employed in fabricating the axial leaded versions. These results should be compared to those presented in Figure 9.16. In this case, the leakage current is much more linear (on the log scale) than the leaded devices, and the amplitude of the leakage current is reduced by an order of magnitude in the surface-mount components.

In both the capacitance-vs.-temperature curves and the leakage current-vs.-temperature curves, it should be noted that the MLV devices are very well-behaved to at least 225°C. The capacitance variations in this temperature range remain well within levels normally tolerable in most circuit applications. In addition, there is only a modest increase in leakage current through the devices in this temperature window. Figure 9.19 shows a composite screen dump obtained from an HP54602A digital oscilloscope and one of the axial leaded 5.6 VDC components. The composite is composed of two pairs of traces. The upper pair represents the voltage across the 10Ω series resistor (and hence may be read as current through the varistor

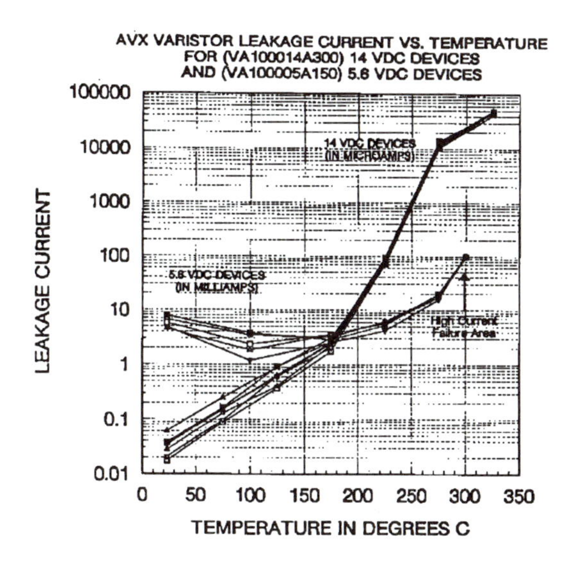

Figure 9.16 Leakage current-vs.-temperature data.

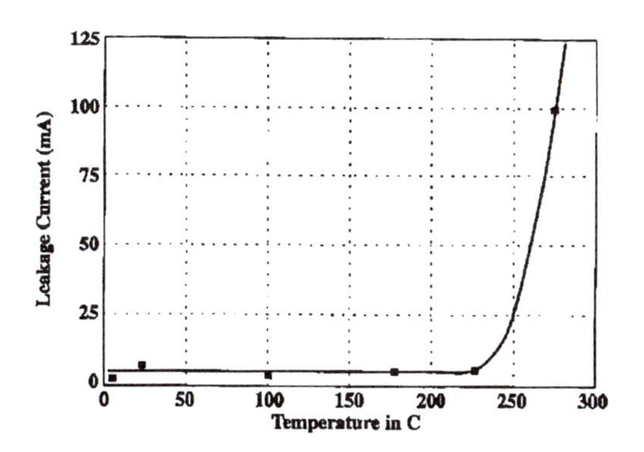

Figure 9.17 Typical curve fit for Leakage current-vs.-temperature data.

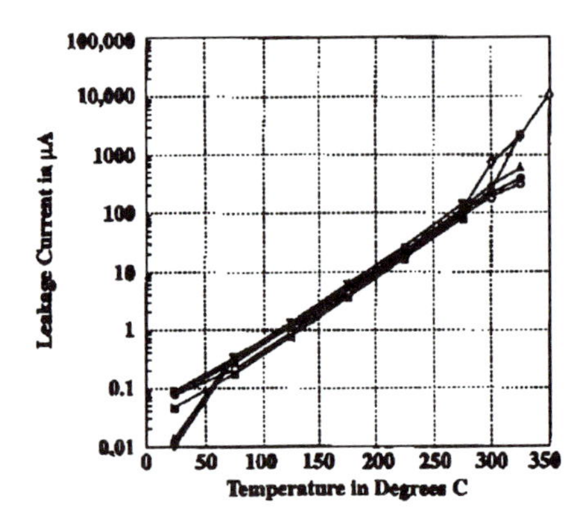

Figure 9.18 Typical curve fit for leakage current-vs.-temperature data for surface-mount devices.

directly on a 500 mA/division scale) at room temperature and at 375°C. The lower pair of traces represents the voltage across the varistor at room temperature and at 375°C. In this figure, the increased leakage current through the device and the decreasing ability of the varistor to sustain a constant clamping voltage are evident. In general, however, the ability of these devices to handle transient energy is only modestly impaired through most of the temperature excursion. Finally, Figure 9.20 shows an identically arranged 14 VDC leaded device tested to a final evaluation temperature of 400°C. This component was specially constrained in a clamping fixture that prevented the failing packaging materials from moving. Similar results are obtained because the intrinsically superior component survives beyond the packaging materials.

This example has projected the steps necessary to fixture, harness, and characterize 50 AVX VA100005A150 and VA100014A300 MLV varistors over a temperature range from room temperature to as high as 400°C. The characterization data enabled the development of modeling equations for the capacitance, C, and the leakage current, and therefore R_P, the required parameters which were obtained via curve fit for typical device responses. These equations can then be used within circuit simulation programs, such as SPICE (as described in [Grzybowski, 1991]) to accurately include the effects of temperature on the circuit being modeled.

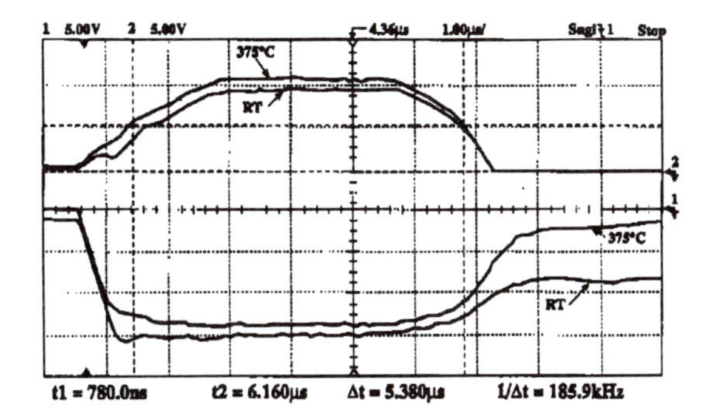

Figure 9.19 Composite screen dump obtained from an HP54602A digital oscilloscope - axial leaded 5.6 VDC device.

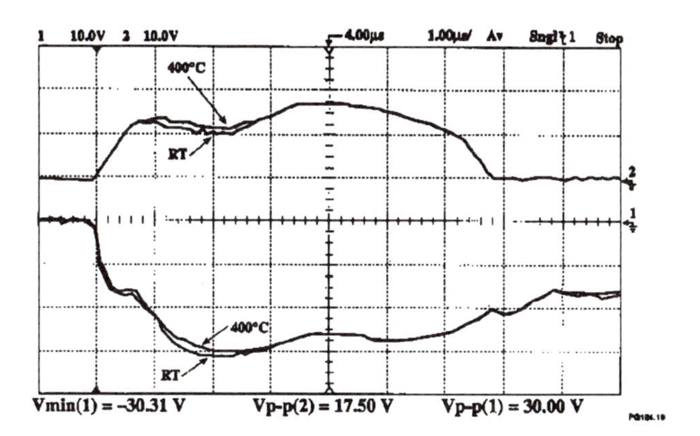

Figure 9.20. Composite screen dump obtained from an HP54602A digital oscilloscope - axial leaded 14 VDC device.

9.4 Summary

This chapter outlines the fabrication steps necessary to build test fixturing and packaging that has worked well at UTRC for the accelerated testing of electronic components to temperatures as high as 600°C. To be as inclusive and pragmatic as possible, it also considers a passive component accelerated testing example. The example assumes that the high-temperature electrical characteristics for incorporation in a SPICE circuit simulation program. The example demonstrates fixturing the devices, routing all of the cabling, obtaining data, curve-fitting the results, and deriving the model for the simulation. However, high-temperature electronics is still in its infancy. We are still working toward a complete solution for high-temperature circuit applications. This means that concomitant issues, such as high-temperature wirebonding and die attach, are still being evaluated and developed. These very issues are being worked on at UTRC for applications utilizing or accelerating the testing of silicon carbide integrated circuits to temperatures as high as 750°C.

REFERENCES

Antonetti, V.W., Oktay, S. and Simons., and R.E. Heat Transfer in Electronic Packages, Microelectronics Packaging Handbook, R.R. Tummala and E.J. Rymaszewski, Eds., Van Nostrand Reinhold, Chapter 4, pp. 168, (1989).

Ablestik Laboratories, Adhesives for Electronics, Company Literature, Rancho Dominguez, CA, (1995).

Adams, S., Severt, C., Leonard, J., Liu, S., and Smith, S. R. *Trans. 2nd Intl. High Temp. Electronics Conf.*, D. B. King and F. V. Thome, Eds., Vol. 1, pp. XIII-9, (1994).

Aday, J., Johnson, R. W., Evans, J. L., and Romanczuk, C. Thick Film Silver Multilayers for Under-the-Hood Automotive Applications, *Proc. Int. Symp. on Microelectronics ISHM*, pp. 126-131, (Nov. 1993).

Alekseenko, M. V., Zabrodskii, A. G., and Timofeev, M. P. Influence of the Degree of Doping and of Compensation on the Activation Energy of e_1 Conduction in 6H-SiC:N, *Soviet Physics Semiconductors*, 21, 494-500, (1982).

Alloys Digest, Engineering Alloys Digest, Inc., Upper Montclair, NJ , (1982).

Ambaum K. and Aderhold, J. Verification of Latchup Design Precautions for 1 Micron Junction Isolated CMOS ASICs Operating at Temperatures up to 520 K, *Trans. Second Int. High-Temp. Electron. Conf.*, Charlotte, NC, pp. P-15 to P-20, (June 1994).

Ames, I., d'Heurle, F. M., and Horstman, R. E. Reduction of Electromigration in Aluminum Films by Copper Doping, *IBM J. Res. Dev.*, 14 (4), pp. 461-463, (1970).

Amoco Performance Products, Inc., *Torlon Engineering Polymers Design Manual,* Atlanta, GA, (1990).

Andrews, J. M. and Phillips, J. C. *Phys. Rev. Lett.*, 35, 56, (1975).

Anikin, M. M., Rastegaeva, M. G., Syrkin, A. L., and Chuiko, I. V. *Amorphous and Crystalline Silicon Carbide III, Springer Proc. in Physics*, Springer-Verlag Berlin, Heidelberg, Vol. 56, pp.183, (1992).

Antler, M. Gold Plated Contacts: Effect of Heating on Reliability, *Plating 57* (6), 615-618, (1970).

Arthur, S.D., Temple, V.A.K., and Watrous, D.L. Forward Blocking Comparison of P and N MCTs, *Proc. IEEE Ind. Appl. Soc. Meet.*, pp. 1144-1149, (1992).

ASM Metals Handbook, Vol. 2, 10th ed., ASM International, Materials Park, OH, (1990).

ASM Handbook, Welding, Brazing, and Soldering, 10th ed., ASM International, Materials Park, OH, vol. 6, (1993).

Atsumi, K., et. al. Ball Bonding Technique for Copper Wire, *36th Proc. IEEE Electron. Compon. Conf.* pp. 312, (1986).

Auberton, A.J., Aspar, B., and Pelloie, J.L. SOI Substrates for Low-Power LSIs, *Solid State Technology*, pp. 89, (July 1994).

Auberton, A.J., Aspar, B., and Pelloie, J.L. SOI Substrates for Low-Power SLIs, *Solid State Technology*, pp.87, (March 1995).

Bader, W. G. Lead Alloys for High Temperature Soldering of Magnet Wire, *Welding Journal*, 154 (10), 370s-375s, (1975).

Baliga, B. J. High-Voltage Junction-Gate FET with Recessed Gates, *IEEE Trans. Electron Dev.*, ED-29, 1560, 1982.

Bar-Cohen, A. Fundamentals of Nucleate Pool Boiling of Highly-Wetting Dielectric Liquids, *Proc. Adv. Study Inst. Cool. Electron. Sys.*, pp. 518-558, Izmir, Turkey, (1993).

Barrett, D. L., Seidensticker, R. G., Gaida, W., Hopkins, R. H., and Choyke, W. J. Sublimation Vapor Transport Growth of Silicon Carbide, *Amorphous and Crystalline Silicon Carbide III and Other Group IV-IV Materials, Springer Proc. Phys. 56*, Springer, Berlin, 33-40, (1990).

Barrett, D. L., McHugh, J. P., Hobgood, H. M., Hopkins, R. H., McMullin, P. G., Clarke, R. C., and Choyke, W. J. Growth of Large SiC Single Crystals, *J. Crystal Growth*, 128, 358-36, (1993).

Bauer, R. E. and Clelland, I. W. Performance of a Multilayer Film Capacitor in Severe Applications, *Proc. Capacitor and Resistor Technol. Symp. (CARTS 1990)*, San Francisco, CA, pp. 128-133, (1990).

Bauer, K. A., Gerson, S., Griffith., D. M., and Trika, A. K. Woods Fly-by-Light/Power-by-Wire Integrated Requirements Analysis and Preliminary Design, NASA Contractor Report CR-4590, NASA Langley Research Center, Hampton, VA, (1993).

Beasom, J. D. and Patterson, R.B. Process Characteristics and Design Methods for a 300°C Quad Operational Amplifier, *IEEE Trans. Ind. Electron.*, IE-29(2), 112-117, (May 1982).

Bencuya, I., Cogan, A. I., Butler, S. J., and Regan, R. J. Static Induction Transistor Optimized for High-Voltage Operation and High Microwave Power Output, *IEEE Trans. Electron. Dev.* ED-32, 1321-1327, (1985).

Berger, H. H. *Solid-State Electron*, 15, 145, (1972).

Bergles, A.E. and Bar-Cohen, A. Direct Liquid Cooling of Microelectronic Components, *Proc. Adv. Study Inst. Cool. Electron. Sys.*, pp. 328-410, Izmir, Turkey, (1993).

Bhatnagar, M., McLarty, P. K., and Baliga, B. J. *IEEE Elec. Dev. Lett.*, 13 (10), 501, (1992).

Biedermann, E. The Optical Absorption Bands and their Anisotropy in the Various Modifications of SiC, *Solid State Commun.*, 3, 343-346, (1965).

Black, J. R. Electromigration - A Brief Survey and Some Recent Results, *IEEE Trans. Electron. Dev.*, ED-16(4), 338-347, (1969).

Borom, M. and Giddings, R. Glass/Metal Compression Seals Designed to Withstand Thermal Excursions, General Electric Company, Document # 74CRD330, (1975).

Bowers, M.B. and Mudawar, I. Two-Phase Electronic Cooling Using Mini-Channel and Micro-Channel Heat Sinks, *Proc. ASME Int. Electron. Packaging Conf.*, 2, 693-712, (1993).

Braun, C.G. and Podlesak, T.F. Power Electronics for the All-Electric Tank, Rep. # ARL-TR-67, Army Research Laboratory, Fort Monmouth, NJ , (August 1993).

Britton, S. C. Spontaneous Growth of Whiskers on Tin Coatings: 20 Years of Observation, *Trans. Inst. Metal Finishing* 52 (1974).

Bromstead, J.R. et al. Performance of Power Semiconductor Devices at High Temperature, *Trans. First Int. High-Temp. Electron. Conf.* Albuquerque, NM, pp. 27-35, (June 1991).

Bromstead, J. R., Weir, G. B., Johnson, R. W., Jaeger, R. C. and Baumann, E. D. Performance of Power Semiconductor Devices at High Tem. *Proc. First Int. High Temp. Electron. Conf.*, Albuquerque, NM, pp. 27-35, (June 1991).

Brown, D. M., Ghezzo, M., Kretchmer, J., Downey, E., Gorczyca, T., Saia, R., Edmond, J., Palmour, J., Carter, C. H. Jr., Gati, G., Dasgupta, S., Pimbley, J., and Chow, P. SiC Electronics for High Temperature Control Systems, *1991 Digest of Government Microcircuit Applications Conference,* Fort Monmouth, NJ, 89-92, (1991).

Brown, G.A., Hosack, H.H., and Joyner, K. Integrity of Gate Oxides Formed on SIMOX Wafers, *Proc. 1994 IEEE Int. SOI Conf.,* pp. 73, (October 1994).

Brush-Wellman, Inc., *Update-Beryllium Copper Design Conference Notes*, Cleveland, OH, (1991).

Brusius, P., Gingerich, B., Liu, S.T., Ohme, B., and Swenson, G. Reliable High Temperature SOI Process, 2nd International High Temperature Electronics Conference Charlotte, pp. II-15, (June 1994).

Bullard, G. L. Solid State Double-Layer Capacitor, *Proc. Workshop High Temp. Radiation-Hardened Electron.*, F.V. Thome, D. B. King, and C. W. Severt Eds. Sandia National Laboratories, pp. C21-C25, (1989).

Buritz, R. S. Advanced Capacitor Development, *Proc. Workshop High Temp. Radiation-Hardened Electron.*, F. V. Thome, D. B. King, and C. W. Severt Eds. Sandia National Laboratories, pp. C3-C7, (1989).

Burk, A. A. Jr., Barrett, D. L., Hobgood, H. M., Siergiej, R. R., Braggins, T. T., Clarke, R. C., Eldridge, G. W., Brandt, C. D., Larkin, D. J., Powell, J. A., and Choyke, W. J. SiC Epitaxial Growth on A-Axis SiC Substrates, *Silicon Carbide and Related Materials, Institute of Physics Conference Series 137*, Institute of Physics Publishing, Bristol, 29-32, (1994).

Burk, A. A. Jr. and Rowland, L. B. The Role of Excess Silicon and In situ Etching on 4H and 6H-SiC Epitaxial Layer Morphology, to be published in *J. Crystal Growth*, (1996).

Butler, S. J. and Regan, R. J. High-Voltage HF/VHF Power Static Induction Transistor Amplifiers, *Proc. RF Technol. Exp.,* (1986).

Campbell, R. B. Whatever Happened to Silicon Carbide., *IEEE Trans. Ind. Electron.*, IE-29, 124-128, (1982).

Capote, M. A., Todd, M., Gandhi, P., Carr, C., Walters, W., and Viajar, H. Multilayer Printed Circuits from Revolutionary Transient Liquid Phase Inks, *Proc. National Electron. Packaging Prod. Conf. West '93*, pp. 1709-1715, (1993).

Carter, C. H., Jr. and Tsvetkov, V. F. Recent Progress in SiC Boule Growth, *Tech. Digest Int. Conf. on Silicon Carbide Mater.,* Kyoto, Japan, (1995).

Celanese, Hoechst Engineering Plastics Division, Advanced Materials Group, *Vectra Liquid Crystal Polymer Design Manual, VC-10,* Chatham, NJ, (1990).

Celanese, H. Engineering Plastics Division, Advanced Materials Group, Designing with Fortron: Polyphenylene Sulfide, Design Guide FN-10, Chatham, NJ, (1991).

Chaddha, A.K., Parsons, J. D., and Kruaval, G. B. *Appl. Phys. Lett.* 66 (6), 760, (1995).

Chang, D. D., Fulton, J. A., Ling, H. C., Schmidt, M. B., Sinitiski, R. E., and Wong, C. P. Accelerated Life Test of Z-Axis Conductive Adhesives, *IEEE Comp. Hybrids Manufactur. Technol.*, 16, pp. 836-842, (1993).

Chang, H. C., LeMay, C. Z., and Wallace, L. F. Use of Silicon Carbide in High Temperature Transistors, Silicon Carbide: A High Temperature Semiconductor, J. R. O'Connor, and J. Smiltens, Eds., Pergamon, Oxford, 496-507, (1960).

Charles, H. K. and Clatterbaugh, G. V. Packaging Considerations for High Temperature Operation, *Proc. Second Int. High Temp. Electron. Conf.,* Charlotte, NC, pp. IX-3, (1994).

Chaudhry, M. I., Berry, W. B., and Zeller, M. V. *Int. J. Electrons*, 71, 439, (1991).

Chen, C., Matloubian, M., Sundaresan, R., Mao, B., Wei, C., and Pollack, G. Single-Transistor Latch in SOI MOSFET's, *IEEE Electron Devices*, Lett. EDL-9(12), 636, (1988).

Chien, I. Y., and Nguyen, M. N. Low Stress Polymer Die Attach Adhesive for Plastic Packages, *Proc. Electron. Comp. and Technol. Conf.* (May 1994).

Cho, H. J., Hwang, C. S., Bang, W. and Kim, H. J. *Silicon Carbide and Related Materials, Proc. 5th Conf. SiC Matler.*, M.G. Spencer, R. P. Devaty, J. A. Edmond, M. A. Khan, R. Kaplan, and M. Rahman, Eds., Institute of Physics Publishing, Bristol and Philadelphia, pp. 663, (1994).

Chow, T. P. and Stecki, A. J. Plasma Etching of Refractory Gates for VLSI Application, *J. Electronochem. Soc.* 131 (10), 2325-2335, (1984).

Clarke, R. C., Siergiej, R. R., Agarwal, A. K., Brandt, C. D., Burk, A. A. Jr., Morse, A., and Orphanos, P. A. Ò30W VHF 6H-SiC Power Static Induction Transistor, *Proc. IEEE /Cornell Conf. Adv. Concepts High Speed Semiconductor Dev. Circuits*, 47-56, (1995).

Clemen, L. L., Yoganathan, M., Choyke, W. J., Devaty, R. P., Kong, H. S., Edmond, J. A., Larkin, D. J., Powell, J. A., and Burk, A. A. Jr. Observation of Doping Dependence of Epitaxially Grown 6H-SiC for Various CVD Growth Directions, Silicon Carbide and Related Materials, *Inst. Phys. Conf. Ser. 137*, Institute of Physics Publishing, Bristol, 251-254, (1994).

Cogan, A., Regan, R., Bencuya, I., Butler, S., and Rock, F. High Performance Microwave Static Induction Transistors, *IEDM Technical Digest*, 221-224, (1983).

Cohen, S.S. and Gildenblat, G.S. *VLSI Electronics Microstructure Science*, Vol. 13, Academic Press, Orlando, FL, pp.47, (1987).

Colinge, J.P. Problems and Issues in SOI CMOS Technology, *Proc.1991 IEEE Int. SOI Conf.* pp. 126, (October 1991a).

Colinge, J.P. Silicon-on-Insulator Technology: Materials to VLSI, Kluwer Academic Publishers, (1991b).

Collinge, J.P. and Tack, M. *Proc.1989 IEEE Int. SOI Conf.* (1989).

Condra, L. Reliability Improvement with Design of Experiments, Marcel Dekker, New York, (1993).

Coors Ceramic Company, *Structural Division,* Materials Properties Data, Golden, CO, (1993).

Crane, J., Breedis, J. F., and Fritzsche, R. M. Lead Frame Materials, *Electronic Materials Handbook, Vol.1 Packaging,* ASM International, Materials Park, OH, pp. 483-492, (1989).

Crofton, J., McMullin, P. G., Williams, J. R., and Bozack, M. J. *Trans. 2nd Intl. High Temp. Electronics Conf.*, King, D. B. King and F. V. Thome, Eds., Vol. 1, pp. XIII-15, (1994).

Crofton, J., Ferrero, J. M., Barnes, P. A., Williams, J. R., Bozack, M. J., Tin, C. C., Ellis, C. D., Spitznagel, J. A., and McMullin, P. G. *Amorphous and Crystalline Silicon Carbide IV, Springer Proc. in Physics*, Vol. 71, Springer-Verlag Berlin, Heidelberg, pp. 176, (1992).

Crofton, J., McMullin, J., Williams, P.G., and Bozack, J.R. *Appl. Phys.* 77, 1317, (1995).

Crofton, J., Williams, J. R., Bozack, M. J., and Barnes, P. A. Silicon Carbide and Related Materials, *Proc. 5th Conf. SiC Mater.*, M. G. Spencer, R. P. Devaty, J. A. Edmond, M. A. Khan, R. Kaplan, and M. Rahman, Eds., Institute of Physics Publishing, Bristol and Philadelphia, pp. 719, (1994).

Crofton, J., Barnes, P. A., Williams, J. R., and Edmond, J. A. *Appl. Phys. Lett.* 62 (4), 384, (1993).

Cygan, S. and McLarney, J. Multilayer Ceramic Capacitors Characterization for High Temperature Applications, *Proc. First Int. High Temp. Electronics Conf.*, Albuquerque, NM, pp. 85-92, (1991).

Daimon, H., Yamanaka, M., Sakuma, E., Misawa, S., and Yoshida, S. *Jap. J. Appl. Phys.* 25 (7), L592, (1986).

Dally, J. W. Packaging of Electronic Systems, McGraw-Hill, (1993).

Dance, D. European SOI Comes of Age, *Semiconductor International*, pp. 83, (November 1993).

Dasgupta, A. Thermomechanical Analysis and Design, Handbook of Electronic Package Design, Marcel Dekker, New York, (1991).

Demcko, R. et al. Specialized Characteristics of Multilayer Glass Dielectric Capacitors, *Proceedings Capacitor and Resistor Technol. Symp.* Amsterdam, Holland, (1989).

Demcko, R. Multilayer Varistor Transient Protection, *EMC Test & Design*, p. 21, (Oct. 1993).

Dikman, C.T., Dogan, N.S., and Osman, M. High Temperature Behavior and Modeling of Integrated Circuit Bipolar Junction Transistors, *Trans. Second Int. High-Temp. Electron. Conference*, Charlotte, NC, pp. P-33 to P-38, (June 1994).

DM Data, Inc., System Design Guidelines for Reliability, 96 (1990).

Dmitriev, V. A., Irvine, K., and Spencer, M. *Appl. Phys. Lett.* 64 (3), 318, (1994).

Dmitriev, V. A., Fekade, K., and Spencer, M. G. *Amorphous and Crystalline Silicon Carbide IV*, C.Y. Yang, M.M. Rahman, and G.L. Harris, Eds., Spring-Verlag Berlin, Heidelberg, pp. 352, (1992).

Doerbeck, F. H., Duncan, W. M., McLevige, W. J., and Yuan, H. T. Fabrication and High-Temperature Characteristics of Ion-implanted GaAs Bipolar Transistors and Ring Oscillators, *IEEE Trans. Ind. Electron.*, IE-29(2), 136-139, (1982).

Doosan Electro-Materials Co., Ltd., *Copper Clad Epoxy Laminates,* Chungbuk, Korea, (1995a).

Doosan Electro-Materials Co., Ltd., *Materials for Multilayer Printed Circuit Boards*, Chungbuk, Korea, (1995b).

Draper B. L. and Palmer, D. W. Extension of High-Temperature Electronics, *IEEE Trans. Components, Hybrids, and Manufacturing Technol.*, Vol. CHMT-2, No. 4, pp. 399-404, (December 1979).

Dreike, P. L., Fleetwood, D. M., King, D. B., Sprauer, D. C., and Zipperian, T.E. An Overview of High-Temperature Electronic Device Technologies and Potential Applications, *IEEE Trans. Components, Packaging and Manufacturing Technol.- Part 1*, 17(4), pp. 594-609, (1994).

E. I. duPont deNemours, Inc., *Guide to Thermount,* (1995).

Edelstein, D. C. Advantages of Copper Interconnects, *Proc. Twelfth Int. VLSI Multilevel Interconnection Conf.*, pp. 301-307, (1995).

Eggermont, J.P., Gentinne, B., Flandre, D., Jespers, P.G.A., and Colinge, J.P. SOI CMOS Operational Amplifiers for Applications Up to 300°C, 2nd International High Temperature Electronics Conference, Charlotte, p II-21, (June 1994).

Eicke, D.M. and Hodge, E.S. Advanced Secondary Power System Study, Contract F33657-84-C-0247, CDRL 3402, Fort Worth Division, (1992).

Elsby, T. W. Microelectronic Packaging of High-Temperature Electronics, Final Report on Workshop on High Temperature Electronics (Metalization and Packaging), June 1989, pp. C53-C62, Sandia Labs Report SAND91-0370, (1989).

Erskine, J. C., Carter, R. G., Hearn, J. A., Fields, H. L. and Himelick, J. M. High Temperature Automotive Electronics: Trends and Challenges, *Trans. Second Int. High Temp. Electron. Conf.*, Vol. I, Charlotte, NC, pp. I-9 - I-17, (June 1994).

Evans, J. L., Romanczuk, C., Johnson, R. W., and Aday, J. Using MCM-L Technology for Under-the-Hood Automotive Environmental Conditions, *Proc. Int. Symp. Microelectronics ISHM*, pp. 584-588, (Nov. 1993).

Evwaraye, A., Smith, S. R., Skowronski, M., and Mitchel, W. C. J. *Appl. Phys.* 74, 5269.

Feinstein, L. Die Attachment Methods, *Electronic Materials Handbook, Vol. 1 Packaging,* ASM International, Materials Park, OH, (1989).

Ferry, D. K., Ed. Gallium Arsenide Technology, Howard W. Sams & Co., Indianapolis, IN, (1985).

Frear, D. R., Jones, W. B., and Kinsman, K. R. Solder Mechanics: A State of the Art Assessment, *TMS Publications*, Warrendale, PA, (1991).

Frear, D. R., Grivas, D., McCormack, M., Tribula, D., and Morris, J. W. Fatigue and Thermal Fatigue of Pb-Sn Solder Joints, Effect of Load and Thermal Histories on Mechanical Behavior, *TMS Spring Conference,* Denver, CO, pp. 113, (1987).

Frear, D. R., Burchett, S. N., Morgan, H. S., and Lau, J. H. The Mechanics of Solder Alloy Interconnections, Van Nostrand Reinhold, New York, NY, (1994).

Frear, D. R. and Vianco, P. T. Intermetallic Growth and Mechanical Behavior of Low Melting Temperature Solder Alloys, Metallurgical Trans. A, 25A, 1509, (1994).

Fricke, K., Hartnagel, H. L., Lee, W. Y., Pirling, T., Shuessler, M., and Wuerfl, J. A Highly Reliable Ohmic Contact on GaAs Based on Pd/In, *Trans. Second Int. High-Temp. Electron. Conf.*, Charlotte, NC, pp. P-197 to P-202, (June 1994).

Fricke, K., Hartnagel, H. L., Lee, W., and Wuerfl, J. AlGaAs/GaAs HBT for High-Temperature Applications, *IEEE Trans. Electron Devices*, 39 (9), 1977-1981, (1982).

Fricke, K., Hartnagel, H. L., Schuetz, R., Schweeger, G., and Wuerfl, J. A New GaAs Technology for Stable FET's at 300°C, *IEEE Electron Device Lett.,* 10(12), 577-579, (1989).

Frost, M. S., Riches, M., and Kerr, T. A p-n-p AlGaAs Heterojunction Bipolar Transistor for High-Temperature Operation, *J. Appl. Phys.*, 60(6), 2149-2153, (1986).

Ganley, G. A. Concorde Propulsion - Did We Get It Right? The Rolls-Royce/Snecma Olympus 593 Engine Reviewed, SAE Technical Paper Series 912180, Aerospace Technology Conference and Exposition, Long Beach, CA, 26 (Sept. 1991).

Gaynes, M. A., Lewis, R. H., Saraf, R. F., and Roldan, J. M. Evaluation of Contact Resistance for Isotropic Electrically Conductive Adhesives, *IEEE-Components Hybrids Manufacturing Technol.*, 18, 299-304, (1995).

Geffken, R. An Overview of Polyimide Use in Integrated Circuits and Packaging, *Proc. 3rd Int. Symp. Large Scale Integration Science and Technology*, (1991).

Geib, K. M., Mahan, J. E., and Wilmsen C. W. *Amorphous and Crystalline Silicon Carbide and Related Materials II, Springer Proc. in Physics*, Vol. 43, Springer-Verlag Berlin, Heidelberg, pp. 224, (1989).

Gersch, G.R. Assessment of Benefits in Utilization of Advanced Thermal Management Packaging Materials, *Proc. Second Int. High Temp. Electron. Conf.*, 2, pp. P-129, P-140, (1994).

Glass, R. C., Spellman, L. M., Tanaka, S., and Davis, R.F. *J. Vac. Sci. Technol.* A 10 (4), 1625, (1992).

Glass, R. C., Spellman, L. M., and Davis, R. F. *Appl. Phys. Lett.* 59 (22), 2868, (1991).

Godbold, C. V., Hudgins, J. L., Braun, C., and Portnoy, W. M. Temperature Variation Effects in MCTs, IGBTs, and BMFETs, *1993 IEEE Power Electronics Specialists Conference Record*, Boston, MA, pp. 93-98 , (June 1992).

Goetz, G. G. unpublished data, (August 1996).

Goetz, G. G. and Johnson, G. Tungsten-based Metallization for High-Temperature Conductors and Contacts, *Trans. Second Int. High Temp. Electron. Conf.*, pp. IX33-IX40, (1994).

Goldberg, C. and Ostroski, J. W. Silicon Carbide Rectifiers, Silicon Carbide: A High Temperature Semiconductor, J. R O'Connor, and J. Smiltens, Eds., Pergamon, Oxford, 453-461, (1960).

Goodnan, S. Handbook of Thermoset Plastics, Noyes Publications, Park Ridge, NJ, (1986).

Gray, P.R., and Meyer, R.G. Analysis and Design of Analog Integrated Circuits, 3rd ed., John Wiley & Sons, Inc., New York, (1993).

Grove, A.S. Physics and Technology of Semiconductor Devices, Wiley, New York, (1967).

Grzybowski, R. R. and Beckman, T. E. Characterization and Modeling of Ceramic Multilayer Capacitors to 500°C and Their Comparison to Glass Dielectric Devices, *Proc.13th Capacitor and Resistor Technol. Symp. (CARTS 1993)*, Tucson, AZ, pp. 157-162, (1993).

Grzybowski, R. Characterization and Modeling of Ceramic Multilayer Capacitors to 500°C and Their Comparison to Glass Dielectric Devices, *Proc. Thirteenth Capacitor and Resistor Technol. Symp.(CARTS)*, pp. 157 - 162, (1993).

Grzybowski, Richard R. "Characterization and Modeling of Ceramic Multilayer Capacitors to 500°C and Their Comparison to Glass Dielectric Devices," Capacitor and Resistor Technology Symposium (CARTS) 1992)," Tucson, AZ, March 16-19, 1992, pp. 114-120.

Grzybowski, R.R., and Tyson, S.M. High Temperature Testing of SOI Devices to 400°C, *Proc. 1993 IEEE International SOI Conference*, pp. 176, (October 1993).

Grzybowski, R. R. Characterization and Modeling of Glass and Glass-K Capacitors to 450°C, *Proceedings of the 12th Capacitor and Resistor Technology Symposium (CARTS 1992*), Tucson, AZ, pp. 114-120, (1992).

Grzybowski, R. R., Kerwin, D. B. High Temperature Reliability of Al/Au and Au/Au Wire Bonds, *Proc. Second Int. High Temp. Electron. Conf.,* Charlotte, NC, pp. VIII-9, (June 1994).

Grzybowski, R. and Tyson, S. M. High Temperature Testing of SOI Devices to 400°C, *1993 IEEE Int. SOI Conf. Proc.*, pp. 176 - 177, (October 5-7, 1993).

Grzybowski, R. R. Development of 600°C Test Fixturing, *Proc. First Int. High Temp. Electron. Conf.,* Albuquerque, NM, pp. 149-155, (1991a).

Grzybowski, Richard R. Characterization and Modeling of Glass and Glass-K Capacitors to 450°C, *Trans. First Int. High Tem. Electron. Conf.*, pp. 99-104, (June, 1991).

Grzybowski, R. R., Gericke, M. 500°C Electronics Packaging and Test Fixturing, *Proc. Second Int. High Temp. Electron. Conf.,* Charlotte, NC, pp. IX-41, (June 1994).

Guide Brush-Wellman Inc., update-beryllium Copper Design Conference Notes, Cleveland, Ohio, (1991).

Guttenplan, J. D. and Violette, D. R. Corrosion Related Problems Affecting Electronic Circuitry, *Mater. Perfor.*, pp. 76-81, (April 1990).

Hall, P. M. et al. Strength of Gold-Plated Copper Leads on Thin Film Circuits under Accelerated Aging, *IEEE Trans. Parts, Hybrids, and Packaging, PHP-11* (3), pp. 202, (1975).

Hamburgen, W.R. and Fitch, J.S. Packaging a 150 W Bipolar ECL Microprocessor, Research Report 92/1, Digital Western Research Laboratory, pp. 20, Palo Alto, CA, (1992).

Hamilton, D. R. Preparation and Properties of Pure Silicon Carbide, Silicon Carbide: A High Temperature Semiconductor, J. R. O'Connor, and J. Smiltens, Eds., Pergamon, Oxford, (1960, 43-51; Barrett, D. L., Evaluation of Trace Impurities in the Preparation of High-Purity Silicon Carbide, *J. Electrochem. Soc.*, 113(11), 1215-1218, 1996.

Hammoud, A. N., and Suthar, J. L. Characterization of Polybenzimidazole (PBI) Film at High Temperatures, *NASA Technical Memorandum 189174*, (1992).

Hammoud, A. N. et. al. High Temperature Power Electronics for Space, *NASA Technical Memorandum 104375*, (1991).

Hammoud, A. N. and Myers, I. T. Evaluation of High Temperature Capacitor Dielectrics, *NASA Technical Memorandum 105622*, (1989).

Harkness, J. C., Spiegelberg, W. D., and Cribb, W. R. Beryllium Copper and Other Beryllium Containing Alloys, Metals Handbook, 10th ed., Vol. 2, (1991).

Harman, G. G. Reliability and Yield Problems of Wirebonding in Microelectronics: The Application of Materials and Interface Science, *A Technical Monograph of the Int. Soc. for Hybrid Microelectronics,* (1989).

Harper, C. A. Handbook of Materials and Processes for Electronics, McGraw-Hill, New York, (1970).

Harris, J. O. Energy Density Capacitor Technology at Sandia National Laboratories, *Proc. Workshop on High Temp. and Radiation Hardened Electronics*, F. V. Thome, D. B. King and C. W. Severt. Sandia National Laboratories, pp. C8-C14, (1989).

Heller, P. Applications of Metal Matrix Composite Materials in SEM-E Cores and Multi Chip Modules, *Proc. Nat. Electron. Packaging and Production Conf. NEPCON West,* pp. 2026-2031, (March 1993).

Hinoda, K. Owade, N., Nishida, T., and Mukal, K. Stress-induced Grain Boundary Fractures in Al-Sl Interconnects, *J. Vacuum Sci. and Technol.* B, 5 (2), pp. 518-522, (1987).

Ho, C. Y., Ackerman, M. W., Wu, K. Y., Havill, T. N., Bogaard, R. H., Matula, R. A., Oh, S. G., and James, H. M. Electrical Resistivity of Ten Selected Binary Alloy Systems, *J. Phys. and Chemistry Reference Data, 12* (2), pp. 183, (1983).

Hobgood, H. M., Barret, D. L., McHugh, J. P., Clarke, R. C., Sriram, S., Burk, A. A., Jr., Greggi, J., Brandt, C. D., Hopkins, R. H., and Choyke, W. J. Large Diameter 6H-SiC for Microwave Dev. Appl., *J. Crystal Growth,* 137, 181-186, (1994).

Hobgood, H. M., Glass, R. C., Augustine, G., Hopkins, R. H., Jenny, J., Skowronski, M., Mitchel, W. C., and Roth, M. Semi-insulating 6H-SiC Grown by Physical Vapor Transport, Applied Phys. Lett., 66(11), 1364-1366, (1995).

Holman, J. P. Heat Transfer, McGraw-Hill, New York, (1986).

Hovel, H., Freeouf, J., Beyer, K., Sadana, D., and Chu, S. Non-Destructive Characterization Techniques for SOI Substrates, *Proc. 1993 IEEE Int. SOI Conf.*, pp. 40, (October 1993).

Hsu, T. S. Water and Moisture Absorption in Laminates, *Printed Circuit Fabrication*, (July 1991).

Hu, J. M., Pecht, M. G., and Dasgupta, A. Design of Reliable Die Attach,

Huang, J.S.T., and Kueng, J.S. An Analytical Model for Snapback in n-Channel SOI MOSFET's, *IEEE Trans. Electron Devices*, Vol. 38, No. 9, pp. 2082, September (1991).

Hudgins, J. L., Menhart, S., Portnoy, W. M., and Sankaran, V. A. Temperature Variations Effects on the Switching Characteristics of MOS-Gate Devices, *Symp. on Mater., and Devices for Power Electronics (EPE-MADEP 1991)*, pp. 262-266, (September 1991).

Hvims, H. L. Conductive Adhesives for SMT and Potential Applications, *IEEE-Compon. Hybrids Manufacturing Technol.*, Vol. 18, pp. 284-291, (1995).

Hwang,Y., Tang, L., and Yu, E. Failure Mechanism Analysis and Thermal Modeling for Lamp Dimmer in Boeing 747-700, Final term project report, ENME 808-D, Department of Mechanical Engineering, University of Maryland, College Park, MD, (1995).

Inayoshi, H., Nishi, K., Okikawa, S., and Wakashima, Y. Moisture-Induced Aluminum Corrosion and Stress on the Chip in Plastic Encapsulated LSIs, Proc. 17th Ann. Int. Reliability Phys. Symp., pp. 113-117, (1979).

Incropera, F.P., and Ramadhyani, S. Single-Phase, Liquid Jet Impingement Cooling of High-Performance Chips, *Proc. Advanced Study Institute on Cooling of Electronic Systems*, pp. 277-327, Izmir, Turkey, (1993).

Incropera, F.P., and Ramadhyani, S. Application of Channel Flows to Single-Phase Liquid Cooling of Chips and Multi-Chip Modules, Proc. Advanced Study Inst. Cooling of Electron. Sys., pp. 460-491, Izmir, Turkey, (1993).

Intel Corporation, *Components Quality/Reliability Handbook,* Santa Clara, CA, (1990).

International Rectifier IR2110 high-voltage bridge-driver data sheets, International Rectifier, El Segundo, CA.

Ioannou, D. E., Papanicolaou, N. A., and Nordquist, P. E., Jr. *IEEE Trans. on Electron Devices*, ED-34 (8), 1694, (1987).

Jacob, C., Nishino, S., Mehregany, M., Powell, J. A., and Pirouz, P. Silicon Carbide and Related Materials, *Proc. of the 5th Conf. on SiC and Related Mater.*, M. G. Spencer, R. P. Devaty, J. A. Edmond, M. A. Khan, R. Kaplan, and M. Rahman, Eds., Institute of Physics Publishing, Bristol and Philadelphia, pp. 247, (1994).

James, K. Reliability Study of Wirebonds to Silver Plated Surfaces, *IEEE Trans. Parts, Hybrids, and Packaging, PHP-13*, pp. 419, (1977).

Jellison, J. L. Susceptibility of Microwelds in Hybrid Microcircuits to Corrosion Degradation, *13th Ann. Proc. Reliability Phys. Symp.*, pp. 70, (1975).

Jenkins, W.C., and Liu, S.T. Hot Electron Lifetime of 0.8 m CMOS Transistor Fabricated in SIMOX, *Proc.6th Int. Symp. Silicon-on-Insulator Technol. and Devices*, PV94-11. The Electrochemical Society, Inc., S. Cristoloveanu, K. Izumi, and Hosack, H. Eds., (1994).

Jenkins, K.A., Sun, J.Y.C., and Pelloie, J.L. Measurement of SOI MOSFET I-V Characteristics Without Self-Heating, *Proc. 1994 IEEE Int. SOI Conf.*, pp. 121, (October 1994).

Johnson, R. W., Weir, G. B., and Bromstead, J. R. 200°C Operation of Semiconductor Power Devices, *IEEE Transactions on Components, Hybrids, and Manufacturing Technol.*, 16, (7), 759-764, (November 1993).

Johnson, R. W., Thomas, E. L., Duren, R. M., Curington, D. W., and Lippincott, A. C. Insulated Metal Substrates for the Fabrication of a Half-Bridge Power Hybrid, *IEEE Trans. Components, Hybrids, and Manufacturing Technol.*, 14, 886-893, (1991).

Johnson, E. O. Physical Limitations on Frequency and Power Parameters of Transistors, RCA Review 26, 163-177, (1965).

Joshi, Y and Kelleher, M.D. Liquid Immersion Cooling of Electronic Equipment, *Naval Res. Rev.*, XLIV, (1), 35-42, (1992).

Kane, M. G. and Frey, R. The PSIFET Emerges as a New Contender, *Microwave Syst. News*, 14, (10), (Sept. 1984).

Katscher, H., Sangster, R., and Schröder, F., Eds. Gmelin Handbook of Inorganic Chemistry, Silicon Supplement, Vol. B3, Springer, Berlin, 62, (1986).

Kearny, K. M. Trends in Die Bonding Materials, Semiconductor Int., pp. 84-89, (1988).

Keyes, R. W. Figure of Merit for Semiconductors for High-Speed Switches, *Proc. IEEE*, 60, 225, (1972).

Khan, M., et al. Effect of High Thermal Stability Mold Material on the Gold-Aluminum Bond Reliability in Epoxy Encapsulated VLSI Devices, *IEEE International Reliability Phys. Symp.*, 40, (1988).

Khatchatourian, Z. Ablestik Laboratories, Inc., private communication, (1995).

Kim, H. J., and Shin, D. W. Growth of 6H-SiC Single Crystals by the Modified Sublimation Method, *Amorphous and Crystalline Silicon Carbide III and Other Group IV-IV Materials, Springer Proceedings in Physics 56*, Springer, Berlin, 23-28, (1990).

Koga, K., Fujikawa, Y., Ueda, Y., and Yamaguchi, T. Growth and Characterization of 6H-SiC Bulk Crystals by the Sublimation Method, *Amorphous and Crystalline Silicon Carbide IV, Springer Proceedings in Physics*, Springer, Berlin, 96-100, (1992).

Krishnamurthy, V., Brown, D. M., Ghezzo, M., Kretchmer, J., Hennessy, W., Downey, E., and Michon, G. Planar Depletion-mode 6H-SiC MOSFETs, Silicon Carbide and Related Materials, *Institute of Physics Conference Series 137*, Institute of Physics Publishing, Bristol, 483-486, (1994).

Krull, W.A. and Lee, J.C. Demonstration of Benefits of SOI for High Temperature Operation, IEEE SOI/SOS Technology Workshop. St. Simons's Island, GA, pp. 69, (1988).

Krumbein, S. J. Metallic Electromigration Phenomena, *IEEE Trans. Comp. Hybrids, Manuf. Tech., CHMT-11* (1), 5 (1988).

Kuphal, E. *Solid-State Electron.*, 24, 69, (1981).

Lambregts, A. A. and McCorkle, R. D. The Boeing Condor Unmanned Air Vehicle, Presentation to SAE Aerospace Control and Guidance Systems Committee (No. 67), Denver CO, 1991.

Larkin, D. J., Neudeck, P. G., Powell, J. A., and Matus, L. G. Site competition epitaxy for controlled doping of CVD silicon carbide, *Silicon Carbide and Related Materials, Inst. Phys. Conf. Ser. 137*, Institute of Physics Publishing, Bristol, 51-54, (1994).

Lau, J. H., Erasmus, S. J., and Rice, D. W. Overview of Tape Automated Bonding Technology, *Electronic Material Handbook,*Vol. 1, pp. 274-295, (1991).

Learn, A. J. Effect of Structure and Processing on Electromigration-induced Failure in Anodize Aluminum, *J. Appl. Phys.*, 44 (3), pp. 1251-1262, (1973).

Lee, R., Trombley, G., Johnson, B., Reston, R., Havasy, C., Mah, M. and Ito, C. Low Leakage GaAs MESFET Devices Operating at 350°C Ambient Temperature, *Trans. Second Int. High-Temp. Electron. Conf.*, Charlotte, NC, pp. V-3 to V-8, (June 1994).

Leksina, L. E., and Novikova, S. I. Thermal Expansion of Copper, Silver, and Gold Within a Wide Range of Temperature, *Soviet Physics-Solid State, 5* (4), (1963).

Lely, J. A. Darstellung von einkristallen von silicium carbid und beherrschung von art und menge der eingebautem verunreingungen, *Berichte Deutche Keramik Geselshaft*, 32, 229-236, (1955).

Li, L., Lizzul, C., Kim, H., Sacolick, I., and Morris, J. E. Electrical, Structural and Processing Properties of Electrical Conductive Adhesives, *IEEE-Components Hybrids Manufacturing Technol.*16, 843-851, (1995).

Liu, S., Reinhardt, K., Severt, C., and Scofield, J. *paper presented at the Workshop on High Temperature Power Electronics for Vehicles*, Fort Monmouth, NJ, USA, April 26-27, (1995a).

Liu, S., Reinhardt, K., Severt, C., and Scofield, J. paper presented at the 6th Intl Conf. on SiC and Related Mater., Kyoto, Japan, (Sept. 18-21, 1995b).

Liu, J., Ljungkrona, L., Lai, Z. Development of Conductive Adhesive Joining for Surface-Mounting Electron., Manufacturing, *IEEE-Components Hybrids Manufacturing Technol.,* 18, pp. 313-319, (1995c).

Liu, K. H., Oruganti, R., and Lee, F. C. Y. Quasi-Resonant Converters - Topologies and Characteristics, IEEE Trans. Power Electron., PE-2, No. 1, pp. 62-71, (January 1987).

Liu, S., Reinhardt, K., Severt, S., Scofield, J., Ramalingam, M. and Tunstall, C. To be published.

Loo, M. C. and Su, K. Die Attach of Large Dice with Ag/glass in Multilayer Packages, *Hybrid Circuits, 11*, pp. 8, (1986).

Lundberg, N. and Östling, M. *Appl. Phys. Lett.* 63 (22), 3069, (1993).

Magistralli, F., Tedesco, C., and Zanoni, E. Failure Mechanisms of GaAs MESFETs and Low-Noise HEMTs, *Semiconductor Device Reliability*, A. Christou Eds.,and B. A. Unger, pp. 211-267, (1990).

Mahefkey, T. Thermal Management of High Temperature Power Electronics - Limitations and Technology Issues, paper presented at the Second Int. High Tem. Electron. Conf., Charlotte, NC, June 5-10, (1994).

Mahefkey, T. Thermal Management of High Temperature Power Electronics Limitations and Technology, *Proc. Second Int. High Temp. Electron. Conf.*, pp. IX-21-IX-2, (1994).

Maher, G. H. et. al. Accelerated Life Testing and Reliability of a X7R Multilayer Ceramic Capacitor with a PLZT Dielectric, Proc.6th Capacitor and Resistor Technol. Symp., New Orleans, LA, (1986).

Maldonado, M.A., Shah, N.M., Cleek, K.J., and Walia, P.S. MADMEL Program Status, paper presented at 30th Intersociety Energy Conversion Engineering Conference (IECEC), Orlando, FL, IECEC Paper No. AP-63, (July 30, 1995).

Manko, H. Solders and Soldering, 3rd ed., McGraw-Hill, New York, pp. 116, (1992).

Manko, H. H. Solders and Soldering, 2nd ed., McGraw-Hill, New York, (1979).

Mapham, N. An SCR Inverter with Good Regulation and Sine-Wave Output, *IEEE Trans. Ind. Gen. Appl.*, IGA - 3, (2), 176-187, (March/April 1967).

Mark, H. F., Bikales, N. M., Overberger, C. G. And Menges, G. Encyclopedia of Polymer Science and Engineering, John Wiley and Sons, New York, (1985).

Marshall, A. Operating Power IC's at 200°, *Proc. 23rd IEEE Power Electron. Conf.*, pp. 1033-1039, (1992).

Marshall, A. Operating Power IC's at 200°, *1992 IEEE Power Electron. Specialists Conf. Record*, Vol. II, Boston, MA, pp. 1033 - 1039, (1992).

Matisoff, B. Wiring and Cable Designer's Handbook, TAB Books, Blue Ridge Summit, PA, (1987).

Matloubian, M. Smart Body Contact for SOI MOSFET's, *Proc. 1989 IEEE Int. SOI Conf.*, pp. 128, (October 1989).

Mattox, D. M. Ceramics, Glasses, and Diamond, Electronic Materials and Processes Handbook, 2nd ed., McGraw-Hill, New York, (1994).

Matula, R. A. Electrical Resistivity of Copper, Gold, Palladium, and Silver, *J. Phys. Chem. Ref. Data*, pp. 1147-1298, (1979).

McDaid, L.J., Hall, S., Mellor, P.H., and Eccleston, W. *Electron Lett*, 25, 827 (1989).

McMullin, P. G., Spitznagel, J. A., Szedon, J. R., and Costello, J. A. *Amorphous and Crystalline Silicon Carbide III*, *Springer Proc. in Physics*, 56, Springer-Verlag Berlin, Heidelberg, pp. 275, (1992).

Menhart, S., Hudgins, J. L., Godbold, C. V., and Portnoy, W. M. Temperature Variation Effects on the Switching Characteristics of Bipolar Mode FETs (BMFETs), *Conf. Rec. 1992 IEEE Ind. Appl. Soc. Ann. Meet.*, I, 1122-1125, (October 1992).

Migitaka M. and Kurachi, K. Silicon Integrated Injection Logic Operating Up to 454°C and Its Applications, *Trans. Second Int. High-Temp. Electron. Conf.*, Charlotte, NC, pp. II-27 to II-32, (June 1994).

MIL (1990) *Military Handbook - Reliability Prediction of Electronic Equipment*, MIL HDBK-217F, (2 January 1990).

Modern Plastics Encyclopedia `95, Vol. 71, McGraw-Hill, New York, (1994).

Moghadan, F. K. Development of Adhesive Die Attach Technol., in CERDIP Packages: Materials Issues, *Solid State Technology* 27, pp. 149, (1984).

Mogro-Campero, A. Simple Estimate of Electromigration Failure in Metallic Thin Films, J. Appl. Phys., 53 (2), 1224 and 1225, (1982).

Mohan, N., Undeland, T. M., and Robbins, W. P. Power Electronics: Converters, Applications, and Design, John Wiley & Sons, New York, (1989).

Mönch, W. Surface Science, 299/300, pp. 928, (1994).

Mroczowski, R. C. AMP Inc., private communication, (1995).

Mroczowski, R. S. Interassembly and Intersystem Interconnect, Physical Architecture of VLSI Systems, John Wiley & Sons, New York, (1994).

Mroczowski, R. S. AMP, Inc., Personal Communication, (1995).

Muller, R.S. and Kamins, T.I. Device Electronics for Integrated Circuits, 2nd ed., John Wiley & Sons, New York, (1986).

Murakami, M. *Matl. Sci. Reports* 5, pp. 273, (1990).

Murakami, M. Shih, Y., Price, W. H., White, E. L., Childs, K. D., and Parks, C. C. Thermally Stable Ohmic Contacts to n-type GaAs. III GeInW and NiInW Contact Metals, *J. App. Phys.*, 64(4), 1974-1982 , (1988).

Nakata, T., Koga, K., Matsushita, Y., Ueda, Y., and Nijna, T. Single Crystal Growth of 6H-SiC by a Vacuum Sublimation Method, and Blue LEDs, *Amorphous and Crystalline Silicon Carbide and Related Materials II, Springer Proc. Phys. 43*, Springer, Berlin, 26-34, (1989).

Nakayama, W. Thermal Management of Electronic Equipment: A Review of Technology and Research Topics, Advances in Thermal Modeling of Electronic Components and Systems, Chapter 1, pp. 1-78, Volume 1, A. Bar-Cohen and A.D. Kraus, Eds., Hemisphere Publishing Corporation, pp. 1-78, (1988).

Nelson, W. Accelerated Testing, John. Wiley & Sons, New York, (1990).

Newsome, J. L. et al. Metallurgical Aspects of Aluminum Wirebonds to Gold Metallization, *14th Ann. Proc. IEEE Electron. Components Conf.*, pp. 63, (1976).

Nishino, S., Kojima, Y., and Saraie, J. Growth and Morphology of 6H-SiC Prepared by the Sublimation Method, *Amorphous and Crystalline Silicon Carbide III and Other Group IV-IV Materials, Springer Proc. Phys. 56*, Springer, Berlin, 15-22, (1990).

Nishizawa, J., Terasaki, T., and Shibata, J. Field-effect Transistor Versus Analog Transistor (Static Induction transistor), *IEEE Trans. Electron Devices*, ED-22, 185-197, (1975).

Nordwall, B. D. Air Force Links Radar Problems to Growth of Tin Whiskers, *Aviation Week and Space Technology*, pp.65, (30 Jun 1986).

Ochoa, A., Sexton, F., Wrobel, T., Hash, G., and Sokel, R. Snapback: A Stable Regenerative Breakdown of MOS Devices, IEEE Trans. Nuclear Sci.,, NS-30, pp. 4127-4130, (December 1983).

Olson, D. R. and Berg, H. M. Properties of Die Bond Alloys Relating to Thermal Fatigue, *IEEE Transactions on Components, Hybrids, and Manufacturing Technol., CHMT-2* (2), pp. 193, (1979).

Olson, R. Johnson-Matthey, Inc., private communication, (1995).

Onuki, J. et al. Investigation of Reliability of Copper Ball Bonds to Aluminum Electrodes, *IEEE Trans. Components, Hybrids, and Manufacturing Technol., CHMT-10,* pp. 550, (1987).

Overton, E. et. al. Thermal Aging Effects on the Electrical Properties of Film and Ceramic Capacitors, *Proc. Electrical/Electronics Insulation Conf.*, Chicago, IL, pp. 201-205, (1993).

O'Connor, J. R. and Smiltens, J., Eds. Silicon Carbide, A High Temperature Semiconductor, Pergamon, Oxford, (1960).

Pal, D. and Joshi, Y. Application of Phase Change Materials to Thermal Control of Electronic Modules: A Computational Study, *Proc. Int. '95*, pp. 1307-1315, Hawaii, (1995).

Palkuta, L. J,. Prince, J. L., and Glista, A. S. Integrated Circuit Characteristics at 260 IC for Aircraft Engine-Control Applications, *IEEE Transactions on Components, Hybrids, and Manufacturing Technol.*, CHMT-2, (4), 405-412, (December 1979).

Palmer, D.W. and Heckman, R.C. Extreme Temperature Range Electronics, *IEEE Transactions on Components, Hybrids and Manufacturing Technol.*, CHMT-1, pp. 333-340, (1978).

Palmour, J. W., Kong H. -S., and Carter, C. H. Jr. Field-Effect Transistors in 6H-Silicon Carbide, *Proc.1991 Int. Semiconductor Device Res. Symp.,* Charlottesville VA, 491-494, (1991).

Palmour, J. W. , Kong, H. -S., and Carter, C. H. Jr. Vertical Power Devices in Silicon Carbide, Silicon Carbide and Related Materials, *Inst. Phys. Conf. Series 137,* Institute of Physics Publishing, Bristol, 499-502, (1994a).

Palmour, J. W., Tsvetkov, V. F., Lipkin, L. A., and Carter, C. H. Jr. Silicon Carbide Substrates and Power Devices, *Compound Semiconductors 1994, Inst. Phys. Conf. Series 141*, Institute of Physics Publishing, Bristol, 377-382, (1994b).

Papanicolaou, N. A., Christou, A., and Gipe, M. L. *J. Appl. Phys.* 65 (9), 3526, (1989).

Papanicolaou, N. A., Jones, S. H., Jones J. R., and Anderson, W. T. All-Refractory GaAs FET Using Amorphous $TiWSi_x$ Source/Drain Metallization and Graded-$In_x Ga_{1-x} As$ Layers, IEEE Electron Device Lett., 15(1), 7-9, (1994).

Partridge, J. H. Glass-to-Metal Seals, Society of Glass Technology, Sheffield, UK (1949).

Paulson, W. M. "Further studies on the reliability of thin film nickel-chromium resistors," 11th Annual Proc. Of the IEEE Reliability Physics Symposium, (1973).

Pecht, M. G., Nguyen, L. T., and Hakim, E. B. Plastic Encapsulated Microelectronics, John Wiley & Sons, New York, (1995).

Pecht, M. G. Handbook of Electronic Package Design, Marcel Dekker, New York, (1991).

Pecht, M. and Lall, P. Resistors, in Chapter 1: Passive Components of The Electrical Engineering Handbook, R.C. Dorf, ed., CRC Press, Boca Raton, FL, p. 5- 15, (1994).

Pecht, M., Hu, J., and Dasgupta, A. A Probabilistic Approach For Predicting Thermal Fatigue Life of Wirebonding in Microelectronics, ASME *J. Electron. Packaging*, 113, pp. 275, (September 1991).

Pecht, M., Lall, P., and Dasgupta, A. A Failure Prediction Model for Wire Bonds, *Proc. 1989 Int. Symp. Hybrid Microelectron.*, pp. 607-613, (1989).

Pecht, M. G. and Agarwal, R. K. Electronic Packaging Materials and Their Properties, Physical Architecture of VLSI Systems, John Wiley & Sons, New York, (1994).

Pecht, M. Integrated Circuit, Hybrid, and Multichip Module Package Design Guideline, John Wiley & Sons, (1994).

Pelletier, J., Gervais, D., and Pomot, C. *J. Appl. Phys.* 55 (4), pp. 994 , (1984).

Peterson, G.P. An Introduction to Heat Pipes, Modeling, Testing and Applications, pp. 306, John Wiley & Sons, New York, (1994).

Petit, J. B., Neudeck, P. G., Salupo, C. S., Larkin, D. J., and Powell, J.A. Silicon Carbide and Related Materials, Proc. 5th Conf. on SiC and Related Mater., M. G. Spencer, R. P. Devaty, J. A. Edmond, M. A. Khan, R. Kaplan, and M. Rahman, Eds., Institute of Physics Publishing, Pristol and Philadelphia, pp. 679, (1994).

Petit, J. B. and Zeller, M. V. *Mat. Res. Symp. Proc.* 242, pp. 567, (1992).

Philosky, E. Design Limits When Using Gold-Aluminum Bonds, *9th Ann. Proc. Int. Reliability Phys. Symp.,* Las Vegas, NV, pp. 11-16, (1971).

Philosky, E. Purple Plague Revisited, *8th Ann. Proc. Int. Reliability Phys. Symp.* Las Vegas, NV, pp. 177-185, (1970).

Pitt, V. A. and Needes, C. R. S. Thermosonic Gold Wirebonding to Copper Conductors, *IEEE Trans. Components, Hybrids, and Manufacturing Technol., CHMT-5,* pp. 435, (1982).

Porter, L. M., Glass, R. C., Davis, R. F., Bow, J. S., Kim, M. J., and Carpenter, R. W., *Mat. Res. Soc. Symp. Proc.* 282, 471, (1993).

Porter, L. M., Davis, R. F., Bow, J. S., Kim, M. J., and Carpenter, R. W. *Silicon Carbide and Related Materials, Proc. 5th Conf. SiC Related Mater.,* M. G. Spencer, R. P. Devaty, J. A. Edmond, M. A. Khan, R. Kaplan, and M. Rahman, Eds., Institute of Physics Publishing, Bristol and Philadelphia, pp. 581, (1994).

Porter, L. M., Davis, R. F., Bow, J. S., Kim, M. J., and Carpenter, R. W. *Trans. 2nd Intl. High Temp. Electronics Conf.,* D. B. King, and F. V. Thome, Eds., Vol. 1, pp. XIII-3, (1994).

Powell, J. A., Larkin, D. J., Matus, L. G., Choyke, W. J., Bradshaw, J. L., Henderson, L., Yoganathan, M., Yang, J. and Pirouz, P. Growth of High Quality 6H-SiC Epitaxial Films on Vicinal (0001) 6H-SiC Wafers, *Appl. Phys. Lett.,* 56, 1442-1444, (1990).

Powell, J. A. and Matus, L. Silicon Carbide, a Semiconductor for Space Power Electronics, Proc. Eighth Symp. Space Nucl. Power Syst., pp. 954-959, (Jan. 1991).

Prince, J.L., Draper, B.L., Rapp, E.A., Kronberg, J.N., and Fitch, L.T. Performance of Digital Integrated Circuit Technologies at Very High Temperatures, *IEEE Trans. Components, Hybrids and Manufacturing Technol.,* CHMT-3, pp. 571-579, (December 1980).

Product & Applications Handbook 1994 Unitrode Integrated Circuits, Merrimack, NH, (1993-1994).

Przybylko, S.J. Developments in SiC for Aircraft Propulsion System Applications, paper presented at AIAA/SAE/ASME/ASEE 29th Joint Propulsion Conference, Monterey, CA, (June 28, 1993).

Quality and Reliability Assurance Manual, TQS Specification #QUA.011, Rev B, Triquint Semiconductor, Beaverton, OR, (Oct. 1992).

Quigley, R.E. More Electric Aircraft, USAF Wright Lab. Aerospace Power Division internal report, (1993).

Rahn, A. The Basics of Soldering, John Wiley & Sons, New York, (1993).

Rastegaeva, M. G. and Syrkin, A. L. *Sensors and Actuators,* A, 33, pp. 95, (1992).

Reeves, G. K. *Solid-State Electron.,* 23, pp. 487, (1980).

Regan, R., Butler, S., Bulat, E., Varallo, A., Abdollahian, M., and Rock, F. HF/VHF/UHF Power Static Induction Transistor Performance, paper presented at the RF Exposition East, (Nov. 1986).

Regan, R. Static Induction Transistors - An Emerging Technology, Proc. Southwest Semiconductor Electron. Expo., (Oct. 1986).

Regan, R. J. and Butler, S. J. High-Voltage UHF Power Static Induction Transistors, *Proc. RF Technol. Expo.,* (1985).

Regan, R. J., Bencuya, I., Butler, S. J., Stites, S., and Harrison, W. New UHF Power Transistor Operates at High Voltage, *Microwaves & RF*, 24, (April 1985).

Regan, R., Cogan, A., Butler, S., Bencuya, I., and Haugsjaa, P. Improved Performance of High Voltage, Microwave Power, Static Induction Transistors, paper presented at the 14th European Microwave Conference, (Aug. 1984).

Robock, P. V. and Nguyen, L. T. Plastic Packaging, Microelectronics Packaging Handbook, R. R. Tummala and E. J. Rymaszewski, Eds., Van Nostrand Reinhold, New York, (1989).

Romanczuk, C.S., Burcham, S.W., Evans, J.L., Knight, R.W., and Johnson, R.W. Finite Element Analysis of the Thermal Characteristics of an Automotive Powertrain Controller, *ISHM '93 Proceedings*, pp. 650-655, (1993).

Rörgren, R. S., and Liu, J. Reliability Assessment of Isotropically Conductive Adhesive Joints in Surface Mount Applications, *IEEE-Components Hybrids Manufacturing Technology*, Vol. 18, pp. 305-312, (1995).

Rosler, R. K. Rigid Epoxies, *Electronic Materials Handbook, Vol.1 Packaging,* ASM International, Materials Park, OH, (1989).

Rusanen, O. Lenkkeri, J. Reliability Issues of Replacing Solder with Conductive Adhesives in Power Modules, *IEEE-Components Hybrids Manufacturing Technol.*, 18, 320-325, (1995).

Sabate, J. A., Vlatkovic, V., Ridley, R. B., Lee, F. C., and Cho, B. H. Design Considerations for High-Voltage, High-Power, Full-Bridge, Zero-Voltage-Switched PWM Converter, High-Frequency Resonant and Soft-Switching PWM Converters, Virginia Power Electronics Center, Blacksburg, VA, (1991).

Sable D. M. and Lee, F. C. The Operation of a Full-Bridge, Zero-Voltage-Switched, PWM Converter, High-Frequency Resonant and Soft-Switching PWM Converters, Virginia Power Electronics Center, Blacksburg, VA, (1991).

Sakamoto, N., Kanai, T., and Ohkawa, K. Thermal Design and Structure of Thick Film Hybrid IC Based on Insulated Aluminum Substrate, *Proc. Ninth IEEE Semi-Therm Symp.*, pp. 186-193, (Feb. 1993).

Sarma, K.R. and Liu, S.T. Silicon-on-Quartz for Low Electronic Applications, *Proc. 1994 IEEE Int. SOI Conf.*, pp. 117, (October 1994).

Schneider, R. P. and Lott, J. A. InAlP/InAlGaP Distributed Bragg Reflectors for Visible Vertical Cavity Surface-Emitting Lasers, Appl. Phys. Lett., 62, 2748-2750, (1993).

Schroder, D. K. Semiconductor Material and Device Characterization, John Wiley & Sons, Inc., New York, pp.101, 130, 507 119, (1990).

Schroder, K. CRC Handbook of Electrical Resistivities of Binary Metal Alloys, CRC Press , Boca Raton, FL, (1983).

Sclater, N. Wire and Cable for Electronics: A User's Handbook, McGraw-Hill, New York, (1991).

Scott, A. W. Cooling of Electronic Equipment, John Wiley & Sons, New York, (1974).

Seraphim, D. P., Lee, L.C., Appelt, B. K., and Marsh, L. L. An Overview of Materials Science in Printed Circuit Packaging, *Mater. Res. Soc. Symp. Proc.*, 40, 21-48, (1985).

Shah, N. M. More Electric F-18 Cost Benefits Study, WL-TR-91-2093, Northrop Corporation, Aircraft Division, (1992).

Shatzkes, M. and Lloyd, J. R. A Model for Conductor Failure Considering Diffusion Concurrently with Electromigration Resulting in a Exponent of 2, *J. Appl. Phys.*, 59 (11), 3890-3893, (1986).

Shor, J. S. And Weber, R. A. Provost L. G., Goldstein, D. and Kurtz, A. D. *Mat. Res. Soc. Symp. Proc*. 242, 573, (1992).

Shoucair F. S. and Ojala, K. High-Temperature Electrical Characteristics of GaAs MESFETs (25-400 °C*), IEEE Trans. Electron Devices*, 39(7), 1551-1557, (1992).

Shoucair, F.S. Potential and Problems of High-Temperature Electronics and CMOS Integrated Circuits (25 to 250 °C) - An Overview, Microelectronics J., 22, 39-54, (1991).

Sims, P. E. and DiNetta, L. C. High Temperature Performance of AlGaP/GaP P-N Junction Photodiodes, *Trans. Second Int. High-Temp. Electronics Conf.*, Charlotte, NC, pp. P-247 to P-251, (June 1994).

Sinclair, I. R. Passive Components: A User's Guide, 2nd Edition, Butterworth-Heinemann Ltd., Oxford, 1991.

Smith, B. L. "Failure Mechanisms in Passive Devices," Electronic Materials Handbook, Vol. 1, Packaging, ASM International, Materials Park, OH, pp. 994- 1005 (1989).

Smithells, C. J. Materials Reference Book, 2nd ed., Vol. 2, Butterworth's Scientific Publications, London, (1995).

Sokolich, M., Yu, K. K., Chiang, M., Lee, H. M., and Shih, Y. C. Performance and Reliability of GaAs Refractory Gate X-Band Power Amplifiers at Elevated Temperatures, Trans. First Int. High-Temp. Electronics Conf., Albuquerque, NM, pp. 302-312, (June 1991).

Spellman, L. M., Glass R. C., Davis, R. F., Humphreys, T. P., Nemanich, R. J., Das, K., and Chevacharoenkul, S. *Amorphous and Crystalline Silicon Carbide IV, Springer Proc. Phys.*, vol. 71, Springer-Verlag Berlin, Heidelberg, pp. 176, (1992).

Sponger, J. W., Gaul, S. J., and Heedley, P. L. An Evaluation of Op Amp Performance up to 300 IC Using Dielectric Isolation and Bonded Wafer Material Technologies, *Proceedings of the First Int. High Temp. Electronics Conf.*, Albuquerque, NM, pp. 281-290, (June 1991).

Sriram, S., Clarke, R. C., Burk, A. A. Jr., Hobgood, H. M., McMullin, P. G., Orphanos, P. A., Siergiej, R. R., Smith, T. J. Brandt, C. D., Driver, M. C., and Hopkins, R. H. RF Performance of SiC MESFETs on High Resistivity Substrates, *IEEE Electron Device Lett.,* 15, 458-459, (1994a).

Sriram, S., Clarke, R. C., Hanes, M. H., McMullin, P. G., Brandt, C. D., Smith, T. J.,

Burk, A. A. Jr., Hobgood, H. M., Barrett, D. L., and Hopkins, R. H. SiC Microwave Power MESFETs, Silicon Carbide and Related Materials, *Inst. Phys. Conf. Series 137*, Institute of Physics Publishing, Bristol, 491-498, (1994b).

Steckl, A. J. and Su, J. N. *IEDM*, pp. 695, (1993).

Steckl, A. J., Su J. N., Yih, P.H., Yuan, C., and Li, J. P. Silicon Carbide and Related Materials, Proc. 5th Conf. SiC Related Mater., M. G. Spencer, R. P. Devaty, J. A. Edmond, M. A. Khan, R. Kaplan, and M. Rahman, Eds. Institute of Physics Publishing, Bristol, p 653, (1994).

Stevenson, R. A. New Hermetically Sealed Ceramic-Cased Axial/Radial Monolithic Ceramic Capacitor Designed for 260°C Operation, *Trans. First Int. High Temp. Electronics Conf.*, Albuquerque, NM, pp. 93-98, (1991).

Stevenson, R. A. New Hermetically Sealed Ceramic-Cased Monolithic Ceramic Capacitor, *Proc. the 11th Capacitor and Resistor Technol. Symp. (CARTS 1991)*, Las Vegas, NV, pp. 210-213, (1991).

Streetman, B. G. Solid State Electronic Devices, 3rd ed., Prentice Hall, Englewood Cliffs, New Jersey, (1990).

Suehle, J.S., Chaparala, P., and Messick, C. High Temperature Reliability on Thin Film SiO_2, Trans. Second Int. High Temp. Electronics Conf., Charlotte, N.C., pp. VIII-15, (June 1994).

Sunayama, T., Kawakami H., and Migitaka, M. Silicon Integrated Injection Logic Operating Above 350°C, *Proc. Int. Electron Devices Meet.*, pp. 161-164, (1991).

Suthar, J. L., et. al. Dielectric Films for High Temperature High Power Electronics Applications, *Trans. First Int. High Temp. Electronics Conf.*, Albuquerque, NM, pp. 105-110, (1991).

Swirhun, S., Hanka, S., Nohava, J., Grider, D., and Bauhahn, P. Refractory Self-Aligned-Gate GaAs FET Based Circuit Technology for High Ambient Temperatures, *Trans. First Int. High-Tem. Electron. Conf.*, Albuquerque, New Mexico, pp. 295-300, (June 1991).

Sze, S. M. Physics of Semiconductor Devices, John Wiley & Sons, New York, pp.182, 246, 261, 274 (1981).

Sze, S.M. Semiconductor Devices Physics and Technology, John Wiley & Sons, New York, pp. 169, (1985).

Tairov, Yu. M. and Tsvetkov, V. F. Investigations of Growth Processes of Ingots of Silicon Carbide Single Crystals, *J. Crystal Growth*, 43, 209-212, (1978).

Takeda, E. and Suzuki, N. *IEEE Electron Dec. Lett.* ED-4(4), pp. 111, (1983).

Takei, W. J. and Francombe, M. H. Measurement of Diffusion-Induced Strains at Metal Bond Interfaces, *Solid State Electronics 11,* Pergamon Press, pp. 205-208, (1968).

Tummala, R. R. and Rymaszewski, E. J. Microelectronic Packaging Handbook, Van Nostrand Reinhold, New York, (1989).

Terry, L. E. and Wilson, R. W. *Proc. IEEE*, 5 (9), 1580, (1969).

The National Technology Roadmap for Semiconductors, Semiconductor Industry Association, pp. 134-135, (1994).

Thompson, C. V. and Kahn, H. Effects of Microstructure on Interconnect and Via Reliability: Multimodel Failure Statistics, J. Electron. Mater., 22 (6), 581-587, (1993).

Thurmond, C.D. The Standard Thermodynamic Function of the Formation of Electrons and Holes in Ge, Si, GaAs and GaP, *J. Electrochem. Soc.*, 122, 1133-1141, (1975).

Thwaites, C. J. Optimum Reliability in Printed Circuit Soldering through Quality Assurance, Welding J., pp. 702, (Oct 1972).

Torri, T. Delco Electronics, Private Communication, (1995).

Touloukian, Y. S., Powell, R. W., Ho, C. Y., and Klemens, R. G. Thermophysical Properties of Matter, Vol. 1, Thermal Conductivity: Metallic Elements and Alloys, IFI Plenum, New York, (1970).

Tyson, S.M. High Temperature Characteristics of Silicon-on-Insulator and Bulk Silicon Devices to 500°C, *Trans. Second Int. High-Temp. Electronics Conf.*, Charlotte, NC, pp. P-9 to P-14, (June 1994).

Tyson, S.M. and Grybowski, R.R. High Temperature Characteristics of Silicon-on-Insulator and Bulk Silicon Devices to 500 °C, 2nd International High Temperature Electronics Conference, Charlotte, NC, pp. 9, (June 1994).

Ueda, T., Nishino, H., and Matsunami, H. Crystal Growth of SiC by Step-Controlled Epitaxy, *J. Crystal Growth* 104, 695-700, (1990).

Valdya, S., Sheng, T. T., and Sinha, A. K. Linewidth Dependence of Electromigration in Evaporated Al-0.5% Cu, *Appl. Phys. Lett.*, 36 (6), 464-466, (1980).

Veneruso, A.F. Ed., Sourcebook on High-Temperature Electronics and Instrumentation, Report# SAND81-2112, Sandia National Laboratories, Albuquerque, NM, (October 1981).

Vianco, P. T. General Soldering, Metals Handbook, 10th ed., Vol 6, Welding, Brazing, and Soldering, ASM International, Materials Park, OH, (1993).

Waldrop, J.R. and Grant, R.W. *J. Appl. Phys.* 72 (10), 4757, (1992).

Waldrop, J.R. and Grant, R.W. *Appl. Phys. Lett.* 56 (6), 557, (1990).

Waldrop, J.R. and Grant, R.W. *Appl. Phys. Lett.* 62 (21), 2685, (1993).

Watanabe, Y. and Nishizawa, J. Japanese patent 205 068, published No. 28-6077, application date, (Dec. 1950).

Weast, R. C., Ed. CRC Handbook of Chemistry and physics, 60th ed., CRC Press, Boca Raton, FL, (1979).

Weichold, M. H., Eknoyan, O., and Kao, Y-C. A GaP MESFET for High Temperature Applications, *IEEE Trans. Components, Hybrids and Manufacturing Technol.*, CHMT-5(4), 342-344, (1982).

Weimer, J.A. Power Management and Distribution for the More Electric Aircraft, paper presented at 30th Intersociety Energy Conversion Engineering Conference (IECEC), Orlando, FL, IECEC Paper No. AP-385, (July 30, 1995).

Wilcoxon Research, private communication, (1995).

Williams, R. Modern GaAs Processing Methods, 2nd ed., Artech House, Boston, MA, (1990).

Wilson, C. D. and O'Neill, A. G. GaAs Heterojunction Devices for Integrated Circuit Technology at Elevated Temperatures, *Trans. Second Int. High-Temp. Electron. Conf.*, Charlotte, NC, pp. V-15 to V-20, (June 1994).

Wong, C. P. An Overview of Integrated Circuit Device Encapsulants, *ASME J. Electron. Packaging 111*, pp. 97, (1989).

Woodall, J. M., Freeouf, J. L., Pettit, G. D., Jackson, T., and Kirchner, P. Ohmic Contacts to n-GaAs Using Graded Band Gap Layers of $Ga_{1-x}In_xAs$ Grown by Molecular Beam Epitaxy, *J. Vacuum Sci. Technol.*, 19(3), 626-627, (1981).

Wright, C. The Effect of Solid State Reactions upon Solder Lap Shear Strength, *27th IEEE Electron. Components Conf.*, (1977).

Yallup, K. SOI Provides Total Dielectric Isolation, *Semiconductor Int.*, pp. 134, (July 1993).

Yallup, K. and Creighton, O. Growing Reliable Gate Oxides on Thick Film SOI Substrates, *Pro. 1993 IEEE Int. SOI Conf.*, pp. 78, (October 1993).

Yeh, H. L., and Strickman, S. Pd Enhanced Dry Soldering Process, *IEEE-Electronic Components Tech. Conf.*, Vol. 42, pp. 492-501, (1992).

Yoshida, S., Sasaki, K., Sakuma, E., Misawa, S., and Gonda, S. *Appl. Phys. Lett.* 46 (8), 766, (1985).

Yost, F. G., Hosking, and F. M., and Frear, D. R. The Mechanics of Solder Alloy Wetting and Spreading, Van Nostrand Reinhold, New York, (1993).

Yost, F. G., Karnowsky, M. M., Drotning, W. D., and Gieske, J. H. Thermal Expansion and Elastic Properties of High Gold-Tin Alloys, *Met. Trans. A 21A*, pp. 1885-1889, (1990).

Yue, J., Liu, S.T., Fechner, P., Gardner, G., Wircraft, W., and Finn, C. An Effective Method to Screen SOI Wafers for Mass Production, *Proc. 1994 IEEE Int. SOI Conf.*, pp. 13, (October 1994).

Zarlingo, S. P. and Scott J. R. Leadframe Materials for Packaging Semiconductors, *First Ann. Int. Electron. Packaging Soc. Conf.*, Cleveland, OH, (1981).

Ziegler, G., Lanig, P., Theis, D., and Weyrich, C. Single Crystal Growth of SiC Substrate Material for Blue Light Emitting Diodes, *IEEE Trans. Electron Devices*, ED-30, 277-281, (1983).

Zipperian, T.E. A Survey of Materials and Device Technologies for High Temperature (T>300°C) Power Semiconductor Electronics, *Proc. Power Conversion Intelli.*, pp. 353-365, October (1986).

Zipperian, T. E., Chaffin, R. J. and Dawson, L.R. Recent Advances in Gallium Phosphide Junction Devices for High-Temperature Electronic Applications, *IEEE Trans. Ind. Electronics,* IE-29(2), pp. 129-136, (1982a).

Zipperian, T. E., Dawson, L. R., and Barnes, C. E. Evaluation of $Al_xGa_{1-x}As$ for High-Temperature Electronic Junction Device Applications, *Institute of Physics Conference Series No. 65: International Symposium on GaAs and Related Compounds*, Chap. 6, pp. 523-528, (1982b).

Index

Materials